T0138447

OF THE PLURALITY OF WORLDS

☞ PUBLISHER'S NOTE ☜

This volume contains facsimiles of the first edition of William Whewell's *Of the Plurality of Worlds: An Essay;* the typeset pages that Whewell excised from the first edition just before the book went to press; and "A Dialogue on the Plurality of Worlds," reproduced here from the second edition, which Whewell wrote as a response to his critics. We have added a new introduction by Michael Ruse, which begins on page 1. We have reproduced the original pages of Whewell's material, and for the convenience of our readers we have added consecutive page numbers to the entire volume; these added numbers appear at the foot of each page in the outside margin.

OF

THE PLURALITY OF WORLDS

A FACSIMILE OF THE FIRST EDITION OF 1853;
PLUS PREVIOUSLY UNPUBLISHED MATERIAL
EXCISED BY THE AUTHOR JUST BEFORE
THE BOOK WENT TO PRESS; AND
WHEWELL'S DIALOGUE
REBUTTING HIS
CRITICS,

REPRINTED FROM THE
SECOND EDITION

WILLIAM WHEWELL

Edited and with New Introductory Material by
MICHAEL RUSE

THE UNIVERSITY OF CHICAGO PRESS
CHICAGO & LONDON

William Whewell (1794–1866) was professor of mineralogy (1828–32) and moral philosophy (1835–55) at the University of Cambridge. He was the author of many books, including *History of the Inductive Sciences; The Elements of Morality, Including Polity;* and *History of Scientific Ideas.*

Michael Ruse is professor of philosophy at Florida State University. He is author of more than a dozen books, including *Can a Darwinian Be a Christian?* and *The Darwinian Revolution,* Second Edition, the latter published by the University of Chicago Press.

The University of Chicago Press, Chicago 60637
The University of Chicago Press, Ltd., London

Printed in the United States of America

10 09 08 07 06 05 04 03 02 01 1 2 3 4 5
ISBN: 0-226-89435-5 (cloth)
ISBN: 0-226-89436-3 (paper)

Library of Congress Cataloging-in-Publication Data
Whewell, William, 1794–1866
Of the plurality of worlds / William Whewell ; edited and with new introductory material by Michael Ruse.
p. cm.
Originally published: London : J. W. Parker, 1853.
"A facsimile of the first edition of 1853; plus previously unpublished material excised by the author just before the book went to press; and Whewell's dialogue rebutting his critics, reprinted from the second edition."
ISBN 0-226-89435-5 (alk. paper) — ISBN 0-226-89436-3 (pbk. : alk. paper)
1. Plurality of worlds. I. Ruse, Michael. II. Title.
QB54 .W47 2001
576.8'39—dc21

00-067667

∞

The paper used in this publication meets the minimum requirements of the American National Standard for Information Sciences—Permanence of Paper for Printed Library Materials, ANSI Z39.48-1992.

CONTENTS

A note to the reader: For convenience, the pages of this book are numbered consecutively, even though the pagination of the original Whewell material is also retained. The consecutive page numbers appear in the margins at the bottom of each page.

INTRODUCTION

Michael Ruse

"ARE YOU A WHEWELLITE or a Brewsterite, or a t'other-manite, Mrs. Bold?" said Charlotte, who knew a little about everything, and had read about a third of each of the books to which she alluded.

"Oh!" said Eleanor; "I have not read any of the books, but I feel sure that there is one man in the moon at least, if not more."

"You don't believe in the pulpy gelatinous matter?" said Bertie.

"I heard about that," said Eleanor; "and I really think it's almost wicked to talk in such a manner. How can we argue about God's power in the other stars from the laws which he has given for our rule in this one?"

"How indeed!" said Bertie. "Why shouldn't there be a race of salamanders in Venus, and even if there be nothing but fish in Jupiter, why shouldn't the fish there be as wide awake as the men and women here?"

"That would be saying very little for them," said Charlotte. "I am for Dr. Whewell myself; for I do not think that men and women are worth being repeated in such countless worlds. There may be souls in other stars, but I doubt their having any bodies attached to them. But come, Mrs. Bold, let us put our bonnets on and walk round the close. . . ."

This exchange, just before a moonlight walk, comes in Anthony Trollope's novel *Barchester Towers,* written in 1856. Needing a wife for her spendthrift brother Bertie, Charlotte Stanhope is plotting to capture Eleanor Bold, a pretty young widow with a nice little fortune. Trollope paints Charlotte, and indeed her whole family, as being without a great deal of moral worth, but much fun to be with and certainly one who would know all of the latest gossip and scandal and controversy. For all that there was now a war going on in the faraway Crimea and that tension was building on the Indian subcontinent—tension soon to explode into the Mutiny—the Stanhopes knew that the great topic of the day was the dreadful clash between William Whewell, professor of moral philosophy and Master of Trinity College at the University of Cambridge, and Sir David Brewster, Scottish man of science, noted Presbyterian, and biographer of Sir Isaac Newton. Are we alone in the universe, as Whewell claimed, or is the whole of creation teeming with intelligence, as Brewster responded? This debate was sparked three years earlier by the quasi-anonymous publication of Whewell's *Of the Plurality of Worlds: An Essay*. It is this volume which is reprinted here before you. Together with this are included some never-before-published pages that Whewell removed from his (already typeset) manuscript just before publication. For us today, knowing much about the Victorian science/religion interface to which Whewell was contributing, these make fascinating reading. Furthermore, concluding this reprint, there is the dialogue that Whewell penned and added to subsequent editions of his *Essay*, a place where he took on the criticisms of his readers and where he tried to show the strength of the case he was making.

Here I will introduce you to the whole debate about extraterrestrials, to put Whewell's *Essay* in context. As so often with these sorts of questions, it is appropriate to begin with the Greeks.

Life in the Universe

Typically we find that the Greeks took diametrically opposed positions on the possibility of the existence of life, especially intelligent life, elsewhere in the universe than here on Earth (Dick

1982). Equally typically, the positions taken were based less on observation and calculation and more on *a priori* speculation and prejudged philosophical commitments. The atomists believed that everything in the universe is the result of small particles buzzing around randomly in a void of infinite size and infinite duration. Normally, this all means nothing: chance encounters by these particles and nothing more. But sometimes the atoms cohere, they stick together, and sometimes, very rarely, this cohesion leads to functioning units. Given the never-ending dimensions of space and time, inevitably there will be a working globe like our own—and even more inevitably, this working globe (or things much like it) will repeat again and again throughout the universe. Infinity means just that. Most of what exists is, was, and will be, nothing but inert nonfunctioning space containing meaningless matter. But throughout this space lies the infinite possibility of living beings, some perhaps like us and some perhaps different.

The greatest of the Greek philosophers, Aristotle, would have none of this. For him, infinity did not counter the need for purpose and form. He believed in one tight, finite universe, with the Earth at the center and with a series of concentric shells ever circulating around our world, holding the sun, moon and planets in place, ending finally with the outer sphere of the stars. The Earth alone is the place of becoming, of change, of decay—and it alone has the potential for life. The heavens are perfect but sterile, and simply could not serve as the abodes of living beings. Aristotle's physics—embracing the four elements of earth, water, air, and fire—precluded any chance of functioning life on heavenly bodies. The heaviest of the elements (earth) makes naturally for the center of the universe—that is why bodies fall downward when released; whereas the lightest of the elements (fire) rises naturally from the center—that is why smoke goes up rather than down. A Martian, for instance, would at once fall to the Earth, save it was encased (as is Mars itself) in an invisible crystalline shell that keeps the planet constantly moving around our central domain.

Both visions—atomists' and Aristotle's—were passed down through antiquity. The great Roman poet Lucretius, in his *De rerum natura,* picked up and endorsed the atomists' philosophy. Aristotle's ideas, long since known to the Arabs, came back to the West through the writings of Saint Thomas Aquinas in the thirteenth century. Given the incorporation into Christianity of the Jewish story of origins, with its total focus on humankind and on our privileged place here on Earth, there was no doubt in anyone's mind but that we here on Earth are unique—there can be no material life elsewhere in the universe, and humans alone have intelligence and freedom. We are made in God's image. To think otherwise would be heresy—and, more than that, just plain silly. This stance of course changed dramatically in 1543, the year Nicholas Copernicus published his *De revolutionibus orbium coelestium* (*On the Revolutions of the Heavenly Spheres*), in which he put the sun at the center of the universe and the planets—including the Earth—forever circling around it. Copernicus himself did not want to challenge Christian orthodoxy or to populate the universe with new beings, but it was not long before some of his supporters started to speculate in this way, notoriously Giordano Bruno, who was burnt at the stake in 1600 for just such flights of fancy.

It was Johannes Kepler, the great astronomer who realized that planetary motion could not be circular, who forcibly posed the dilemma then facing the Copernican. We have great tension between revealed theology—the belief in God based on faith, revelation, and the word of spiritual authority (the Church for Catholics and, for both Catholics and Protestants, the Bible)—and natural theology—the belief in God based on physical evidence and reason. On the one hand, intelligent beings not of this earth seem to threaten our special relationship with God. As Kepler pondered, "For if their globes are nobler, we are not the noblest of rational creatures. Then, how can all things be for man's sake? How can we be the masters of God's handiwork?" (Kepler 1965, 43). On the other hand, worlds without denizens seem a pointless exercise in the Almighty's power: "Our moon exists for us on the earth, not the other globes. Those four little moons exist for Jupiter, not for us. Each planet in turn, together with its occupants, is served by its own satellites. From this line of reasoning we deduce with the highest degree of probability that Jupiter is inhabited" (Kepler [1610] 1965, 42).

Notwithstanding the difficulties and the threats to traditional religion, with the increasing success of modern science, there was a corresponding, increasing enthusiasm for a fully populated universe (Crowe 1986). Widely read and immensely popular were the witty and informative (in some respects, too informative) dialogues of the Cartesian, Bernard de Fontenelle. Purportedly eavesdropping on the encounters between a philosopher and a cultured noblewoman, *Entretiens sur la pluralité des mondes* (*Conversations on the Plurality of Worlds*) did not hesitate to postulate life throughout the heavens, nor to fill out details for the curious. Venusians, we learn, are "little black people, scorch'd with the sun, witty, full of fire, very amourous," whereas the distant Jupiterians are "Flegmatik: They no [sic] not what 'tis to laugh, they take a day's time to answer the least question. . . ." (Marsak 1970, 121). Humans, expectedly given our relative position among the planets, come somewhere in between, with "no fix'd or determined character." Fantasy stuff but not wilder than the eighteenth-century speculations of that most sober of philosophers, Immanuel Kant, according to whom the ultimate law of morality (the Categorical Imperative) binds the denizens of Mercury and Saturn no less than those of Earth. Kant believed that intelligence increases as we move from the sun, "From the one side [Mercury and Venus] we saw thinking creatures among whom a man from Greenland or a Hottentot would be a Newton, and on the other side [Saturn and Jupiter] some others would admire him [as if he were] an ape" (Kant [1755] 1981, 190). Kant was here echoing the English poet Alexander Pope:

> Superior beings, when of late they saw
> A mortal man unfold all Nature's law,
> Admir'd such wisdom in an earthly shape,
> And shew'd a Newton as we shew an Ape.
> (*Essay on Man,* Epistle II, ll. 31–34)

Focusing now on Britain, the nineteenth-century scene was set by Thomas Chalmers, Presbyterian divine, charismatic preacher, and a man of immeasurable moral and spiritual strength, who in 1843 was to become the leader of those who could no longer abide the chafing restrictions of the Church of Scotland and who broke loose to form the Free Church. Called to the pulpit of Tron Church in Glasgow, in 1815 Chalmers delivered an electrifying series of lunchtime sermons—one needed to appear four hours early to find a seat—on the subject of Christianity and life elsewhere in the universe. Published in 1817, Chalmers's *Astronomical Discourses* became an instant best seller, and he convinced virtually all of his countrymen that God's universe teems with life—intelligent life. Although far from threatening Christianity, the belief in extraterrestrials is virtually demanded by the *Discourses.* Learned in modern science, Chalmers brought the findings and theories of natural philosophy to his deeply felt biblical commitment in a brilliant fusion of faith and fact.

Asked the psalmist: "When I consider the heavens, the work of thy fingers, the moon and the stars, which thou has ordained: What is man, that thou are mindful of him? and the son of man, that thou visitest him?" Answered Chalmers: no one sensitive to modern science can doubt that the whole of creation is filled with living beings. The analogies between Earth and objects elsewhere are altogether too strong. Were our world destroyed tomorrow, there are many others on which life would continue. "Is it presumptuous to say," Chalmers asked, "that the moral world extends to these distant and unknown regions? that they are occupied with people? that the charities of neighbourhood and home flourish there? that the praises of God are there lifted up, and his goodness rejoiced in?" (Chalmers [1817] 1906, 44). Does this then mean that we are worthless and that God has no interest in us? Worthless we may most certainly seem, but lack of interest most certainly not! The very opposite is the case. The "divine condescension" is so great that He stretches out to us, sinners as we are, on that insignificant speck of dust that we call home. Indeed, according to Chalmers, it is precisely because we are so insignificant that we can start to comprehend the strength and glory of God's love and grace.

> But is it not adding to the bright catalogue of his other attributes to say, that while magnitude does not overpower him, minuteness cannot escape him, and variety cannot bewilder him; and at the very time while the mind of the Deity is abroad over the whole vastness of creation, there is not one particle of matter, there is not one individual principle of rational or of animal existence, there is not one single world in that expanse which teems with them, that his eye does not discern them as constantly, and his hand does not guide as unerringly, and his Spirit does not watch and care for as vigilantly, as if it formed the one and exclusive object of his attention? (95)

We are as nothing and naught that we do can count for anything, Chalmers proclaimed. It is only through God's unearned grace that we have hope of salvation, which, according to his Presbyterian beliefs, we can pray for but cannot earn. Stressing the plurality of worlds served two ends. It underlined God's magnificence and the fact that He does nothing by chance or without reason. It underlined our insignificance and the wonder that God would care so much for such creatures as we humans. It was against this background that, some forty years later, Whewell wrote and published his *Essay*.

William Whewell

Born at the end of the eighteenth century in the northern town of Lancaster, William Whewell was the son of a carpenter. His formidable talents were recognized and, thanks to free grammar school training, he moved up to Trinity College where he was to spend his whole academic and professional career. Shining at his mathematical studies, he became a fellow of the college, then a tutor, and finally in 1841 he was appointed master (principal), a position he was to hold until his death. His first university post was that of professor of mineralogy, after which he was elected professor of moral philosophy. Whewell was one of a group of university teachers (which included John Henslow, professor of botany, and Adam Sedgwick, professor of geology) who were determined to upgrade the quality of instruction in science at the older institutions (Ruse 1979). To this end, Whewell wrote several highly regarded and much-used mathematical texts. As a practicing scientist, he was, as he himself was the first to admit, not of the first rank. He devoted much attention to the movements of the tides, hoping to find general patterns or laws which govern the waters from country to country. Although he did sterling work of empirical collation, true theoretical insight evaded him. Typically, however, Whewell interested himself in a much broader range of scientific issues, and today there is quite some considerable admiration for brief forays he made into crystallography and mineralogy, as well as in the theoretical foundations of economics. Had he pursued these real insights, he might be better known as a scientist *per se* (Ruse 1991).

Whewell, however, needs no defense or apology, especially today. As part of the general move to upgrade science, Whewell was much concerned in defining and elaborating the bounds of science (it was Whewell indeed who invented the term "scientist") and to articulate the norms and practices of good science. To this end, long anticipating the mid-twentieth-century claim of Thomas Kuhn in his *The Structure of Scientific Revolutions* that the key to understanding the science of today lies in understanding the science of yesterday, Whewell wrote the massive three-volume *History of the Inductive Sciences* in 1837. In 1840 Whewell added a two-volume analysis, the *Philosophy of the Inductive Sciences.* Much influenced by the thinking of Kant, Whewell argued that science must be seen as a combination of *a priori* concepts or ideas applied to empirical discoveries and generalizations. Taking as his guide the science of Isaac Newton—the most illustrious of all the previous fellows of Trinity College—Whewell provided a full picture of scientific inquiry: a picture that today garners increasing interest and respect. Idealist though he may have been, Whewell's feel for the texture of real science has never been surpassed and rarely equalled.

As a fellow of an Oxbridge college, Whewell was ordained in the established Church and became an Anglican priest. As a man much immersed in science, his interests naturally inclined more to natural than revealed theology, and so it was appropriate when, at the end of the 1820s, Whewell was asked to contribute a volume to a series of works in natural theology, the *Bridgewater Treatises.* Given his specific expertise, he fitted comfortably with the topic of astronomy and how and why it is conducive to a belief in God. Whewell's treatise was the first in the series to appear, and general opinion regarded it as the most brilliant. This in itself is somewhat of a paradox because, in respects, astronomy seems the least promising of topics for support of natural theology. British natural theology of the day centered on the argument from design, and the traditional version of that argument—first proposed by Plato, endorsed by Saint Thomas Aquinas, criticized by David Hume and defended by William

Paley—centers on the living world. The eye is like a telescope, telescopes have telescope makers, and therefore the eye must have an eye maker—the Great Optician in the sky, otherwise known as God. General opinion was that the nonorganic world does not show such tight and obvious design, and indeed the great philosopher of science Francis Bacon (also a Trinity man) had, in the early seventeenth century, quipped that the invocation of design—of final causes—in physics was akin to the status of the Roman Vestal Virgins—decorative but sterile (Bacon [1605] 1915, 97). How then could Whewell, who was certainly not about to deprecate the memory of Bacon, show God's attention and care in the realm of astronomy?

In part, Whewell did this by bringing astronomy around to the living world. He offered an anticipation of what today is known as the anthropic principle, the claim that, were the laws of nature any different from how they are, life would be impossible. Whewell pointed out that living things are dependent on the effects of the heavens in motion—the tides for instance, and the lengths of days and seasons. Were these not as they are, chaos and disaster would follow. Thus having detailed the need of living things for sleep, and sleep of a certain duration daily, Whewell asked: "Now how should such a reference be established at first in the constitution of man, animals, and plants, and transmitted in them from one generation to another?" Then he answered his own question: "If we suppose a wise and benevolent Creator, by whom all the parts were fitted to their uses and to each other, this is what we might expect and can understand. On any other such a supposition such a fact appears altogether incredible and inconceivable" (Whewell 1833, 41).

It is not for me to comment here on the validity of this inference, although I will note that at least one reader—a young Charles Darwin just entering what was to prove a distinguished scientific career—commented (in one of his very private notebooks) on the "arrogance" of Whewell's assumption that the world was made for man rather than man for the world. What I will remark is that, to his biological inferences, Whewell offered a supplementary argument, based on the very existence of law itself, which made much of our human trials here on Earth. These trials were apparently not just moral but intellectual, for it is our task to trace the motions of the heavens, whether or not they have any direct utilitarian purpose: "The contemplation of the material universe exhibits God to us as the author of the laws of material nature; bringing before us a wonderful spectacle, in the simplicity, the comprehensiveness, the mutual adaptation of these laws, and in the vast variety of harmonious and beneficial effects produced by their mutual bearing and combined operation" (Whewell 1833, 251).

The Anglican mind of William Whewell was very distant from the Presbyterian mind of Thomas Chalmers. The God of the Church of England was very different from the God who ruled north of the border in Scotland. For Whewell, God is a Greek intellectual—or, more probably, a sometime fellow of Trinity College—who thinks rationally and who expects His favored creation to do likewise. We are made in God's image to serve the end of producing good science. For Chalmers, God is a Jewish prophet, little interested in reason as such and waiting on our supplication for His grace. In the tradition of Calvin himself, Chalmers was interested in science, but not as training for our intellectual well being. Rather, the end of science is to attest to God's glory and magnificence.

In the early 1830s, however, Whewell was prepared to accept the Chalmersian conclusion about extraterrestrial life. He allowed that some stars may "have planets revolving round them, and these may, like our planet, be the seats of vegetable and animal and rational life" (Whewell 1833, 270). There was nothing odd about this concession. The oddity would have been if Whewell had argued otherwise, and yet this was precisely what he was to do some twenty years later! Two things in particular brought about the reversal. First was Whewell's own personal change of status and consequent redirection of interest. By the 1850s, he was no longer a mere tutor, textbook writer, and second-class researcher, but an important figure of influence and authority with responsibility for the moral and spiritual welfare of young people. Whewell had been writing more on ethics and standards of right behavior, and was much interested in educating the young. As an ordained member of a church institution, Whewell was pushed more toward issues of revealed religion rather than the purely natural variety. Immortal souls were at stake, including his own (as Whewell was growing older) and that of his wife (who was dying). For Whewell, the direct teachings and consolations of Christianity loomed larger.

The second factor changing Whewell's views was that, in the 1840s, Whewell and his circle had been shocked and battered by the anonymous publication of the evolutionary tract, *Vestiges of the Natural History of Creation,* by the Scottish publisher Robert Chambers. Starting with the so-called nebular hypothesis, a theory that the whole universe developed in an evolutionary fashion from clouds of gas, Chambers argued that the world of life is one of change and development. The cycle of life began in the world of chemicals, which produced primitive life forms, and so climbed progressively up the chain of being until finally it arrived at humankind. Indeed, the message of *Vestiges* was so wide and successful that it was taken up by none less than the poet and former student of Whewell, Alfred, Lord Tennyson, who made the message of upward evolution the saving theme at the conclusion of the poem *In Memoriam* (which in 1850 was to gain him the title of Poet Laureate).

Such a doctrine was anathema to Whewell who, although no biblical literalist, could not accept anything other than a miraculous origin for living beings, especially humanlike living beings. Collecting passages from his earlier writings (specifically his contribution to the *Bridgewater Treatises,* and the *History* and *Philosophy of the Inductive Sciences*), Whewell published a relatively small book, *Indications of the Creator,* which countered the vile heresies of *Vestiges.*

These two factors—change of Whewell's personal status and the threat that evolution posed—made Whewell radically rethink the plurality of worlds issue. On the one hand, there was the need to preserve the unique status of humankind. It is we here on Earth who have the special relationship with God, it is we who are made in His image, and it is for our sins alone that He died on the cross that we might be saved. The Crucifixion, the conquering of sin, and the Resurrection must be for our ends only. We cannot have Jesus dying on the cross sequentially through time and space. On the other hand, as Whewell saw it, a full-bloodied plurality begs for—practically necessitates—a background of evolutionism. Instead of organisms appearing providentially here on Earth—coming through the miraculous interventions of a benevolent Creator—they arrive apparently to order, whenever and wherever the space is available. If this is not to come by material law, in an evolutionary fashion, it is certainly to come by something suspiciously similar. Such a horrendous possibility must be stopped in its tracks.

Hence, never one to let a good thought go unpublished, in late 1853 Whewell put out a slim (for him) volume, in which he argued against the possibility of life elsewhere in the universe. Knowing that his position would be much contested, Whewell published this volume anonymously—it would be unseemly for one in his position to arouse public controversy. Unlike Chambers however, Whewell made little effort to conceal his authorship. Before long, it became public knowledge that Whewell had published his argument, a position which led one wag to quip that Whewell set out to prove that "through all infinity, there is none so great as the Master of Trinity" (Stephen 1921–22, 1370).

The Essay

Whewell's scientific biographer, Isaac Todhunter, called the *Essay* the "cleverest of the many works of the author," and the judgment continues to stand, even though it is the *History* and the *Philosophy* which are the major works which most justly burnish Whewell's reputation (Todhunter 1876, 100). The arguments are laid out clearly, in language which can be understood by the general reader, without in any sense trivializing their importance for the professional. First, Whewell sets up the problem, making much use of Chalmers's celebrated arguments. On the one hand, the evidence of modern science seems to point to a multiplicity of inhabited worlds throughout the universe—worlds which may even contain intelligent life. More than this, Whewell presents the Scotsman's ingenious argument that if one worried how God could care for so many different spheres of activity—spheres unknown to us—he should consider that the microscope shows us a whole dimension of life and action which was completely unknown until a century or two previous. Such discoveries "shew us that all the notions which our knowledge, hitherto, had enabled us to form of the powers and attributes of the Creator and Preserver of all living things, are vastly, are immeasurably below the real truth of the case" (Whewell 1853, 26).

On the other hand, this would all seem flatly irreconcilable with the special status of humankind in the Christian scheme of things. Man is a spiritual creature, with unique intellectual abilities, bearing the hope of eternal life. More than this, the doctrine of Christ's Atonement shows that this world of ours is no ordinary place. Whewell argued that, "The earth, thus selected as the theatre of such a scheme of Teaching and Redemption, can not, in the eyes of anyone who accepts this Christian faith, be regarded as being on a level with any other domiciles. It is the Stage of the Great Drama of God's Mercy and Man's Salvation;

the Sanctuary of the Universe; the Holy Land of Creation; the Royal Abode, for a time at least, of the Eternal King" (44). In short, things on Earth have to be special. Having stated the problem, Whewell moves to answer it in a three-stage argument.

First, he turns to geology. Although he was not a professional geologist, Whewell was well qualified for this task. From his early years he had taken a keen interest in movements in the science. Adam Sedgwick was a close friend and a fellow member of Trinity; Whewell had often reviewed major works in the subject (including Charles Lyell's seminal *Principles of Geology*); and at one point Whewell had even served as president of the Royal Geological Society in London. Simply put, Whewell noted that the history of this globe shows that God did not solely intend it to be, for each and every moment of time, the habitation of intelligent life. Our human existence is but an instant of the globe's full span. Why therefore should we expect, and insist, that humanlike life (or life of any kind) fill the rest of space? It shouldn't be needed, for the full history of our own earth proves a major exception:

> Let the difficulty be put in any way the objector pleases. Is it that it is unworthy of the greatness and majesty of God, according to our conceptions of Him, to bestow such peculiar care on so small a part of His creation? But, we know, from geology, that He has bestowed upon this small part of His creation, mankind, this special care;—He has made their period, though only a moment in the ages of animal life, the only period of intelligence, morality, religion. If then, to suppose that He has done this, is contrary to our conceptions of His greatness and majesty, it is plain that our conceptions are erroneous; they have taken a wrong direction. (103)

As Whewell stated in the *History,* geology says nothing, but it points upwards. Everything that geology teaches us is that man is something different from the creation which went before.

With our intelligence, our morality, our sociality, our spirituality, we are of an order different from any other creature. We are different in kind and not just dimension. Whatever you may think about other organisms, our origin seems to spell "miracle." The advent of humankind is out of the course of nature, and for that reason alone there is simply no ground for expecting such events to be repeated elsewhere in the universe.

Second, there is the argument from astronomy. Whewell gives a detailed rebuttal to all of those (including Chalmers, of course) who think that there really is an analogy between our Earth and the rest of creation, or that modern thinking on developments and causes points to repetition of Earth-life on other potential domains of existence. For a start, Whewell denied that nebulae are clusters of stars, possible places of habitation. As like as not, they are simply massive clouds of gas, unfit for supporting life. It is doubtful that the stars themselves are suns like our own, and certainly doubtful that they are suns with their own complete planetary systems. And even if such systems do exist, the knowledge that we have of the planets of our own system make it highly unlikely that there is anything living on their surfaces. Jupiter appeared to be little more than solid ice and Saturn not much better, and of course, being so big, gravity would be colossal on these planets. "For such reasons, then, as were urged in the case of Jupiter, we must either suppose that it has no inhabitants; or that they are aqueous, gelatinous creatures; too sluggish, almost, to be deemed alive, floating in their ice-cold waters, shrouded for ever by their humid skies" (185–86).

Third, Whewell turns to theological and philosophical arguments. The fullest of these, given in the *Essay* as it was actually published, focused on the argument from design, and for a number of reasons, both internal and external, it is a most interesting chapter. Whewell, as a Kantian, had always stressed the teleological nature of the living world—the fact that organisms' parts are adapted toward specific ends. In the *Philosophy*, Whewell followed Kant so far as to argue that a necessary condition of studying biology is to assume a teleological stance. Final causes are not optional. The ground for this position had been prepared in the *History,* where the merits were sung of the great French biologist Georges Cuvier, who made much of such causes in his science. Whewell leveled much of his criticism against those who downplayed final causes in favor of other factors, notably in favor of the isomorphisms between organisms, phenomena now known as homologies. Ostensibly the main focus of Whewell's attack was Cuvier's fellow French biologist, Geoffroy Saint-Hilaire, but truly aimed for were the German morphologists, the *Naturphilosophen,* who made much of homology and bound it all up in a developmental, quasi-evolutionary picture. It did not help that such a view had been endorsed by the German idealist philosopher G. F. W. Hegel, already in bad odor for his criticism of Newton.

But even in Britain homology could not be denied, and in the 1840s it was championed by the leading anatomist Richard Owen, a close friend and former schoolmate of Whewell. In 1848 Owen developed the notion of the archetype, a kind of blueprint or pattern on which groups of organisms were formed, and pressed on Whewell the necessity of considering the repeated patterns—patterns without any direct utilitarian function—that one finds throughout living nature. The forelimb of the horse, the arm of man, the paw of the mole, the flipper of the whale, the wing of the bird, are all used for very different ends but united by a shared pattern of bone structure. Turning this to good account, Whewell in the *Essay* makes much of such exceptions to direct function and suggests that there is no longer the need to assume that all of astronomical creation has a direct function. In a somewhat flowery metaphor, Whewell writes:

> Instead of manufacturing a multitude of worlds on patterns more or less similar, He has been employed in one great work, which we cannot call imperfect, since it includes and suggests

> all that we can conceive of perfection. It may be that all of the other bodies, which we can discover in the universe, show the greatness of this work, and are rolled into forms of symmetry and order, into masses of light and splendour, by the vast whirl which the original creative energy imparted to the luminous element. The planets and the stars are the lumps which have flown from the potter's wheel of the Great Worker;—the shred-coils which in the working, sprang from His mighty lathe;—the sparks which darted from His awful anvil when the solar system lay incandescent thereon;—the curls of vapour which arose from the great cauldron of creation when its elements were separated. (243)

But there are dangers lurking here, and Whewell knew that full well. If homologies do not exist for the purpose of direct function—what biologists speak of as "adaptive" ends—then what does one say about their existence, and indeed, about other organic nonfunctional phenomena like rudimentary nipples on males? Clearly they were the product of God's laws, and as we shall see in a moment, Whewell could turn this to his advantage. What Whewell knew only too well, however, was that the producing laws might be taken to include laws of development and evolution. After all, this was precisely the move made by Chambers in *Vestiges* and earlier by Geoffroy, and this was certainly hinted at by the *Naturphilosophen*. Nonfunctional homology points to common origins, and how more easily can one get from these origins than by evolutionary change? Indeed, Owen himself was running into trouble from vigilantes like Sedgwick for skating too close to transmutationary speculations. (In fact, Owen was privately moving towards evolution, and had sent a letter of praise to Chambers.) Whewell had to stop this line of argument dead in its tracks:

> The evidences of design in the anatomy of man are not less striking than they were, when no such gradation was thought of. And what is more to the purpose of our argument, the evidences of the peculiar nature and destination of man, as

> shown in other characters than his anatomy,—his moral and intellectual nature, his history and capacities,—stand where they stood before; nor is the vast chasm which separates man, as a being with such characters as these latter . . . at all filled up or bridged over. (216–17)

The workings of law—the task set by Whewell's Anglican God for men of intelligence and diligence—had, of course, been the argument of the *Bridgewater Treatises.* Twenty years later, God again makes His appearance, as Whewell turns now to argue that law itself is a mark of God's activity and magnificence, and that the discernment of this law is our task here on earth. As it happens, the argument of law as it appears in the *Essay* is somewhat truncated, for while he was reading proof Whewell was sending the material to his friend Sir James Stephen (Regius Professor of Modern History at Cambridge), and on Stephen's advice Whewell compounded five of the theological/philosophical chapters into one. These five chapters are now published in their original form, for the first time ever, in this present volume. From a literary point of view, this editorial advice was probably sound, and the chief lines of the argument from law are not lost altogether. As the original, extended version shows even more clearly, however, Whewell takes the position that because the universe works according to unbroken law, this is all that is needed to justify God's actions, and (at least as important to us) to explain how we humans are the favored creation of God:

> If God have placed upon the earth a creature who can so far sympathize with Him, if we may venture upon the expression;—who can raise his intellect into some accordance with the Creative Intellect; and that, not once only, nor by a few steps, but through an indefinite gradation of discoveries, more and more comprehensive, more and more profound; each, an advance, however slight, towards a Divine Insight;—then, so far as intellect alone (and we are here speaking of intellect alone) can make Man a worthy object of all the vast

> magnificence of Creative Power, we can hardly shrink from believing that he is so. (253)

And so with a brief concluding chapter about the future, and the need of man as a social being to form international societies (Whewell's will provided for a professorship to promote global harmony), the *Essay* comes to an end.

Responses

Sir James Stephen put his finger right on the big problem: Say what you like, was it truly plausible that so vast a universe should have been created empty, except for one scrap of dust off the center? He did not think so:

> I am almost afraid to add that your conclusion has one characteristic from which many will shrink;—myself among the number. Can it really be that this world is the best product of omnipotence, guided by omniscience, and animated by Love?—that the Deity has called into existence one race of rational Beings only, and that one race corrupt from the very dawn of its appearance?—that of this solitary family "many are called but few are chosen"?—that for the vast majority of them, as far as we can judge, it had been infinitely better that they had never been born?—that this is not an exceptional case, a dark spot on the fair face of creation, but the single spot from which can arise any thoughts or any affections tending towards the creator as their object and their end? that the measureless shoreless ocean of space enveloping us, embraces no one world sinless, wise, holy, happy? no region in which the all seeing eye can rest with complacency, or of which the Divine voice can once again declare that "it is very good"? (Stephen, J., 1853)

Stephen was a friendly critic but he was perceptive. He knew where the objections would come and he was not wrong. The *Essay* occasioned a huge controversy, the biggest scandal between Chambers's *Vestiges* of 1844 and Darwin's *Origin of Species* of 1859. By the time Darwin published the *Origin,* there had been over twenty books and at least fifty articles and reviews responding to Whewell's *Essay*—and while not all of them were hostile, most were opposed to Whewell's conclusions. As Charlotte Stanhope knew full well, Sir David Brewster led the list. He and Whewell had been old foes ever since Brewster wrote critical reviews of the *Bridgewater Treatises,* the *History,* and the *Philosophy* decades earlier. There were many reasons for their differences: from science (Brewster backed the Newtonian corpuscular theory of light, whereas Whewell boosted the newly favored wave theory), through nationality (Brewster thought Whewell's writings belittled Scotsmen), to religion (Whewell the Anglican pitted against Brewster the Presbyterian). Ultimately, it was a clash of personalities. Brewster was thin skinned and thinking himself always maligned. Whewell tended to be pompous and bullying despite, or perhaps because of, his own humble origins.

Brewster tore into Whewell with a long, critical article in the *North British Review,* and followed up with a full-length book with the glorious title, *More Worlds than One: The Creed of the Philosopher and the Hope of the Christian.* Brewster left nothing in his universe to chance and left no possible world unpopulated. You might think it would be enough for the sun and the moon to exist for the sake of Earth's denizens, but not Brewster! Where Kant saw possible forms of life as becoming more intelligent the further out they live in the solar system, Brewster believed the opposite, with the most intelligent beings found closer to the sun and the "highest orders of intelligence" found on the sun.

Design, purpose, and function—that is the story of creation, and Brewster preached it at the top of his lungs:

> The chariots of flame and the horses of fire that bore Elijah from his star of earth, and surrounded Elisha in the mountains of Syria, and the wheels of amber and of fire that

> were exhibited to the captive prophet on the banks of the Chebar, become, in the poet's eye, the vehicle from planet to planet, and from star to star, in which the heavenly host is to survey the wonders and glories of the Universe. (1854, 261)

Like Chalmers before him, the very insignificance of man was the glorious starting point of Brewster's Calvinist theology. We are as nothing and deserve nothing—that God will extend His grace to such worms as us is in itself the greatest mark of His love and care.

No one else was quite so vitriolic and unbalanced as Brewster (who, if a psychological answer is needed, was going through a time of doubt and depression following the death of his wife). Not much sympathy was shown, though, by others for Whewell's position, and few of Whewell's arguments were left untouched. Yet, as is often the case, one learns more about the critics themselves from their objections than the critics might suspect. Thomas Henry Huxley, for instance, who was soon to be renowned as Darwin's great champion (his "bulldog," as it were), was still a very junior scientist making his way by attacking the high and mighty church authorities, clearing the ground for a secular scientific community. After savaging Richard Owen, Huxley turned to Whewell. Huxley was less interested in addressing Whewell's arguments than he was to show that Whewell had misquoted sources, including Owen himself. If Huxley could sow discord among his opponents, so much the better (Huxley 1854a, b). The Reverend Baden Powell—Oxford professor and liberal church man—was a similar critic, though perhaps more striking because he wrote from within the Anglican community. Like Huxley, he had little time for the human-centered miracle-monger of Whewell or the stern grace-and-favor God of Brewster. Unlike Huxley, however, Powell did not see this as a victory for the secular world. In his 1855 *Essays on the Spirit of Inductive Philosophy,* Powell argued that God's greatness was revealed by the extent to which He could work through unbroken law and, already sympathetic to evolution (he was to praise the *Origin* as soon as it appeared), Powell tied the plurality issue straight into the lawbound, developmental nature of the whole universe:

> All astronomical presumption, taking the truths of geology into account, seems to be in favour of progressive order, advancing from the inorganic to the organic, and from the insensible up to the intellectual and moral in all parts of the material world alike, though not necessarily in all at the same time or with the same rapidity; in some worlds one stage being reached, while in others only a comparatively small advance may have been made. (230)

The support was for Brewster's side of the case but at a cost that the Scotsman would surely not have appreciated.

Whewell was never a man to let a slight go unnoticed or a criticism unanswered. To the second edition of the *Essay* he added a dialogue (reprinted here) in which he took on his critics (who are identified and listed therein). One senses that, apart from Brewster, every critic was having a really good time. The topic was interesting, the underlying issues were important, resolution was impossible, and no quarter was given or expected in the struggle back and forth. Because, officially, the *Essay* was anonymously published, no real personal offence could be taken. Critics could be as pugnacious as they wished, and Whewell could respond with the vigor he had shown in his schooldays when, in playground fights, fairness decreed that only two opponents could take on the burly Whewell at any one time, but common sense decreed that it would be rank foolishness if less than two took on Whewell at once.

Of course, we today have the benefit of hindsight, knowing that Darwin's *Origin* was to appear at the end of the decade. Whewell did not write the *Essay* with the *Origin* in mind, and we should not read it (or the criticisms) in this way. However, we certainly can and should look at the reaction to the *Origin* in light of the controversy that Whewell occasioned, and there is surely one major conclusion that can be drawn (Ruse 1979). Darwin's ideas had much success, and in both the scientific community and the general public, evolution conquered practically overnight. Moreover, although the mechanism of natural selection was less popularly accepted, there was growing agreement that the causes of change had to be natural in some wise.

It would be ungracious to pretend that the *Origin*'s success was not due in major part to Darwin's own genius, as well as to the network of friends and supporters Darwin had gathered around him by the time he published. The *Origin*'s success, however, was also due to the troubles among the other side, if somewhat tendentiously we may so characterize the theological and other opponents of evolutionism. Revealed theology, natural theology, metaphysics, philosophy, anatomy, astronomy, and more had become an ever-increasingly uncomfortable melange, and Whewell's *Essay* and the reactions it provoked show this. If you promoted revealed theology, then you threatened natural theology. If you stood by natural theology, then you opened the way for evolution. If you denied evolution, then you were back to excusing God for His "pointless" creation. And so the dreadful circle continued. Every time you patched up one point, another piece fell out. If you downplayed adaptation, then homology—the best support for evolution—became important. If you belittled homology, you went against the cutting edge of modern science. If you tried to mediate, you would find yourself promoting uncomfortably idealistic notions like Owen's archetypes. Without pretending that the Darwinian revolution fits nicely the somewhat abrupt nature of scientific change pictured in Thomas Kuhn's aforementioned *Structure of Scientific Revolutions* (and I do not think it does), Kuhn is surely right when he draws one of his major messages from the history of science: change comes from the collapse of the old as well as of the success of the new.

Science is like love: you do not start looking for something new until you are tired of what you have.

The Evolutionary Paradox

Whewell *malgré lui* was preparing the way for the *Origin*. What Whewell could not know, and perhaps for his peace of mind it is just as well, was that although he made few converts at the time of his writing of the *Essay*, posterity has given him supporters that would have surprised (and shocked) him as much as they would have surprised (and shocked) his critics. For all that Whewell feared that a plurality of worlds was the other side of the metaphysical coin to evolution—an assumption we have just seen reinforced and confirmed by Baden Powell—in the past century and a half, down to and including the present, it has been those very evolutionists who have almost alone agreed with Whewell that life, intelligent life certainly, is not probable through the universe!

I need hardly say how much of a paradox lies in this support. The twentieth century saw virtually unbroken enthusiasm for the idea of extraterrestrials (Dick 1996). As distant world after distant world was uncovered by ever more powerful methods of investigation, the fullness of creation has become more and more apparent. Galaxies, stars, and even planetary systems—these are the stuff of existence. Surely, many people think, in the vastness of space, on some of the billions and billions of extraterrestrial domains, life has begun and grown and flourished. There has to be another Shakespeare out there, another Einstein, another Mother Teresa, and alas another Hitler. Science tells us this, and one novel after another, one blockbuster film following the next, makes the imagination of Bernard de Fontenelle look positively impoverished. *Close Encounters of the Third Kind*, *ET*, and *Star Wars* are all fiction, but surely there has to be something to some of this. Certainly, government agencies have swallowed the line. Huge sums are spent on sending probes into space, on testing for life in our solar system, on listening for words from without. The failure to uncover a single thing is usually a spur to renewed effort and yet more demands on the taxpayer.

The naysayers are those very people Whewell most feared would enthusiastically embrace a plurality and in turn use it as support for their own vile doctrines—the evolutionists! This is not to say that these supporters have shared Whewell's theological concerns. Modern evolutionists are often indifferent (if not outright hostile) to religion and its foundations, although in the early twentieth century Alfred Russel Wallace, the codiscoverer with Darwin of the mechanism of natural selection, argued for the uniqueness of planet Earth. He did so because, as a spiritualist, he found a much-populated universe to be theologically offensive. Evolutionists of the twentieth century (and continuing on into the twenty-first century), however, have their own secular reasons to doubt the existence of beings—intelligent beings certainly—in outer space (Ruse 2001). Understanding a "humanoid" to be an intelligent humanlike being, the paleontologist G. G. Simpson authored a classic exercise in scepticism. Down the line, his attitude was extremely negative: "There are four successive probabilities to be judged: the probability that suitable planets do exist; the probability that life has arisen on them; the probability that life has evolved in a predictable way; and the probability that such evolution would lead eventually to humanoids. . . ." (Simpson 1964, 258–59). Simpson argued that the first probability is fair, the second much lower although still possible, the third vanishingly small, and the fourth effectively zero. "Each of these probabilities depends on that preceding it, so that they must be multiplied together to obtain the probability of the over-all probability of the final event, the emergence of humanoids. The product of these probabilities, each a fraction, is probably not significantly greater than zero" (Simpson 1964, 258–59).

What made Simpson—and indeed all true followers of Darwin—different from the evolutionists of Whewell's day, and different from the evolution that Whewell himself feared, was his denial of any built-in progressionism driving evolution upwards. Chambers, to take the specific example that Whewell abominated most, thought that once you start on the process of developmental change through time, the human or something humanlike was bound to emerge from so upwardly progressive a process. For the Darwinian, however, depending on a relativistic mechanism like natural selection, one which is working on the random changes of genes (today identified with strips of molecular DNA), it is chance and contingency all the way. You simply cannot bank on humans or humanlike organisms coming out at the end. Simpson observed, "Both the course followed by evolution and its processes clearly show that evolution is not repeatable. No species or any larger group has ever evolved, or can ever evolve, twice. Dinosaurs are gone forever. Nothing very like them occurred before them or will occur after them" (Simpson 1964, 267). Simpson saw this as absolutely true of humans also. We have arrived here on Earth. There is absolutely no reason to think—and many reasons not to think—that humans will ever appear again on earth or on any other planets:

> There is a more or less random element in evolution involved in mutation and recombination, which are stochastic, technically speaking. Repetition is virtually impossible for nonrandom actions of selection on what is there in populations. It becomes still less probable when one considers that duplication of what are, in a manner of speaking, accidents is also required. This essential nonrepeatability of evolution on earth obviously has a decisive bearing on the chances that it has been repeated or closely paralleled on any other planet. (267)

Humans simply will not be repeated, nor did Simpson think it very likely that humans will continue to evolve for us to expect there to be superintelligent humanoids throughout the universe. Using a version of the argument which haunted many an evolutionist, Simpson pointed out that the poor and stupid tend to breed and have large families, whereas the rich and intelligent practice birth control and hence (in a way highly counterproductive from a Darwinian perspective) have much smaller numbers of offspring. Superintelligence is possible, but our experience is that it is not likely. "Future evolution could raise man to superb heights as yet hardly glimpsed, but it will not automatically do so. As far as can now be foreseen, evolutionary degeneration is at least as likely in our future as is further progress" (Simpson 1964, 285).

One has to be careful about just how far one is to take this denial as absolute. No one yet has shown exactly how life might have arisen from nonlife, ultimately from inert chemicals. But there is little doubt in the minds of today's evolutionists that such an event did occur. Moreover, this event was natural in the sense of being ruled by the laws of physics and chemistry, nothing else. Likewise, few would absolutely deny that there may be some form of life in other parts of the universe, perhaps even fairly complex forms of life. If sexuality evolved once, perhaps it evolved repeatedly. After all, something as clever as the eye seems to have come about many times independently. As Simon Conway Morris states in his *The Crucible of Creation:*

> The point I wish to stress is that again and again we have evidence of biological form stumbling on the same solution to a problem. Consider animals that swim in water. It turns out that there are only a few fundamental methods of propulsion. It hardly matters if we choose to illustrate the method of swimming by reference to water beetles, pelagic snails, squid, fish, newts, ichthyosaurs, snakes, lizards, turtles, dugongs, or whales; we shall find that the style in which the given animal moves through the water will fall into one of only a few basic categories. (Conway Morris 1998, 204–5)

But how much further one can go is the big question. Can one apply this necessitarian line of argument to thinking and to culture and all of the things we take to be distinctively human? Some, especially those who have roots in a pre-Darwinian, nonselective form of evolutionism think it may be possible. Paleontologist Stephen Jay Gould claims, "I can present a good argument from 'evolutionary theory' against the repetition of anything like a human body elsewhere; I cannot extend it to the general proposition that intelligence in some form might pervade the universe" (Dick 1996, 395). Others are far from sure. Natural selection just does not work that way. They would very much doubt that anything to be encountered would strike us as particularly intelligent. To put matters in terms that Whewell would understand, conventional Darwinians would think it unlikely that one could find creatures anywhere else in the universe made in the image of God for whom the sacrifice of Jesus on the cross—here on Earth or elsewhere in the universe—was needed to ensure their eternal salvation.

None of this makes Whewell right—or wrong, for that matter. In a way, it is probably a mistake to insist on answers to questions about truth and falsity. Whewell was writing at another time, in another context. He was not a twenty-first-century evolutionist wrestling with the confident assertions of astronomers. He was a mid-nineteenth-century Victorian wrestling with issues of faith, in the light of the modern science of his day. Let us read him in this way—learning where we can, criticizing where we think appropriate, and for the rest, simply enjoying a rattling good argument. Why should we have less fun than Whewell, his critics, and the fictional characters of Trollope's novel?

Bibliography

Bacon, F. [1605] 1915. *The advancement of learning.* London: Everyman.

Brewster, D. 1854. *More worlds than one: The creed of the philosopher and the hope of the Christian.* London: Camden Hotten.

Chalmers, T. [1817] 1906. *A series of discourses on the Christian revela-*

tion, viewed in connection with the modern astronomy. New York: American Tract Society.

Chambers, R. 1844. *Vestiges of the natural history of creation.* London: Churchill.

Conway Morris, S. 1998. *The crucible of creation: The Burgess shale and the rise of animals.* Oxford: Oxford University Press.

Crowe, M. J. 1986. *The extraterrestrial life debate, 1750–1900: The idea of a plurality of worlds from Kant to Lowell.* Cambridge: Cambridge University Press.

Dick, S. J. 1982. *Plurality of worlds: The origins of the extraterrestrial life debate from Democritus to Kant.* Cambridge: Cambridge University Press.

———. 1996. *The biological universe: The twentieth-century extraterrestrial life debate and the limits of science.* Cambridge: Cambridge University Press.

Douglas, S., ed. 1881. *The life and selections from the correspondence of William Whewell, D.D.* London: Kegan Paul.

Huxley, T. H. 1854a. Review of William Whewell's "Plurality of Worlds." *Westminster Review* 61: 313–15.

———. 1854b. Review of David Brewster's "More Worlds than One." *Westminster Review* 62: 242–46.

Kant, I. [1755] 1981. *Universal natural history and theory of the heavens.* Translated by Stanley L. Jaki. Edinburgh: Edinburgh University Press.

Kepler, J. [1610] 1965. *Kepler's conversation with Galileo's sidereal messenger.* Translated by Edward Rosen. New York and London: Johnson Reprint Corp.

Marsak, Leonard M., ed. 1970. *The achievement of Bernard le Bovier de Fontanelle.* New York: Johnson Reprint Corp.

Owen, R. 1848. *On the archetype and homologies of the vertebrate skeleton.* London: Voorst.

———. 1849. *On the nature of limbs.* London: Voorst.

Powell, B. 1855. *Essays on the spirit of the inductive philosophy.* London: Longman, Brown, Green, and Longmans.

Ruse, M. 1979. *The Darwinian revolution: Science red in tooth and claw.* Chicago: University of Chicago Press.

———. 1991. William Whewell: Omniscientist. In *William Whewell: A composite portrait,* edited by Menachem Fisch and Simon Schaffer. Oxford: Oxford University Press.

———. 2001. *Can a Darwinian be a Christian? The relationship between science and religion.* Cambridge: Cambridge University Press.

Simpson, G. G. 1964. *This view of life.* New York: Harcourt, Brace, and World.

Stephen, J. 1853. Letter to William Whewell, 15 October.
Stephen, L. 1921–22. "William Whewell." *Dictionary of national biography* XX. London: Oxford University Press.
Todhunter, I. 1876. *William Whewell, D.D. Master of Trinity College Cambridge: An account of his writings with selections from his literary and scientific correspondence.* London: Macmillan.
Trollope, Anthony. 1994. *Barchester Towers.* New York: Penguin Classics: 160–61.
Wallace, A. R. 1903. *Man's place in the universe.* London: Chapman and Hall.
Whewell, W. 1833. *Astronomy and general physics* (*Bridgewater Treatises,* 3). London: William Pickering.
———. 1837. *The history of the inductive sciences.* London: Parker.
———. 1840. *The philosophy of the inductive sciences.* London: Parker.
———. 1845. *Indications of the Creator.* London: Parker.
———. 1853. *Of the plurality of worlds: An essay.* London: Parker.

PART ONE

Of the Plurality of Worlds: An Essay

WILLIAM WHEWELL

First edition, 1853

51 MESSIER

H. Adlard, sc.

99 MESSIER

OF

THE PLURALITY OF WORLDS:

An Essay.

On Nature's Alps I stand,
And see a thousand firmaments beneath!
A thousand systems, as a thousand grains!
So much a stranger, *and so late arrived,*
How shall man's curious spirit not inquire
What are the natives of this world sublime,
Of this so distant, unterrestrial sphere,
Where mortal, untranslated, never stray'd?

NIGHT THOUGHTS.

LONDON:
JOHN W. PARKER AND SON,
WEST STRAND.

1853.

PREFACE.

ALTHOUGH the opinions presented in the following Essay are put forwards without claiming for them any value beyond what they may derive from the arguments there offered, they are not published without some fear of giving offense. It will be a curious, but not a very wonderful event, if it should now be deemed as blameable to doubt the existence of inhabitants of the Planets and Stars, as, three centuries ago, it was held heretical to teach that doctrine. Yet probably there are many who will be willing to see the question examined by all the light which modern science can throw upon it; and such an examination can be undertaken to no purpose, except the view which has of late been generally rejected have the arguments in its favour fairly stated and candidly considered.

Though Revealed Religion contains no doctrine relative to the inhabitants of planets and stars; and though, till within the last three centuries, no Christian thinker deemed such a doctrine to be required, in order to complete our view of the attributes of the Creator; yet it is possible that at the present day, when the assumption of such inhabitants is very generally made and assented to, many persons have so mingled this assumption with their religious belief, that they regard it as an essential part of Natural Religion. If any such persons find their religious convictions interfered with, and their consolatory impressions disturbed, by what is said in this Essay, the Author will deeply regret to have had any share in troubling any current of pious thought belonging to the time. But, as some excuse, it may be recollected, that if such considerations had prevailed, this very doctrine, of the Plurality of Worlds, would never have been publicly maintained. And if such considerations are to have weight, it must be recollected, on the other hand, that there are many persons to whom the assumption of an endless multitude of Worlds appears difficult to reconcile with the belief of that which, as the Christian Revelation teaches us, has been done for this our World of Earth. In this conflict of religious difficulties, on a point which rather belongs to science than to religion, perhaps philosophical arguments may be patiently listened to, if urged as arguments merely: and in that hope, they are here stated, without reserve and without exaggeration.

All speculations on subjects in which Science and Religion bear upon each other, are liable to one of the two opposite charges;—that the speculator sets Philosophy and Religion at variance; or that he warps Philosophy into a conformity with Religion. It is confidently hoped that no candid reader will bring either of these charges against the present Essay. With regard to the latter, the arguments must speak for themselves. To the Author at least, they appear to be of no small philosophical force; though he is quite ready to weigh carefully and candidly any answers which may be offered to them. With regard to the amount of agreement between our Philosophy and Religion, it may perhaps be permitted to the Author to say, that while it appears to him that some of his philosophical conclusions fall in very remarkably with certain points of religious doctrine, he is well aware that Philosophy alone can do little in providing man with the consolations, hopes, supports, and convictions which Religion offers; and he acknowledges it, as a ground of deep gratitude to the Author of all good, that man is not left to Philosophy for those blessings; but has a fuller assurance of them, by a more direct communication from Him.

Perhaps, too, the Author may be allowed to say, that he has tried to give to the book, not only a moral, but a scientific interest; by collecting his scientific facts from the best authorities, and the most recent discoveries. He would flatter himself, in particular, that the view of the Nebulæ and of the Solar System, which he has here given, may be not unworthy of some attention on the part of astronomers and observers, as an occasion of future researches in the skies.

CONTENTS

OF THE PLURALITY OF WORLDS.

OF THE PLURALITY OF WORLDS.

CHAPTER I.

Astronomical Discoveries.

'WHEN I consider the heavens, the work of thy fingers, the moon and the stars, which thou hast ordained; What is man, that thou art mindful of him? and the son of man, that thou visitest him?'

1 These striking words of the Hebrew Psalmist have been made, by an eloquent and pious writer of our own time, the starting point of a remarkable train of speculation. Dr Chalmers, in his *Astronomical Discourses*, has treated the reflexion thus suggested, in connexion with such an aspect of the heavens and the stars, the earth and the universe, as modern astronomy presents to us. Even from the point of view in which the ancient Hebrew looked at the stars; seeing only their number and splendour, their lofty position, and the vast space which they visibly occupy in the sky; compared with the earth, which lies dark, and mean, and perhaps small in extent, far beneath them, and on which man has his habitation; it appeared wonderful, and scarcely credible, that the maker of all that array of luminaries, the lord of that wide and magnificent domain, should occupy himself with the concerns of men: and yet, without a belief in His fatherly care and goodness to us, thoughtful and religious persons, accustomed to turn their minds constantly to a Supreme

Governor and constant Benefactor, are left in a desolate and bewildered state of feeling. The notion that while the heavens are the work of God's fingers, the sun, moon, and stars ordained by him, He is *not* mindful of man, does not regard him, does not visit him, was not tolerable to the thought of the Psalmist. While we read, we are sure that he believed that, however insignificant and mean man might be, in comparison with the other works of God,—however difficult it might seem to conceive, that he should be found worthy the regards and the visits of the Creator of All,—yet that God *was* mindful of him, and *did* visit him. The question, 'What is man, that this is so?' implies that there is an answer, whether man can discover it or not. '*What* is man, that God is mindful of him?' indicates a belief, unshaken, however much perplexed, that man is *something*, of such a kind that God *is* mindful of him.

2 But if there was room for this questioning, and cause for this perplexity, to a contemplative person, who looked at the skies, with that belief concerning the stars, which the ancient Hebrew possessed, the question recurs with far greater force, and the perplexity is immeasurably increased, by the knowledge, concerning the stars, which is given to us by the discoveries of modern astronomy. The Jew probably believed the earth to be a region, upon the whole, level, however diversified with hills and valleys, and the skies to be a vault arched over this level;—a firmament in which the moon and the stars were placed. What magnitude to assign to this vault, he had no means of knowing; and indeed, the very aspect of the nocturnal heavens, with the multitude of stars, of various brightness, which come into view, one set after another, as the light of day dies away, suggests rather the notion of their being scattered through a vast depth of space, at various distances, than of their being so many lights fastened to a single vaulted surface. But however he

44|

might judge of this, he regarded them as placed in a space, of which the earth was the central region. The host of heaven all had reference to the earth. The sun and the moon were there, in order to give light to it, by day and by night. And if the stars had not that for their principal office, as indeed the amount of light which they gave was not such as to encourage such a belief,—and perhaps the perception, that the stars must have been created for some other object than to give light to man, was one of the principal circumstances which suggested the train of thought that we are now considering;—yet still, the region of the stars had the earth for its center and base. Perhaps the Psalmist, at a subsequent period of his contemplations, when he was pondering the reflexions which he has expressed in this passage, might have been led to think that the stars were placed there in order to draw man's thoughts to the greatness of the Creator of all things; to give some light to his mental, rather than to his bodily eye; to shew how far His mode of working transcends man's faculties; to suggest that there are things in heaven, very different from the things which are on earth. If he thought thus, he was only following a train of thought on which contemplative minds, in all ages and countries, have often dwelt; and which we cannot, even now, pronounce to be either unfounded or exhausted; as we trust hereafter to shew. But whether or not this be so, we may be certain that the Psalmist regarded the stars, as things having a reference to the earth, and yet not resembling the earth; as works of God's fingers, very different from the earth with its tribes of inhabitants; as luminaries, not worlds. In the feeling of awe and perplexity, which made him ask, 'What is man that thou art mindful of him?' there was no mixture of a persuasion that there were, in those luminaries, creatures, like man, the children and subjects of God; and therefore, like man, requiring his care and attention. In asking, 'What is man, that thou visitest him?' there was no latent comparison, to make the question imply, 'that thou visitest *him*, rather than those who dwell in those abodes?' It was the multitude and magnificence of God's works, which made it seem strange that he should care for a *thing* so small and mean as man; not the supposed multitude of God's intelligent creatures inhabiting those works, which made it seem strange that he should attend to every *person* upon this earth. It was not that the Psalmist thought that, among a multitude of earths, all peopled like this earth, man might seem to be in danger of being overlooked and neglected by his Maker; but that, there being only one earth, occupied by frail, feeble, sinful, shortlived creatures, it might be unworthy the regards of Him who dwelt in regions of eternal light and splendour, unsullied by frailty, inaccessible to corruption.

3 This, we can have no doubt, or something resembling this, was the Psalmist's view, when he made the reflexion, which we have taken as the basis of our remarks. And even in this view, (which, after all that science has done, is perhaps still the most natural and familiar,) the reflexion is extremely striking; and the words cannot be uttered without finding an echo in the breast of every contemplative and religious person. But this view is, as most readers at this time are aware, very different from that presented to us by Modern Astronomy. The discoveries made by astronomers are supposed by most persons to have proved, or to have made it in the highest degree probable, that this view of the earth, as the sole habitation of intelligent subjects of God's government; and of the stars, as placed in a region of which the earth is the centre, and yet differing in their nature from this lower world; is altogether erroneous. According to astronomers, the earth is not a level space, but 46| a globe. Some of the stars which we see in the vault of

heaven, are globes, like it; some smaller than the earth, some larger. There are reasons, drawn from analogy, for believing that these globes, the other planets, are inhabited by living creatures, as the earth is. The earth is not at rest, with the celestial luminaries circulating above it, as the ancients believed, but itself moves in a circle about the sun, in the course of every year; and the other planets also move round the sun in like manner, in circles, some within and some without that which the earth describes. This collection of planets, thus circulating about the sun, is the SOLAR SYSTEM: of which the earth thus forms a very small part. Jupiter and Saturn are much larger than the earth. Mars and Venus are nearly as large. If these be inhabited, as the Earth is, which the analogy of their form, movements and conditions, seems to suggest, the population of the earth is a very small portion of the population of the solar system. And if the mere number of the subjects of God's government could produce any difficulty in the application of his providence to them, a person to whom this view of the world which we inhabit had been disclosed, might well, and with far more reason than the Psalmist, exclaim, 'Lord, what is man, that thou art mindful of him? the inhabitant of this Earth, that thou regardest him?'

4 But this is only the first step in the asserted revelations of astronomy. Some of the stars, are, as we have said, planets of the kind just described. But these stars are a few only:—five, or at most six, of those visible to the unassisted eye of man. All the rest, innumerable as they appear, and numerous as they really are, are, it is found, objects of another kind. They are not, as the planets are, opaque globes, deriving their light from a sun, about which they circulate. They shine by a light of their own. They are of the nature of the sun, not of the planets. That they appear mere specks of light, arises from their being at a vast

distance from us. At a vast distance they undoubtedly are; for even with our most powerful telescopes, they still appear mere specks of light;—mere luminous points. They do not, as the planets do, when seen through telescopes, exhibit to us a circular face or disk, capable of being magnified and distinguished into parts and features. But this impossibility of magnifying them by means of telescopes, does not at all make us doubt that they may be far larger than the planets. For we know, from other sources of information, that their distance is immensely greater than that of any of the planets. We can measure the bodies of the solar system;—the earth, by absolutely going round a part of it, or in other ways; the other bodies of the system, by comparing their positions, as seen from different parts of the earth. In this manner we find that the earth is a globe 8000 miles in diameter. In this way, again, we find that the circle which the earth describes round the sun has, in round numbers, a radius about 24,000 times the earth's radius; that is, nearly a hundred millions of miles. The earth is, at one time, a hundred millions of miles on one side of the sun; and at another time, half a year afterwards, a hundred millions of miles on the other side. Of the bright stars which shine by their own light,—the *fixed stars*, as we call them, (to distinguish them from the planets, the *wandering stars*,)—if any one were at any moderate distance from us, we should see it change its apparent place with regard to the others, in consequence of our thus changing our point of view two hundred millions of miles: just as a distant spire changes its apparent place with regard to the more distant mountain, when we move from one window of our house to the other. But no such change of place is discernible in any of the fixed stars: or at least, if we believe the most recent asserted discoveries of astronomers, the 48| change is so small as to imply a distance in the star, of more than two hundred thousand times the radius of the earth's orbit, which is, itself, as we have said, one hundred millions of miles*. This distance is so vastly great, that we can very well believe that the fixed stars, though to our best telescopes they appear only as points of light, are really as large as our sun, and would give as much light as he does, if we could approach as near to them. For since they are thus, the nearest of them, two hundred thousand times as far off as he is, even if we could magnify them a thousand times, which we can hardly do, they would still be only one two-hundredth of the breadth of the sun; and thus, still a mere point.

5 But if each fixed star be of the nature of the sun, and not smaller than the sun, does not analogy lead us to suppose that they have, some of them at least, planets circulating about them, as our sun has? If the Sun is the center of the Solar System, why should not Sirius, (one of the brightest of the fixed stars,) be the center of the *Sirian System?* And why should not that system have as many planets, with the same resemblances and differences of the figure, movements, and conditions of the different planets, as this? Why should not the Sirian System be as great and as varied as the solar system? And this being granted, why should not these planets be inhabited, as men have inferred the other planets of the solar system, as well as the earth, to be? And thus we have, added to the population of the universe of which we have already spoken, a number (so far as we have reason to believe) not inferior to the number of

* It is quite to our purpose to recollect the impression which such discoveries naturally make upon a pious mind.

Oh! rack me not to such extent,
These distances belong to Thee;
The world's too little for Thy tent,
A grave too big for me!

GEORGE HERBERT.

inhabitants of the solar system: this number being, according to all the analogies, very manyfold that of the population of the whole earth?

And this is the conclusion, when we reason from one star only, from Sirius. But the argument is the same, from each of the stars. For we have no reason to think that Sirius, though one of the brightest, is more like our sun than any of the others is. The others appear less bright in various degrees, probably because they are further removed from us in various degrees. They may not be all of the same size and brightness; it is very unlikely that they are. But they may as easily be larger than the sun, as smaller. The natural assumption for us to make, having no ground for any other opinion, is, that they are, upon the average, of the size of our sun. On that assumption, we have as many solar systems as we have fixed stars: and, it may be, six or ten, or twenty times as many inhabited globes; inhabited by creatures of whom we must suppose, by analogy, that God is mindful, if he is mindful of us. The question recurs with overwhelming force, if we still follow the same train of reflexion: 'What is man, that God is mindful of him?'

6 But we have not yet exhausted the views which thus add to the force of this reflexion. The fixed stars, which appear to the eye so numerous, so innumerable, in the clear sky on a moonless night, are not really so numerous as they seem. To the naked eye, there are not visible more than four or five thousand. The astronomers of Greece, and of other countries, even in ancient times, counted them, mapped them, and gave them names and designations. But Astronomy, who thus began her career by diminishing, in some degree, the supposed numbers of the host of heaven, has ended by immeasurably increasing them. The first application of the telescope to the skies discovered a vast 50| number of fixed stars, previously unseen: and every improve-

ment in that instrument has disclosed myriads of new stars, visibly smaller than those which had before been seen; and smaller and smaller, as the power of vision is more and more strengthened by new aids from art; as if the regions of space contained an inexhaustible supply of such objects;—as if infinite space were strewn with stars in every part of it to which vision could reach. The small patch of the sky which forms, at any moment, the field of view of one of the great telescopes of Herschel, discloses to him as many stars, and those, of as many different magnitudes, as the whole vault of the sky exhibits to the naked eye. But the magnifying power of such an instrument only discloses, it does not make, these stars. There appears to be quite as much reason to believe, that each of these telescopic stars is a sun, surrounded by its special family of planets, as to believe that Sirius or Arcturus is so. Here, then, we have again an extension, indefinite to our apprehension, of the universe, as occupied by material structures; and if so, why not by a living population, such as the material structures which are nearest to us support?

7 Even yet we have not finished the series of successive views which astronomers have had opened to them, extending more and more their spectacle of the fulness and largeness of the universe. Not only does the telescope disclose myriads of stars, unseen to the naked eye, and new myriads with each increase of the powers of the instrument; but it discloses also patches of light, which, at first at least, do not appear to consist of stars: *Nebulæ*, as they are called; bright specks, it might seem, of stellar matter, thin, diffused, and irregular; not gathered into regular and definite forms, such as we may suppose the stars to be. Every one who has noticed the starry skies, may understand what is the general aspect of such nebulæ, by looking at the milky way or galaxy, an irregular band of nebulous light, which runs |51

quite round the sky; 'A circling zone, powdered with stars;' as Milton calls it. But the nebulæ of which I more especially speak, are minute patches, discovered mainly by the telescope, and in a few instances only discernible by the naked eye. And what I have to remark especially concerning them at present is, that though, to visual powers which barely suffice to discern them, they appear like mere bright clouds, patches of diffused starry matter; yet that, when examined by visual powers of a higher order, by more penetrating telescopes, these patches of continuous feeble light, are, in many instances at least, distinguishable into definite points: they are found, in fact, to be aggregations of stars; which before appeared as diffused light, only because our telescopes, though strong enough to reveal to our senses the aggregate mass of light of the cluster, were not strong enough to enable us to discern any one of the stars of which the cluster consists. The galaxy, in this way, may, in almost every part, be *resolved* into separate stars; and thus, the multitude of the stars in the region of the sky occupied by that winding stream of light, is, when examined by a powerful telescope, inconceivably numerous.

8 The small telescopic nebulæ are of various forms; some of them may be in the shape of flat strata, or cakes, as it were, of stars, of small thickness, compared with the extent of the stratum. Now if our sun were one of the individuals of such a stratum, we, looking at the stars of the stratum from his neighbourhood, should see them very numerous and close in the direction of the edge of the stratum, and comparatively few and rare in other parts of the sky. We should, in short, see a galaxy running round the sky, as we see in fact. And hence Sir William Herschel has inferred, that our sun, with its attendant planets, has its place in such a stratum; and that it thus belongs to a host of stars which

52| are, in a certain way, detached from the other nebulæ which we see. Perhaps, he adds, some of those other nebulæ are beds and masses of stars not less numerous than those which compose our galaxy, and which occupy a larger portion of the sky, only because we are immersed in the interior of the crowd. And thus, a minute speck of nebulous light, discernible only by a good telescope, may contain not only as many stars as occupy the sky to ordinary vision, but as many as is the number into which the most powerful telescope resolves the milky light of the galaxy. And of such resolvable nebulæ the number which are discovered in the sky is very great, their forms being of the most various kind; so that many of them may be, for aught we can tell, more amply stocked with stars than the galaxy is. And if all the stars, or a large proportion of the stars, of the galaxy, be suns attended by planets, and these planets peopled with living creatures, what notion must we form of the population of the universe, when we have thus to reckon as many galaxies as there are resolvable nebulæ! the stock of discoverable nebulæ being as yet unexhausted by the powers of our telescopes; and the possibility of resolving them into stars being also an operation which has not yet been pursued to its limit.

9 For, (and this is the last step which I shall mention in this long series of ascending steps of multitude apparently infinite,) it now begins to be suspected that not some nebulæ only, but *all*, are resolvable into separate stars. When the nebulæ were first carefully studied, it was supposed that they consisted, as they appeared to consist, of some diffused and incoherent matter, not of definite and limited masses. It was conceived that they were not stars, but Stellar Matter in the course of formation into stars; and it was conceived, further, that by the gradual concentration of such matter, whirling round its centre while it concentrated, not only stars, that is, suns, might be formed, but also systems of planets, circling |53

round these suns; and thus this *Nebular Hypothesis,* as it has been termed, gave a kind of theory of the origin and formation of systems, such as the solar system. But the great telescope which Lord Rosse has constructed, and which is much more powerful than any optical instrument yet fabricated, has been directed to many of the nebulæ, whose appearance had given rise to this theory; and the result has been, in a great number of cases, that the nebulæ are proved to consist entirely of distinct stars; and that the diffused nebulous appearance is discovered to have been an illusion, resulting from the accumulated light of a vast number of small stars near to each other. In this manner, we are led to regard every nebulæ, not as an imperfectly formed star or system, but as a vast multitude of stars, and, for aught we can tell, of systems: for the apparent smallness and nearness of these stars are, it is thought, mere results of the vast distance at which they are placed from us. And thus, perhaps, all the nebulæ are, what some of them seem certainly to be, so many vast armies of stars, each of which stars, we have reason to believe, is of the nature of our sun; and may have, and according to analogy has, an accompaniment of living creatures, such as our sun has, certainly on the earth, probably, it is thought, in the other planets.

10 It is difficult to grasp, in one view, the effect of the successive steps from number to number, from distance to distance, which we have thus been measuring over. We may, however, state them again briefly, in the way of enumeration.

From our own place on the earth, we pass, in thought, as a first step, to the whole globe of the Earth; from this, as a second step, to the Planets, the other globes which compose the Solar System. A third step carries us to the Fixed Stars, as visible to the naked eye; very numerous and immensely distant. The transition to the Telescopic Stars makes a fourth 54| step; and in this, the number and the space are increased, almost beyond the power of numbers to express how many there are, and at what distances. But a fifth step:—perhaps all this array of stars, obvious and telescopic, only make up our Nebula; while the universe is occupied by other Nebulæ innumerable, so distant that, seen from them, our nebula, though including, it may be, stars of the 20th magnitude, which may be 20 times or 2000 times more remote than Sirius, would become a telescopic speck, as their nebulæ is to us.

11 Various images and modes of representation have been employed, in order to convey to the mind some notion of the dimensions of the scheme of the universe to which we are thus introduced. Thus, we may reckon that a cannon-ball, moving with its usual original velocity unabated, would describe the interval between the sun and the earth in about one year. And this being so, the same missile would, from what has been said, occupy more, we know not how much more, than 200,000 years in going to the nearest fixed star: and perhaps a thousand times as much, in going to other stars belonging to our group; and then again, 200,000 times so much, or some number of the like order, in going from one group to another. When we have advanced a step or two in this mode of statement, the velocity of the cannon-ball hardly perceptibly affects the magnitude of the numbers which we have to use.

And the same nearly is the case if we have recourse to the swiftest motion with which we are acquainted; that of Light. Light travels, it is shewn by indisputable scientific reasonings, in about eight minutes from the sun to the earth. Hence we can easily calculate that it would occupy at least three years to travel as far as Sirius, and probably, three thousand years, or a much greater number, to reach to the smallest stars, or to come from them to us. And thus, as Sir W. Herschel remarked, since light is the only vehicle by |55 which information concerning these distant bodies is conveyed to us, we do, by seeing them, receive information, not what they are at this moment, but what they were, as to visible condition, thousands of years ago. Stars may have been created when man was created, and yet their light may not have yet reached him*. Stars may have been extinguished thousands of years ago, and yet may still be visible to our eyes, by means of the light which they emitted previous to their extinction, and which has not yet died away.

12 So vast then are the distances at which the different bodies of the universe are distributed; and yet so numerous are those bodies. In the vastness of their distances, there is, indeed, nothing which need disturb our minds, or which, after a little reflexion, is likely to do so: for when we have said all that can be said, about the largeness of these distances, still there is no difficulty in finding room for them. We necessarily conceive *Space* as being infinite in its extent: however much space the heavenly bodies occupy, there is space beyond them: if they are not there, space is there nevertheless. That the stars and planets are so far from each other, is an arrangement which prevents their disturbing each other with their mutual attractions, to any destructive extent; and is an arrangement which the spacious, the infinite universe, admits of, without any difficulty.

13 But we are more especially concerned with the *Numbers* of the heavenly bodies. So many planets about our sun: so many suns, each perhaps with its family of planets: and then, all these suns making but one group: and

* This thought is, however, older. Young expresses it in his *Night Thoughts*, Night IX., (published in 1744):

How distant some of these nocturnal suns!
So distant (says the sage) 'twere not absurd
To doubt if beams, set out at nature's birth,
Are yet arrived at this so foreign world.

other groups coming into view, one after another, in seemingly endless succession: and all these planets being of the nature of our earth, as all these stars are of the nature of our sun:—all this, presents to us a spectacle of a world—of a countless host of worlds—of which, when we regard them as thus arranged in planetary systems, and as having, according to all probability, years and seasons, days and nights, as we have, we cannot but accept it as at least a likely suggestion, that they have also inhabitants;—intelligent beings who can reckon these days and years; who subsist on the fruits which the seasons bring forth, and have their daily and yearly occupations, according to their faculties. When we take, as our scheme of the universe, such a scheme as this, we may well be overwhelmed with the number of provinces, besides that in which man dwells, which the empire of the Lord of all includes; and, recurring to the words of the Psalmist, we may say with a profundity of meaning immeasurably augmented—'Lord, what is man?'

It was this view, I conceive which Dr Chalmers had in his thoughts, in pursuing the speculations which I have mentioned, in the outset of this Essay.

CHAPTER II.

The Astronomical Objection to Religion.

---

1 SUCH astronomical views, then, as those just stated, we may suppose to be those to which Chalmers had reference, in the argument of his *Astronomical Discourses.* These real or supposed discoveries of astronomers, or a considerable part of them, were the facts which were present to his mind, and of which he there discusses the bearings upon religious truths. This multiplicity of systems and worlds, which the telescopic scrutiny of the stars is assumed to have disclosed, or to have made probable, is the main feature in the constitution of the universe, as revealed by science, to which his reflections are directed. Nor can we say that, in fixing upon this view, he has gone out of his way, to struggle with obscure and latent difficulties, such as the bulk of mankind know and care little about. For in reality, such views are generally diffused in our time and country, are common to all classes of readers, and as we may venture to express it, are the *popular* views of persons of any degree of intellectual culture, who have, directly or derivatively, accepted the doctrines of modern science. Among such persons, expressions which imply that the stars are globes of luminous matter, like the sun; that there are, among them, systems of revolving bodies, seats of life and of intelligence; are so frequent and familiar, that those who so speak, do not seem to be aware that, in using such expressions, they are making any assumption at all; any more than they suppose themselves to be making assumptions, when they speak of the globular form of the earth, or of its motion round the 58| sun, or of its revolution on its axis. It was, therefore, a

suitable and laudable purpose, for a writer like Chalmers, well instructed in science, of large and comprehensive views with regard both to religion and to philosophy, of deep and pervasive piety, and master of a dignified and persuasive eloquence, to employ himself in correcting any erroneous opinions and impressions respecting the bearing which such scientific doctrines have upon religious truth. It was his lot to labour among men of great intellectual curiosity, acuteness, and boldness: it was his tendency to deal with new views of others on the most various subjects, religious, philosophical, and social; and, on such subjects, to originate new views of his own. It fell especially within his province, therefore, to satisfy the minds of the public who listened to him, with regard to the conflict, if a conflict there was, or seemed to be, between new scientific doctrines, and permanent religious verities. He was, by his culture and his powers, peculiarly fitted, and therefore peculiarly called, to mediate between the scientific and the religious world of his time.

2 The scientific doctrine which he especially deals with, in the work to which I refer, is the multiplicity of worlds;—the existence of many seats of life, of enjoyment, of intelligence; and it may be, as he suggests also, of moral law, of transgression, of alienation from God, and of the need, and of the means, of reconciliation to Him; or of obedience to Him and sympathy with Him. That, if there be many worlds resembling our world in other respects, they may resemble it in some of these, is an obvious, and we may say, an irresistible conjecture, in any speculative mind to which the doctrine itself has been conveyed. Nor can it fail to be very interesting, to see how such a writer as I have described deals with such a suggestion; how far he accepts or inclines to accept it; and if so, what aspect such a view leads him to give to truths, either belonging to Natural or |59

to Revealed Theology, which, before the introduction of such a view, were regarded as bearing only upon the world of which man is the inhabitant.

3 The mode in which Chalmers treats this suggestion, is to regard it as the ground of an objection to Religion, either Natural or Revealed. He supposes an objector to take his stand upon the multiplicity of worlds, assumed or granted as true; and to argue that, since there are so many worlds beside this, all alike claiming the care, the government, the goodness, the interposition, of the Creator, it is in the highest degree extravagant and absurd, to suppose that he has done, for this world, that which Religion, both Natural and Revealed, represents him as having done, and as doing. When we are told that God has provided, and is constantly providing, for the life, the welfare, the comfort of all the living things which people this earth, we can, by an effort of thought and reflexion, bring ourselves to believe that it is so. When we are further told that He has given a moral law to man, the intelligent inhabitant of the earth, and governs him by a moral government, we are able, or at least the great bulk of thoughtful men, on due consideration of all the bearings of the case, are able, to accept the conviction, that this also is so. When we are still further asked to believe that the imperfect sway of this moral law over man has required to be remedied by a special interposition of the Governor of the world, or by a series of special interpositions, to make the Law clear, and to remedy the effects of man's transgression of it; this doctrine also,—according to the old and unscientific view, which represents the human race as, in an especial manner, the summit and crown of God's material workmanship, the end of the rest of creation, and the selected theatre of God's dealings with transgression and with obedience,—we can conceive, and, as religious per-60|sons hold, we can find ample and satisfactory evidence to

believe. But if this world be merely one of innumerable worlds, all, like it, the workmanship of God; all, the seats of life, like it; others, like it, occupied by intelligent creatures, capable of will, of law, of obedience, of disobedience, as man is; to hold that this world has been the scene of God's care and kindness, and still more, of his special interpositions, communications, and personal dealings with its individual inhabitants, in the way which Religion teaches, is, the objector is conceived to maintain, extravagant and incredible. It is to select one of the millions of globes which are scattered through the vast domain of space, and to suppose that one to be treated in a special and exceptional manner, without any reason for the assumption of such a peculiarity, except that this globe happens to be the habitation of us, who make this assumption. If Religion require us to assume, that one particular corner of the Universe has been thus singled out, and made an exception to the general rules by which all other parts of the Universe are governed; she makes, it may be said, a demand upon our credulity which cannot fail to be rejected by those who are in the habit of contemplating and admiring those general laws. Can the Earth be thus the centre of the moral and religious universe, when it has been shewn to have no claim to be the centre of the physical universe? Is it not as absurd to maintain this, as it would be to hold, at the present day, the old Ptolemaic hypothesis, which places the Earth in the centre of the heavenly motions, instead of the newer Copernican doctrine, which teaches that the Earth revolves round the Sun? Is not Religion disproved, by the necessity under which she lies, of making such an assumption as this?

4 Such is, in a general way, the objection to Religion with which Chalmers deals: and, as I have said, his mode of treating it is highly interesting and instructive. Perhaps, however, we shall make our reasonings and speculations |61

apply to a wider class of readers, if we consider the view now spoken of, not as an objection, urged by an opponent of religion, but rather, as a difficulty, felt by a friend of religion. It is, I conceive, certain that many of those who are not at all disposed to argue against religion, but who, on the contrary, feel that their whole internal comfort and repose are bound up indissolubly with their religious convictions, are still troubled and dismayed at the doctrines of the vastness of the universe, and the multitude of worlds, which they suppose to be taught and proved by astronomy. They have a profound reverence for the Idea of God; they are glad to acknowledge their constant and universal dependence upon His preserving power and goodness; they are ready and desirous to recognize the working of His providence; they receive the moral law, as His law, with reverence and submission; they regard their transgressions of this law as sins against Him; and are eager to find the mode of reconciliation to Him, when thus estranged from Him: they willingly think of God, as near to them. But while they listen to the evidence which science, as we have said, sets before them, of the long array of groups, and hosts, and myriads, of worlds, which are brought to our knowledge, they find themselves perturbed and distressed. They would willingly think of God as near to them; but during the progress of this enumeration, He appears, at every step, to be removed further and further from them. To discover that the Earth is so large, the number of its inhabitants so great, its form so different from what man at first imagines it, may perhaps have startled them; but in this view, there is nothing which a pious mind does not easily surmount. But if Venus and Mars also have their inhabitants; if Saturn and Jupiter, globes so much larger than the earth, have a proportional amount of population; may not man be 62| neglected or overlooked? Is he worthy to be regarded by the Creator of all? May not, must not, the most pious mind recur to the exclamation of the Psalmist; 'Lord, what is man, that thou art mindful of him?' And must not this exclamation, under the new aspect of things, be accompanied by an enfeebled and less confident belief that God *is* mindful of him? And then, this array of planets, which derive their light from the Sun, extends much further than even the astronomer at first suspected. The orbit of Saturn is ten times as wide as the orbit of the earth; but beyond Saturn, and almost twice as far from the sun, Herschel discovers Uranus, another great planet; and again, beyond Uranus, and again at nearly twice *his* distance, the subtle sagacity of the astronomers of our day, surmises, and then detects, another great planet. In such a system as this, the earth shrinks into insignificance. Can its concerns engage the attention of him who made the whole? But again, this whole Solar System itself, with all its orbits and planets, shrinks into a mere point, when compared with the nearest fixed star. And again, the distance which lies between us and such stars, shrinks into incalculable smallness, when we journey in thought to other fixed stars. And again, and again, the field of our previous contemplation suffers an immeasurable contraction, as we pass on to other points of view.

5 And in all these successive moves, we are still within the dominions of the same Creator and Governor; and at every move, we are brought, we may suppose, to new bodies of his subjects, bearing, in the expansion of their number, some proportion to the expanse of space which they occupy. And if this be so, how shall the earth, and men, its inhabitants, thus repeatedly annihilated, as it were, by the growing magnitude of the known Universe, continue to be anything in the regard of Him who embraces all? Least of all, how shall men continue to receive that special, preserving, pro- |63

vidential, judicial, personal care, which religion implies; and without the belief of which, any man who has religious thoughts, must be disturbed and unhappy, desolate and forsaken?

6 Such are, I conceive, the thoughts of many persons, under the influence of the astronomical views which Chalmers refers to as being sometimes employed against religious belief. Of course, it is natural that the views which are used by unbelievers as arguments against religious belief, should create difficulties and troubles in the minds of believers; at least, till the argument is rebutted. And of course also, the answers to the arguments, considered as infidel arguments, would operate to remove the difficulties which believers entertain on such grounds. Chalmers's reasonings against such arguments, therefore, will, so far as they are valid, avail to relieve the mental trouble of believers, who are perplexed and oppressed by the astronomical views of which I have spoken; as well as to confute and convince those who reject religion, on such astronomical grounds. It may, however, as I have said, be of use to deal with these difficulties rather as difficulties of religious men, than as objections of irreligious men; to examine rather how we can quiet the troubled and perplexed believer, than how we can triumph over the dogmatic and self-satisfied infidel. I, at least, should wish to have the former, rather than the latter of these tasks, regarded as that which I propose to myself.

I shall hereafter attempt to explain more fully the difficulties which the doctrine of the Plurality of Worlds appears to some persons to throw in the way of Revealed Religion: but before I do so, there is one part of Chalmers's answer, bearing especially upon Natural Religion, which it may be proper to attend to.

CHAPTER III.

The Answer from the Microscope.

---

1 IT is not my business, nor my intention, to criticize the remarkable work of Chalmers to which I have so often referred. But I may say, that the arguments there employed by him, so far as they go upon astronomical or philosophical grounds, are of great weight; and upon the whole, such as we may both assent to, as scientifically true, and accept as rationally persuasive. I think, however, that there are other arguments, also drawn from scientific discoveries, which bear, in a very important and striking manner, upon the opinions in question; and which Chalmers has not referred to: and I conceive that there are philosophical views of another kind, which, for those who desire, and who will venture, to regard the Universe and its Creator, in the wider and deeper relations which appear to be open to human speculation, may be a source of satisfaction. When certain positive propositions, maintained as true, while they are really highly doubtful, have given rise to difficulties in the minds of religious persons; other positive propositions, combating these, propounded and supported by argument, that they may be accepted according to their evidence, may, at any rate, have force enough to break down and dissipate such loosely founded difficulties. To present to the reader's mind such speculations as I have thus indicated, is the object of the following pages. They can, of course, pretend to no charm, except for persons who are willing to have their minds occupied with such difficulties and such speculations as I have referred to. Those who are willing to be so employed, may, |65

perhaps, find in what I have to say, something which may interest them. For of the arguments which I have to expound, some, though they appear to me both very obvious and very forcible, have never, so far as I am aware, been put forth, in that religious bearing which seems to belong to them; and others, though aspiring to point out in some degree the relation of the Universe and its Creator, are of a very simple kind; that is, for minds which are prepared to deal with such subjects at all.

2 As I have said, the arguments with which we are here concerned refer both to Natural Religion, and to Revealed Religion: and there is one of Chalmers's arguments, bearing especially upon the former branch of the subject, which I may begin by noticing. Among the thoughts, which, it was stated, might naturally arise in men's minds, when the telescope revealed to them an innumerable multitude of worlds, besides the one which we inhabit, was this: that the Governor of the Universe, who has so many worlds under his management, cannot be conceived as bestowing upon this Earth, and its various tribes of inhabitants, that care which, till then, Natural Religion had taught men that he does employ, to secure to man the possession and use of his faculties of mind and body; and to all animals, the requisites of animal existence and animal enjoyment. And upon this, Chalmers remarks, that just about the time when science gave rise to the suggestion of this difficulty, she also gave occasion to a remarkable reply to it. Just about the same time that the invention of the *Telescope* shewed that there were innumerable worlds, which might have inhabitants, requiring the Creator's care as much as the tribes of this earth do,—the invention of the *Microscope* shewed that there were, in this world, innumerable tribes of animals, which had been, all along, enjoying the benefits of the Creator's care, as much as those kinds with which

66|

man had been familiar from the beginning. The telescope suggested that there might be dwellers in Jupiter or in Saturn, of giant size and unknown structure, who must share with us the preserving care of God. The microscope shewed that there had been, close to us, inhabiting minute crevices and crannies, peopling the leaves of plants, and the bodies of other animals, animalcules of a minuteness hitherto unguessed, and of a structure hitherto unknown, who had been always sharers with us in God's preserving care. The telescope brought into view worlds as numerous as the drops of water which make up the ocean; the microscope brought into view a world in almost every drop of water. Infinity in one direction was balanced by infinity in the other. The doubts which men might feel, as to what God could do, were balanced by certainties which they discovered, as to what he had always been doing. His care and goodness could not be supposed to be exhausted by the hitherto known population of the earth, for it was proved that they had not hitherto been confined to that population. The discovery of new worlds at vast distances from us, was accompanied by the discovery of new worlds close to us; even in the very substances with which we were best acquainted: and was thus rendered ineffective to disturb the belief of those who had regarded the world as having God for its governor.

3 This is a striking reflexion; and is put by Chalmers in a very striking manner: and it is well fitted to remove the scruples to which it is especially addressed. If there be any persons to whom the astronomical discoveries which the telescope has brought to light, suggest doubts or difficulties with regard to such truths of Natural Religion, as God's care for and government of the inhabitants of the earth; the discoveries of the many various forms of animalcular life which the microscope has brought to light are well fitted to

remove such doubts, and to solve such difficulties. We may easily believe that the power of God to sustain and provide for animal life, animal sustenance, animal enjoyment, can suffice for innumerable worlds besides this, without being withdrawn or distracted or wearied in this earth; for we find that it does suffice for innumerable more inhabitants of this earth than we were before aware of. If we had imagined before, that, in conceiving God as able and willing to provide for the life and pleasure of all the sentient beings which we knew to exist upon the earth, we had formed an adequate notion of his power and of his goodness, these microscopical discoveries are well adapted to undeceive us. They shew us that all the notions which our knowledge, hitherto, had enabled us to form of the powers and attributes of the Creator and Preserver of all living things, are vastly, are immeasurably, below the real truth of the case. They shew us that God, as revealed to us in the animal creation, is the Author and Giver of life, of the organization which life implies, of the contrivances by which it is conducted and sustained, of the enjoyment by which it is accompanied,—to an extent infinitely beyond what the unassisted vision of man could have suggested. The facts which are obvious to man; from which religious minds in all ages have drawn their notions and their evidence of the Divine power and goodness, care and wisdom, in providing for its creatures,—require, we find, to be indefinitely extended, in virtue of the new tribes of minute creatures, and still new tribes, and still more minute, which we find existing around us. The views of our Natural Theology must be indefinitely extended on one side; and therefore we need not be startled or disturbed, at having to extend them indefinitely on the other side;—at having to believe that there are, in other worlds, creatures whom God has created, whom he sustains in life, for whom he provides the pleasures

68|

of life, as he does for the long unsuspected creatures of this world.

4 This is, I say, a reflexion which might quiet the mind of a person, whom astronomical discoveries had led to doubt of the ordinary doctrines of Natural Religion. But, I think, it may be questioned, whether, to produce such doubts, is a common or probable effect of an acquaintance with astronomical discoveries. Undoubtedly, by such discoveries, a person who believes in God, in his wisdom, power, and goodness, on the evidence of the natural world, is required to extend and exalt his conceptions of those Divine Attributes. He had believed God to be the Author of many forms of life;—he finds him to be the Author of still more forms of life. He had traced many contrivances in the structure of animals, for their sustentation and well being; his new discoveries disclose to him (for that is undoubtedly among the effects of microscopic researches,) still more nice contrivances. He had seen reason to think that all sentient beings have their enjoyments; he finds new fields of enjoyment of the same kind. But in all this, there is little or nothing to disturb the views and convictions of the Natural Theologian. He must, even by the evidence of facts patent to ordinary observation, have been led to believe that the Divine Wisdom and Power are not only great, but great in a degree which we cannot fathom or comprehend;—that they are, to our apprehension, infinite: his new discoveries only confirm the impression of this infinite character of the Divine Attributes. He had before believed the existence of an intelligent and wise Creator, on the evidence of the marks of design and contrivance, which the creation exhibited: of such design and contrivance, he discovers new marks, new examples. He had believed that God is good, because he found those contrivances invariably had the good of the creature for their object: he finds, still, that this is

the general, the universal scheme of the creation, now when his view of it is extended. He has no difficulty in expanding his religious conceptions, to correspond with his scientific discoveries, so far as the microscope is the instrument of discovery; there is no reason why he should have any more difficulty in doing the same, when the telescope is his informant. It is true, that in this case the information is more imperfect. It does not tell him, even that there are living inhabitants in the regions which it reveals; and, consequently, it does not disclose any of those examples of design which belong to the structure of living things. But if we suppose, from analogy, that there are living things in those regions, we have no difficulty in conceiving, from analogy also, that those living things are constructed with a care and wisdom such as appear in the inhabitants of earth. It will not readily or commonly occur to a speculator on such subjects, that there is any source of perplexity or unbelief, in such an assumption of inhabitants of other worlds, even if we make the assumption. It is as easy, it may well and reasonably be thought, for God to create a population for the planets as to make the planets themselves;—as easy to supply Jupiter with tenants, as with satellites;—as easy to devise the organization of an inhabitant of Saturn, as the structure and equilibrium of Saturn's ring. It is no more difficult for the Universal Creator to extend to those bodies the powers which operate in organized matter, than the powers which operate in brute matter. It is as easy for Him to establish circulation and nutrition in material structures, as cohesion and crystallization, which we must suppose the planetary masses to possess; or attraction and inertia, which we know them to possess. No doubt, to our conception, organization appears to be a step beyond cohesion; circulation of living fluids, a step beyond crystallization of dead
70| masses:—but then, it is in tracing such steps, that we dis-

cern the peculiar character of the Creator's agency. He does not merely work with mechanical and chemical powers, as man to a certain extent can do; but with organic and vital powers, which man cannot command. The Creator, therefore, can animate the dust of each planet, as easily as make the dust itself. And when from organic life we rise to sentient life, we have still only another step in the known order of Creative Power. To create animals, in any province of the Universe, cannot be conceived as much more incomprehensible or incredible, than to create vegetables. No doubt, the addition of the living and sentient principle to the material, and even to the organic structure, is a mighty step; and one which may, perhaps, be made the occasion of some speculative suggestions, in a subsequent part of this Essay: but still, it is not likely that any one, who had formed his conceptions of the Divine Mind from its manifestations in the production and sustentation of animal, as well as vegetable life, on this earth, would have his belief in the operation of such a Mind, shaken, by any necessity which might be impressed upon him, of granting the existence of animal life on other planets, as well as on the earth, or even on innumerable such planets, and on innumerable systems of planets and worlds, system above system.

5 The remark of Chalmers, therefore, to which I have referred, striking as it is, does not appear to bear directly upon a difficulty of any great force. If astronomy gives birth to scruples which interfere with religion, they must be found in some other quarter than in the possibility of mere animal life existing in other parts of the Universe, as well as on our earth. That possibility may require us to enlarge our idea of the Deity, but it has little or no tendency to disturb our apprehension of his attributes.

CHAPTER IV.

Further Statement of the Difficulty.

---

1 WE have attempted to shew that if the discoveries made by the Telescope should excite in any one's mind, difficulties respecting those doctrines of Natural Religion,—the adequacy of the Creator to the support and guardianship of all the animal life which may exist in the universe,—the discoveries of the Microscope may remove such difficulties: but we have remarked also, that the train of thought which leads men to dwell upon such difficulties does not seem to be common.

But what will be the train of thought to which we shall be led, if we suppose that there are, on other planets, and in other systems, not animals only, living things, which, however different from the animals of this earth, are yet in some way analogous to them, according to the difference of circumstances: but also creatures analogous to man;—intellectual creatures, living, we must suppose, under a moral law, responsible for transgression, the subjects of a Providential Government? If we suppose that, in the other planets of our solar systems, and of other systems, there are creatures of such a kind, and under such conditions as these, how far will the religious opinions which we had previously entertained be disturbed or modified? Will any new difficulty be introduced into our views of the government of the world by such a supposition?

2 I have spoken of man as an Intellectual Creature; meaning thereby that he has a Mind;—powers of thought, by 72| which he can contemplate the relations and properties of

things in a general and abstract form; and among other relations, moral relations, the distinction of *right* and *wrong* in his actions. Those powers of thought lead him to think of a Creator and Ordainer of all things; and his perception of right and wrong leads him to regard this Creator as also the Governor and Judge of his creatures. The operation of his mind directs him to believe in a Supreme Mind: his moral nature directs him to believe that the course of human affairs, and the condition of men, both as individuals and as bodies, is determined by the providential government of God.

3 With regard to the bearing of a merely *intellectual* nature on such questions, it does not appear that any considerable difficulty would be *at once* occasioned in our religious views, by supposing such a nature to belong to other creatures, the inhabitants of other planets, as well as to man. The existence of our own minds directs us, as I have said, to a Supreme Mind; and the nature of Mind is conceived to be, in all its manifestations, so much the same, that we can conceive minds to be multiplied indefinitely, without fear of confusion, interference, or exhaustion. There may be, in Jupiter, creatures endowed with an intellect which enables them to discover and demonstrate the relations of space; and if so, they cannot have discovered and demonstrated anything of that kind as true, which is not true for us also: their Geometry must coincide with ours, as far as each goes:—thus shewing how absurdly, as Plato long ago observed, we give to the science which deals with the relations of space, a name, (*geometry*,) borrowed from the art of measuring the earth. The earth with its properties is no more the special basis of geometry, than are Jupiter or Saturn, or, so far as we can judge, Sirius or Arcturus and their systems, with their properties. Wherever pure intellect is, we are compelled to conceive that, when employed upon the

same objects, its results and conclusions are the same. If there be intelligent inhabitants of the Moon, they may, like us, have employed their intelligence in reasoning upon the properties of lines and angles and triangles; and must, so far as they have gone, have arrived, in their thoughts, at the same properties of lines and angles and triangles, at which we have arrived. They must, like us, have had to distinguish between right angles and oblique angles. They may have come to know, as some of the inhabitants of the earth came to know, four thousand years ago, that, in a right-angled triangle, the square on the larger side is equal to the sum of the squares on the other two sides. We can conceive occurrences which would give us evidence that the Moon, as well as the Earth, contains geometers. If we were to see, on the face of the full moon, a figure gradually becoming visible, representing a right-angled triangle with a square constructed on each of its three sides as a base; we should regard it as the work of intelligent creatures there, who might be thus making a signal to the inhabitants of the earth, that they possessed such knowledge, and were desirous of making known to their nearest neighbours in the solar system, their existence and their speculations. In such an event, curious and striking as it would be, we should see nothing but what we could understand and accept, without unsettling our belief in the Supreme and Divine Intelligence. On the contrary, we could hardly fail to receive such a manifestation as a fresh evidence that the Divine Mind had imparted to the inhabitants of the Moon, as he has to us, a power of apprehending, in a very general and abstract form, the relations of that space in which he performs his works. We should judge, that having been led so far in their speculations, they must, in all probability, have been led also to a conception of the Universe, as the field of action of a universal and Divine Mind; that having thus become geometers,

74

they must have ascended to the Idea of a God who works by geometry.

4 But yet, by such a supposition, on further consideration, we find ourselves introduced to views entirely different from those to which we are led by the supposition of mere animal life, existing in other worlds than the earth. For, not to dwell here upon any speculations as to how far the operations of our minds may resemble the operations of the Divine Mind;—a subject which we shall hereafter endeavour to discuss;—we know that the advance to such truths as those of geometry has been, among the inhabitants of the earth, gradual and progressive. Though the human mind have had the same powers and faculties, from the beginning of the existence of the race up to the present time, (as we cannot but suppose,) the results of the exercise of these powers and faculties have been very different in different ages; and have gradually grown up, from small beginnings, to the vast and complex body of knowledge concerning the scheme and relations of the Universe, which is at present accessible to the minds of human speculators. It is, as we have said, probably about four thousand years, since the first steps in such knowledge were made. Geometry is said to have had its origin in Egypt; but it assumed its abstract and speculative character first among the Greeks. Pythagoras is related to have been the first who saw, in the clear light of demonstration, the property of the right-angled triangle, of which we have spoken. The Greeks, from the time of Socrates, stimulated especially by Plato, pursued, with wonderful success, the investigation of this kind of truths. They saw that such truths had their application in the heavens, far more extensively than on the earth. They were enabled, by such speculations, to unravel, in a great degree, the scheme of the universe, before so seemingly entangled and perplexed. They determined, to a very considerable extent, the relative motions of the planets |75

and of the stars. And in modern times, after a long interval in which such knowledge was nearly stationary, the progress again began; and further advances were successively made in man's knowledge of the scheme and structure of the visible heavens; till at length the intellect of man was led to those views of the extent of the Universe and the nature of the stars, which are the basis of the discussions in which we are now engaged. And thus man, having probably been, in the earliest ages of the existence of the species, entirely ignorant of abstract truth, and of the relations which, by the knowledge of such truth, we can trace in nature, (as the barbarous tribes which occupy the greater part of the earth's surface still are;) has, by a long series of progressive steps, come into the possession of knowledge, which we cannot regard without wonder and admiration; and which seems to elevate him in no inconsiderable degree, towards a community of thought with that Divine Mind, into the nature and scheme of whose works he is thus permitted to penetrate.

5 Now the knowledge which man is capable, by the nature of his mental faculties, of acquiring, being thus blank and rudimentary at first, and only proceeding gradually, by the steps of a progress, numerous, slow, and often long interrupted, to that stage in which it is the basis of our present speculations; the view which we have just taken, of the nature of Intellect, as a faculty always of the same kind, always uniform in its operations, always consistent in its results, appears to require reconsideration; and especially with reference to the application which we made of that view, to the intelligent inhabitants of other planets and other worlds, if such inhabitants there be. For if we suppose that there are, in the Moon, or in Jupiter, creatures possessing intellectual faculties of the same kind as those of man; capable of apprehending the same abstract and general truths; able, 76| like man, to attain to a knowledge of the scheme of the

Universe; yet this supposition merely gives the capacity and the ability; and does not include any security, or even high probability, as it would seem, of the exercise of such capacity, or of the successful application of such ability. Even if the surface of the Moon be inhabited by creatures as intelligent as men, why must we suppose that they know any thing more of geometry and astronomy, than the great bulk of the less cultured inhabitants of the earth, who occupy, really, a space far larger than the surface of the Moon; and, all intelligent though they be, and in the full possession of mental faculties, are yet, on the subjects of geometry and astronomy, entirely ignorant;—their minds, as to such knowledge, a blank? It does not follow, then, that even if there be such inhabitants in the Moon, or in the Planets, they have any sympathy with us, or any community of knowledge on the subjects of which we are now speaking. The surface of the Moon, or of Jupiter, or of Saturn, even if well peopled, may be peopled only with tribes as barbarous and ignorant as Tartars, or Esquimaux, or Australians; and therefore, by making such a supposition, we do little, even hypothetically, to extend the dominion of that intelligence, by means of which all intelligent beings have some community of thought with each other, and some suggestion of the working of the Divine and Universal Mind.

6 But, in fact, the view which we have given, of the mode of existence of the human species upon the earth, as being a progressive existence, even in the development of the intellectual powers and their results, necessarily fastens down our thoughts and our speculations to the earth; and makes us feel how visionary and gratuitous it is, to assume any similar kind of existence, in any region occupied by other beings than man. As we have said, we have no insuperable difficulty in conceiving other parts of the Universe to be tenanted by animals. Animal life implies no progress in the |77

species. Such as they are in one century, such are they in another. The conditions of their sustentation and generation being given, which no difference of physical circumstances can render incredible, the race may, so far as we can see, go on for ever. But a race which makes a progress in the development of its faculties, cannot thus, or at least cannot with the same ease, be conceived as existing through all time, and under all circumstances. Progress implies, or at least suggests, a beginning and an end. If the mere existence of a race imply a sustaining and preserving power in the Creator, the progress of a race implies a guiding and impelling power; a Governor and Director, as well as a Creator and Preserver. And progress, not merely in material conditions, not merely in the exercise of bodily faculties, but in the exercise of mental faculties, in the intellectual condition of a portion of the species, still more implies a special position and character of the race; which cannot, without great license of hypothesis, be extended to other races; and which, if so extended, becomes unmeaning, from the impossibility of our knowing what is progress in any other species;—from what and towards what it tends. The intellectual progress of the human species has been a progress in the use of thought, and in the knowledge which such use procures; it has been a progress from mere matter to mind; from the impressions of sense to ideas; from what in knowledge is casual, partial, temporary, to what is necessary, universal and eternal. We can conceive no progress, of the nature of this, which is not identical with this; nothing like it, which is not the same. And therefore, if we will people other planets with creatures, intelligent as man is intelligent, we must not only give to them the intelligence, but the intellectual history of the human species. They must have had their minds unfolded 78| by steps, similar to those by which the human mind has been unfolded; or at least, differing from them, only as the intellectual history of one nation of the earth differs from that of another. They must have had their Pythagoras, their Plato, their Kepler, their Galileo, their Newton, if they know what we know. And thus, in order to conceive, on the Moon or on Jupiter, a race of beings, intelligent like man, we must conceive, there, colonies of men, with histories resembling more or less the histories of human colonies: and indeed resembling the history of those nations whose knowledge we inherit, far more closely than the history of any other terrestrial nation resembles that part of terrestrial history. If we do this, we exercise an act of invention and imagination which may be as coherent as a fairy tale, but which, without further proof, must be as purely imaginary and arbitrary. But if we do not do this, we cannot conceive that those regions are occupied at all by intelligent beings. Intelligence, as we see in the human race, in order to have those characters which concern our argument, implies a history of intellectual development: and to assume arbitrarily a history of intellectual development for the inhabitants of a remote planet, as a ground of reasoning, either for or against Religion, is a proceeding which we can hardly be expected either to assent to or to refute. If we are to form any opinions with regard to the condition of such bodies, and to trace any bearing of such opinions upon our religious views, we must proceed upon some ground which has more of reality than such a gratuitous assumption.

7 Thus the condition of man upon the earth, as a condition of intellectual progress, implies such a special guidance and government exercised, over the race, by the Author of his being, as produces progress; and we have not, so far as we yet perceive, any reason for supposing that He exercises a like guidance and government over any of the other bodies with which the researches of astronomers have made us

acquainted. The earth and its inhabitants are under the care of God in a special manner; and we are utterly destitute of any reason for believing that other planets and other systems are under the care of God in the same manner. If we regarded merely the existence of unprogressive races of animals upon our globe, we might easily suppose that other globes also, are similarly tenanted; and we might infer, that the Creator and Upholder of animal life was active on those globes, in the same manner as upon ours. But when we come to a progressive creature, whose condition implies a beginning, and therefore suggests an end, we form a peculiar judgment with respect to God's care of that creature, which we have not, as yet, seen the slightest grounds to extend to other possible fields of existence, where we discern no indication of progress, of beginning, or of end. So far as we can judge, God is mindful of man; and has launched and guided his course in a certain path, which makes his lot and state different from that of all other creatures.

8 Now when we have arrived at this result, we have, I conceive, reached one of the points, at which the difficulties, which astronomical discovery puts in the way of religious conviction, begin to appear. The Earth and its human inhabitants are, as far as we yet know, in an especial manner, the subjects of God's care and government, for the race is progressive. Now can this be? Is it not difficult to believe that it is so? The earth, so small a speck, only one among so many, so many thousands, so many millions, of other bodies, all, probably, of the same nature with itself, wherefore should it draw to it the special regards of the Creator of all, and occupy his care in an especial manner? The teaching of the history of the human race, as intellectually progressive, agrees with the teaching of Religion, in impressing upon us that God is mindful of man; that he does regard him: but

80| still, there naturally arises in our minds a feeling of perplexity and bewilderment, which expresses itself in the words already so often quoted, What is man, that this should be so? Can it be true that this province is thus singled out for a special and peculiar administration, by the Lord of the Universal Empire?

9 Before I make any attempt to answer these questions, I must pursue the difficulty somewhat further, and look at it in other forms. As I have said, the history of Man has been, in certain nations, a history of intellectual progress, from the earliest times up to our own day. But intellectual progress has been, as I have also said, in a great measure confined to certain nations, thus especially favoured. The greater part of the earth's inhabitants have shared very scantily in that wealth of knowledge, to which the brightest and happiest intellects among men have thus been led. But though the bulk of mankind have thus had little share in the grand treasures of science, which are open to the race, their life has still been very different from that of other animals. Many nations, though they may not have been conspicuous in the history of intellectual progress, have yet not been without their place in progress of other kinds;—in arts, in arms, and, above all, in morals;—in the recognition of the distinction of right and wrong in human actions, and in the practical application of this distinction. Such a progress as this has been far more extensively aimed at, than a progress in abstract and general knowledge; and, we may venture to say, has been, in many nations, and in a very great measure, really effected. No doubt, the imperfection of this progress, and the constant recurrence of events which appear to counteract and reverse it, are so obvious and so common, as to fill with grief and indignation the minds of those who regard such a progress as the great business of the human race: but yet still, looking at the whole history of the human race, the progress is visible; and even the grief and |81

the indignation of which we have spoken, are a part of its evidences. There has been, upon the whole, a moral government of the human race. The moral law, the distinction of right and wrong, has been established in every nation; and penalties have been established for wrong-doing. The notion of right and wrong has been extended, from mere outward acts, to the springs of action, to affection, desire, and will. The course of human affairs has generally been such that the just, the truthful, the kind, the chaste, the orderly portion of mankind have been happier than the violent and wicked. External wrong has been commonly punished by the act of human society. Internal sins, impure and dishonest designs, falsehood, cruelty, have very often led to their own punishment, by their effect upon the guilty mind itself. We do not say that the moral government which has prevailed among men has been such, that we can consider it complete and final in its visible form. We see that the aspect of things is much the contrary; and we think we see reasons why it may be expected to be so. But still, there has existed upon earth a moral government of the human race; exercised, as we must needs hold, by the Creator of man; partly through the direct operation of man's faculties, affections, and emotions; and partly through the authorities which, in all ages and nations, the nature of man has led him to establish. Now this moral progress and moral government of the human race is one of the leading facts on which Natural Religion is founded. We are thus led to regard God as the Moral Governor of man; not only his Creator and Preserver, but his Lawgiver and his Judge. And the grounds on which we entertain this belief, are peculiarly the human faculties of man, and their operation in history and in society. The belief is derived from the whole complex nature of man;—the working of his Affections, Desires, 82| Convictions, Reason, Conscience, and whatever else enters into the production of human action, and its consequences. God is seen to be the Moral Governor of man, by evidence which is especially derived from the character of Man; and which we could not attempt to apply to any other creature than man, without making our words altogether unmeaning. But would it not be too bold an assumption to speak of the Conscience of an inhabitant of Jupiter? Would it not be a rash philosophy to assume the operation of Remorse or Self-approval on the planet, in order that we may extend to it the moral government of God? Except we can point out something more solid than this, to reason from, on such subjects, there is no use in our attempting to reason at all. Our doctrines must be mere results of invention and imagination. Here then, again, we are brought to the conviction that God is, so far as we yet see, in an especial and peculiar manner, the Governor of the earth and of its human inhabitants; in such a way that the like government cannot be conceived to be extended to other planets, and other systems, without arbitrary and fanciful assumptions; assumptions either of unintelligible differences with incomprehensible results; or of beings in all respects human, inhabiting the most remote regions of the universe. And here, again, therefore, we are led to the same difficulty which we have already encountered: Can the earth, a small globe among so many millions, have been selected as the scene of this especially Divine Government?

10 That when we attempt to extend our sympathies to the inhabitants of other planets and other worlds, and to regard them as living, like us, under a moral government, we are driven to suppose them to be, in all essential respects, human beings like ourselves, we have proof, in all the attempts which have been made, with whatever license of hypothesis and fancy, to present to us descriptions and representations of the inhabitants of other parts of the universe. Such repre- |83

sentations, though purposely made as unlike human beings as the imagination of man can frame them, still are merely combinations, slightly varied, of the elements of human being; and thus shew us that not only our reason, but even our imagination, cannot conceive creatures subjected to the same government to which man is subjected, without conceiving them as being men, of one kind or other. A mere animal life, with no interest but animal enjoyment, we may conceive as assuming forms different from those which appear in existing animal races; though even here, there are, as we shall hereafter attempt to shew, certain general principles which run through all animal life. But when in addition to mere animal impulses, we assume or suppose moral and intellectual interests, we conceive them as the moral and intellectual interests of man. Truth and falsehood, right and wrong, law and transgression, happiness and misery, reward and punishment, are the necessary elements of all that can interest us,—of all that we can call *Government.* To transfer these to Jupiter or to Sirius, is merely to imagine those bodies to be a sort of island of Formosa, or new Atlantis, or Utopia, or Platonic Polity, or something of the like kind. The boldest and most resolute attempts to devise some life different from human life, have not produced anything more different, than romance-writers and political theorists have devised *as* a form of human life. And this being so, there is no more wisdom or philosophy in believing such assemblages of beings to exist in Jupiter or Sirius, without evidence, than in believing them to exist in the island of Formosa, with the like absence of evidence.

11 Any examination of what has been written on this subject would shew that, in speculating about moral and intellectual beings in other regions of the universe, we merely make them to be men in another place. With regard to 84| the plants and animals of other planets, fancy has freer

play; but man cannot conceive any moral creature who is not man. Thus Fontenelle, in his *Dialogues on the Plurality of Worlds*, makes the inhabitants of Venus possess in an exaggerated degree, the characteristics of the men of the warm climates of the earth. They are like the Moors of Grenada: or rather the Moors of Grenada would be to them as cold as Greenlanders and Laplanders to us. And the inhabitants of Mercury have so much vivacity that they would pass with us for insane. 'Enfin c'est dans Mercure que sont les Petites-Maisons de l'Univers.' The inhabitants of Jupiter and Saturn are immensely slow and phlegmatic. And though he and other writers attempt to make these inhabitants of remote regions in some respects superior to man, telling us that instead of only five senses, they may have six or ten, or a hundred; still, these are mere words, which convey no meaning; and the great astronomer Bessel had reason to say, that those who imagined inhabitants in the Moon and Planets, supposed them, in spite of all their protestations, as like to men as one egg to another*.

12 But there is one step more, which we still have to make, in order to bring out this difficulty in its full force. As we have said, the moral law has been, to a certain extent, established, developed, and enforced among men. But, as I have also said, looking carefully at the law, and at the degree of man's obedience to it, and at the operation of the sanctions by which it is supported, we cannot help seeing, that man's knowledge of the law is imperfect, his conviction of its authority feeble, his transgressions habitual, their punishment and consequences obscure. When, therefore, we regard God, as the Lawgiver and Judge of man, it will not appear strange to us, that he should have taken some mode of promulgating his Law, and announcing his Judgments, in

* Populäre Vorlesungen über Wissenschaftliche Gegenstände. p. 31.

addition to that ordinary operation of the faculties of man, of which we have spoken. Revealed Religion teaches us that he has done so: that from the first placing of the race of man upon the earth, it was his purpose to do so: that by his dealings with the race of man in the earlier times, and at various intervals, he made preparation for the mission of a special Messenger, whom, in the fulness of time, he sent upon the earth in the form of a man; and who both taught men the Law of God in a purer and clearer form than any in which it had yet been given; and revealed His purpose, of rewards for obedience, and punishments for disobedience, to be executed in a state of being to which this human life is only an introduction; and established the means by which the spirit of man, when alienated from God by transgression, may be again reconciled to Him. The arrival of this especial Message of Holiness, Judgment, and Redemption, forms the great event in the history of the earth, considered in a religious view, as the abode of God's servants. It was attended with the sufferings and cruel death of the Divine Messenger thus sent; was preceded by prophetic announcements of his coming; and the history of the world, for the two thousand years that have since elapsed, has been in a great measure occupied with the consequences of that advent. Such a proceeding shows, of course, that God has an especial care for the race of man. The earth, thus selected as the theatre of such a scheme of Teaching and of Redemption, can not, in the eyes of any one who accepts this Christian faith, be regarded as being on a level with any other domiciles. It is the Stage of the great Drama of God's Mercy and Man's Salvation; the Sanctuary of the Universe; the Holy Land of Creation; the Royal Abode, for a time at least, of the Eternal King. This being the character which has thus been conferred upon it, 86| how can we assent to the assertions of Astronomers, when they tell us that it is only one among millions of similar habitations, not distinguishable from them, except that it is smaller than most of them that we can measure; confused and rude in its materials like them? Or if we believe the Astronomers, will not such a belief lead us to doubt the truth of the great scheme of Christianity, which thus makes the earth the scene of a special dispensation?

13 This is the form in which Chalmers has taken up the argument. This is the difficulty which he proposes to solve; or rather, (such being as I have said the mode in which he presents the subject,) the objection which he proposes to refute. It is the bearing of the Astronomical discoveries of modern times, not upon the doctrines of Natural Religion, but upon the scheme of Christianity, which he discusses. And the question which he supposes his opponent to propound, as an objection to the Christian scheme, is;—How is it consistent with the dignity, the impartiality, the comprehensiveness, the analogy of God's proceedings, that he should make so special and pre-eminent a provision for the salvation of the inhabitants of this Earth, when there are such myriads of other worlds, all of which may require the like provision, and all of which have an equal claim to their Creator's care?

14 The answer which Chalmers gives to this objection, is one drawn, in the first instance, from our ignorance. He urges that, when the objector asserts that other worlds may have the like need with our own, of a special provision for the rescue of their inhabitants from the consequences of the transgression of God's laws, he is really making an assertion without the slightest foundation. Not only does Science not give us any information on such subjects, but the whole spirit of the scientific procedure, which has led to the knowledge which we possess, concerning other planets and other systems, is utterly opposed to our making such assumptions, |87

respecting other worlds, as the objection involves. Modern Science, in proportion as she is confident when she has good grounds of proof, however strange may be the doctrines proved, is not only diffident, but is utterly silent, and abstains even from guessing, when she has no grounds of proof. Chalmers takes Newton's reasoning, as offering a special example of this mixed temper, of courage in following the evidence, and temperance in not advancing when there is no evidence. He puts, in opposition to this, the example of the true philosophical temper,—a supposed rash theorist, who should make unwarranted suppositions and assumptions, concerning matters to which our scientific evidence does not reach;—the animals and plants, for instance, which are to be found in the planet Jupiter. No one, he says, would more utterly reject and condemn such speculations than Newton, who first rightly explained the motion of Jupiter and of his attendant satellites, about which Science *can* pronounce her truths. And thus, nothing can be more opposite to the real spirit of modern science, and astronomy in particular, than arguments, such as we have stated, professing to be drawn from science and from astronomy. Since we know nothing about the inhabitants of Jupiter, true science requires that we say and suppose nothing about them; still more requires that we should not, on the ground of assumptions made with regard to them, and other supposed groups of living creatures, reject a belief, founded on direct and positive proofs, such as is the belief in the truths of Natural and of Revealed Religion.

15 To this argument of Chalmers, we may not only give our full assent, but we may venture to suggest, in accordance with what we have already said, that the argument, when so put, is not stated in all its legitimate force. The assertion that the inhabitants of Jupiter have the same need as 88| we have, of a special dispensation for their preservation from

moral ruin, is not only as merely arbitrary an assumption, as any assertion could be, founded on a supposed knowledge of an analogy between the botany of Jupiter, and the botany of the earth; but it is a great deal more so. There may be circumstances which may afford some reason to believe that something of the nature of vegetables grows on the surface of Jupiter; for instance, if we find that he is a solid globe surrounded by an atmosphere, vapour, clouds, showers. But, as we have already said, there is an immeasurable distance between the existence of unprogressive tribes of organized creatures, plants, or even animals, and the existence of a progressive creature, which can pass through the conditions of receiving, discerning, disobeying, and obeying a moral law; which can be estranged from God, and then reconciled to him. To assume, without further proof, that there are, in Jupiter, creatures of such a nature that these descriptions apply to them, is a far bolder and more unphilosophical assumption, than any that the objector could make concerning the botany of Jupiter; and therefore, the objection thus supposed to be drawn from our supposed knowledge, is very properly answered by an appeal to our really utter ignorance, as to the points on which the argument rests.

16 This appeal to our ignorance is the main feature in Chalmers's reasonings, so far as the argument on the one side or the other has reference to science. Chalmers, indeed, pursues the argument into other fields of speculation. He urges, that not only we have no right to assume that other worlds require a redemption of the same kind as that provided for man, but that the very reverse may be the case. Man may be the only transgressor; and this, the only world that needed so great a provision for its salvation. We read in Scripture, expressions which imply that other beings, besides man, take an interest in the salvation of man. May not this be true of the inhabitants of other worlds, if such |89

inhabitants there be? These speculations he pursues to a considerable length, with great richness of imagination, and great eloquence. But the suppositions on which they proceed are too loosely connected with the results of science, to make it safe for us to dwell upon them here.

17 I conceive, as I have said, that the argument with which Chalmers thus deals admits of answers, also drawn from modern science, which to many persons will seem more complete than that which is thus drawn from our ignorance. But before I proceed to bring forward these answers, which will require several steps of explanation, I have one or two remarks still to make.

18 Undoubtedly they who believe firmly both that the earth has been the scene of a Divine Plan for the benefit of man, and also that other bodies in the universe are inhabited by creatures who may have an interest in such a Plan, are naturally led to conjectures and imaginations as to the nature and extent of that interest. The religious poet, in his Night Thoughts, interrogates the inhabitants of a distant star, whether their race too has, in its history, events resembling the fall of man, and the redemption of man.

> Enjoy your happy realms their golden age?
> And had your Eden an abstemious Eve?
> Or, if your mother fell are you redeemed?
> And if redeemed, is your Redeemer scorned?

And such imaginations may be readily allowed to the preacher or the poet, to be employed in order to impress upon man the conviction of his privileges, his thanklessness, his inconsistency, and the like. But every form in which such reflexions can be put shews how intimately they depend upon the nature and history of man. And when such reflexions are made the source of difficulty or objection in the way of religious thought, and when these difficulties and
90| objections are represented as derived from astronomical dis-

coveries, it cannot be superfluous to inquire whether astronomy has really discovered any ground for such objections. To some persons it may be more grateful to remedy one assumption by another: the assumption of moral agents in other worlds, by the assumption of some operation of the Divine Plan in other worlds. But since many persons find great difficulty in conceiving such an operation of the Divine Plan in a satisfactory way; and many persons also think that to make such unauthorized and fanciful assumptions with regard to the Divine Plans for the government of God's creatures is a violation of the humility, submission of mind, and spirit of reverence which religion requires; it may be useful if we can shew that such assumptions, with regard to the Divine Plans, are called forth by assumptions equally gratuitous on the other side: that Astronomy no more reveals to us extra-terrestrial moral agents, than Religion reveals to us extra-terrestrial Plans of Divine government. Chalmers has spoken of the *rashness* of making assumptions on such subjects without proof; leaving it however, to be supposed, that though astronomy does not supply proof of intelligent inhabitants of other parts of the universe, she yet does offer strong analogies in favour of such an opinion. But such a procedure is more than rash: when astronomical doctrines are presented in the form in which they have been already laid before the reader, which is the ordinary and popular mode of apprehending them, the analogies in favour of 'other worlds,' are (to say the least) greatly exaggerated. And by taking into account what astronomy really teaches us, and what we learn also from other sciences, I shall attempt to reduce such 'analogies' to their true value.

14 The privileges of man, which make the difficulty in assigning him his place in the vast scheme of the Universe, we have described as consisting in his being an *intellectual*, *moral*, and *religious* creature. Perhaps the privileges im- |91

plied in the last term, and their place in our argument, may justify a word more of explanation. Religion teaches us that there is opened to man, not only a prospect of a life in the presence of God, after this mortal life, but also the possibility and the duty of spending this life as in the presence of God. This is properly the highest result and manifestation of the effect of Religion upon man. Precisely because it is this, it is difficult to speak of this effect without seeming to use the language of enthusiasm: and yet again, precisely because it is so, our argument would be incomplete without a reference to it. There is for man, a possibility and a duty of bringing his thoughts, purposes, and affections more and more into continual unison with the will of God. This, even Natural Religion taught men, was the highest point at which man could aim; and Revealed Religion has still more clearly enjoined the duty of aiming at such a condition. The means of a progress towards such a state belong to the Religion of the heart and mind. They include a constant purification and elevation of the thoughts, affections, and will, wrought by habits of religious reflexion and meditation, of prayer and gratitude to God. Without entering into further explanation, all religious persons will agree that such a progress is, under happy influences, possible for man, and is the highest condition to which he can attain in this life. Whatever names may have been applied at different times to the steps of such a progress;—the cultivation of the divine nature in us; resignation; devotion; holiness; union with God; living in God, and with God in us;—religious persons will not doubt that there is a reality of internal state corresponding to these expressions; and that, to be capable of elevation into the condition which these expressions indicate, is one of the especial privileges of man. Man's soul, considered especially as the subject 92| of God's government, is often called his *Spirit:* and that man is capable of such conformity to the will of God, and approximation to Him, is sometimes expressed by speaking of him as a *spiritual creature.* And though the privilege of being, or of being capable of becoming, in this sense, a spiritual creature, is a part of man's religious privileges; we may sometimes be allowed to use this additional expression, in order to remind the reader, how great those religious privileges are, and how close is the relation between man and God, which they imply.

15 We have given a view of the peculiar character of man's condition, which seem to claim for him a nature and place, unique and incapable of repetition, in the scheme of the universe: and to this view, astronomy, exhibiting to us the habitation of man, as only one among many similar abodes, offers an objection. We are, therefore, now called upon, I conceive, to proceed to exhibit the answer which a somewhat different view of modern science suggests to this difficulty or objection.

For this purpose, we must begin by regarding the Earth in another point of view, different from that hitherto considered by us.

CHAPTER V.

Geology.

1 MAN, as I trust has been made apparent to the consciousness and conviction of the reader, is an intelligent, moral, religious, and spiritual creature; and we have to discuss the difficulty, or perplexity, or objection, which arises in our minds, when we consider such a creature, as occupying an habitation, which is but one among many globes, apparently equally fitted to be the dwelling-places of living things;—a mere speck in the immensity of creation;—an atom among such a vast array of material structures;—a world, as we needs must deem it, among millions of other objects, which appear to have an equal claim to be regarded as worlds.

2 The difficulty appears to be great, either way. Can the earth alone be the theatre of such intelligent, moral, religious and spiritual action? On the other hand; Can we conceive such action to go on in the other bodies of the universe? If we take the latter alternative, we must people other planets and other systems with men, such as we are, even as to their history. For the intellectual and moral condition of man implies a *history* of the species: and the view of man's condition, which religion presents, not only involves a scheme of which the history of the human race is a part, but also asserts a peculiar reference had, in the provisions of God, to the nature of man; and even a peculiar relation and connexion between the human and the divine nature. To extend such suppositions to other worlds,

94|

would be a proceeding so arbitrary and fanciful, that we are led to consider whether the alternative supposition may not be more admissible. The alternative supposition is, that man is, in an especial and eminent manner, the object of God's care;—that his place in the creation is, not that he merely occupies one among millions of similar domiciles, provided in boundless profusion by the Creator of the Universe; but that he is the servant, subject and child of God, in a way unique and peculiar: that his being a spiritual creature, (including his other attributes in the highest for the sake of brevity,) makes him belong to a spiritual world, which is not to be judged of merely by analogies belonging to the material universe.

3 Between these two difficulties, the choice is embarrassing, and the decision must be unsatisfactory, except we can find some further ground of judgment. But perhaps this is not hopeless. We have hitherto referred to the evidence and analogies supplied by one science, namely, astronomy. But there are other sciences, which give us information concerning the nature and history of the earth. From some of these, perhaps, we may obtain some knowledge of the place of the earth, in the scheme of creation;—how far it is, in its present condition, a thing unique, or only one thing among many like it. Any science which supplies us with evidence or information on this head, will give us aid, in forming a judgment upon the question under our consideration. To such sciences, then, we will turn our attention.

One science has employed itself in investigating the nature and history of the earth by an examination of the materials of which it is composed; namely, Geology. Let us call to mind some of the results at which this science has arrived.

4 A very little attention to what is going on among the materials of which the earth's surface is composed, suf-

fices to shew us that there are causes of change constantly and effectually at work. The earth's surface is composed of land and water, hills and vallies, rocks and rivers. But these features undergo change, and produce change in each other. The mountain-rivers cut deeper and deeper into the ravines in which they run; they break up the rocks over which they rush, use the fragments as implements of further destruction, pile them up in sloping mounds where the streams issue from the mountains, spread them over the plains, fill up lakes with sediment, push into the sea great deltas. The sea batters the cliffs, and eats away the land; and again, forms banks and islands, where there had been deep water. Volcanos pour out streams of lava, which destroy the vegetation over which they flow; and which again, after a series of years, are themselves clothed with vegetation. Earthquakes throw down tracts of land beneath the sea, and elevate other tracts from the bottom of the ocean. These agencies are every where manifest; and though, at a given moment, at a given spot, their effect may seem to us almost imperceptible, too insignificant to be taken account of; yet in a long course of years, almost every place has undergone considerable changes. Rivers have altered their courses, lakes have become plains, coasts have been swept away or have become inland districts, rich vallies have been ravaged by watery or fiery deluges, the country has in some way or other assumed a new face. The present aspect of the earth is in some degree different from what it was a few thousand years ago.

5 But yet, in truth, the changes of which we thus speak, have not been very considerable. The forms of countries, the lines of coasts, the ranges of mountains, the groups of vallies, the courses of rivers, are much the same now, as they were in ancient times. The face of the earth, since

96| man has had any knowledge of it, may have undergone some change; but the changeable has borne a small proportion to the permanent. Changes have taken place, and are taking place; but they do not take place rapidly. The ancient earth and the modern earth are, in all their main physical features, identical; and we must go backwards through a considerably larger interval than that which carries us back to what we usually term *antiquity*, before we are led, by the operation of causes now at work, to an aspect of the earth's surface very different from that which it now presents.

6 For instance, rivers do, no doubt, more or less alter, in the course of years, by natural causes. The Rhine, the Rhone, the Po, the Danube, have, certainly, during the last four thousand years, silted up their beds in level places, expanded the deltas at their mouths, changed the channels by which they enter the sea; and very probably, in their upper parts, altered the forms of their waterfalls and of their shingle beds. Yet even if we were thus to go backwards ten thousand, or twenty, or thirty thousand years, (setting aside great and violent causes of change, as earthquakes, volcanic eruptions, and the like,) the general form and course of these rivers, and of the ranges of mountains in which they flow, would not be different from what it is now. And the same may be said of coasts and islands, seas and bays. The present geography of the earth may be, and from all the evidence which we have, must be, very ancient, according to any measures of antiquity which can apply to human affairs.

7 But yet the further examination of the materials of the earth carries us to a view beyond this. Though the general forms of the land and the waters of continents and seas, were, several thousand years ago, much the same as they now are; yet it was not always so. We have clear evidence that large tracts which are now dry ground, were

|97

formerly the bed of the ocean: and these, not tracts of the shore, where the varying warfare of sea and land is still going on, but the very central parts of great continents; the Alps, the Pyrenees, the Himalayas. For not only are the rocks of which these great mountain-chains consist, of such structure that they appear to have been formed as layers of sediment at the bottom of water; but also, these layers contain vast accumulations of shells, or impressions of shells, and other remains of marine animals. And these appearances are not few, limited, or partial. The existence of such marine remains, in the solid substance of continents and mountains, is a general, predominant, and almost universal fact, in every part of the earth. Nor is any other way of accounting for this fact admissible, than that those materials really have, at some time, formed bottoms of seas. The various other conjectures and hypotheses, which were put forward on this subject, when the amount, extent, multiplicity, and coherence of the phenomena were not yet ascertained, and when their natural history was not yet studied, cannot now be considered as worthy of the smallest regard. That many of our highest hills are formed of materials raised from the depths of ocean, is a proposition which cannot be doubted, by any one, who fairly examines the evidence which nature offers.

8 If we take this proposition only, we cannot immediately connect it with our knowledge respecting the surface of the earth in its present form. We learn that what is now land, has been sea; and we may suppose (since it is natural to assume that the bulk of the sea has not much changed,) that what is now sea was formerly land. But, except we can learn something of the manner in which this change took place, we cannot make any use of our knowledge. Was the change sudden, or gradual; abrupt, or successive; brief, or long continuing?

98|

9 To these questions, the further study of the facts enables us to return answers with great confidence. The change or changes which produced the effects of which we have spoken—the conversion of the bottom of the ocean into the center of our greatest continents and highest mountains,—were undoubtedly gradual, successive, and long continued. We must state very briefly the grounds on which we make this assertion.

10 The masses which form our mountain-chains, offer evidence, as I have said, that they were deposited as sediment at the bottom of a sea, and then hardened. They consist of successive layers of such sediment, making up the whole mass of the mountain. These layers are, of course, to a certain extent, a measure of the time during which the deposition of sediment took place. The thicker the mass of sediment, the more numerous and varied its beds, and the longer period must we suppose to have been requisite for its formation. Without making any attempt at accurate or definite estimation, which would be to no purpose, it is plain that a mass of sedimentary strata five thousand or ten thousand feet thick, must have required, for its deposit, a long course of years, or rather, a long course of ages.

11 But again: on further examination it is found, that we have not merely one series of sedimentary deposits, thus forming our mountains. There are a number of different series of such layers or strata, to be found in different ranges of hills, and even in the same range, one series resting upon another. These different series of strata are distinguishable from one another by their general structure and appearance, besides more intimate characters, of which we shall shortly have to speak. Each such series appears to have a certain consistency of structure within itself; the layers of which it is composed being more or less parallel; but the successive

|99

series are not thus always parallel, the lower ones being often highly inclined and irregular, while the upper ones are more level and continuous; as if the lower strata had been broken up and thrown into disorder, and then a new series of strata had been deposited horizontally on their fragments. But in whatever way these different sedimentary series succeeded each other, each series must have required, as we have seen, a long period for its formation; and to estimate the length of the interval between the two series, we have, at the present stage of our exposition, no evidence.

12 But the mechanical structure of the strata, the result, as it seems, of aqueous sedimentary deposit, is not the only, nor the most important evidence, with regard to the length of time occupied by the formation of the rocky layers which now compose our mountains. As we have said, they contain shells, and other remains of creatures which live in the sea. These they contain, not in small numbers, scattered and detached, but in vast abundance, as they are found in those parts of the ocean which is most alive with them. There are the remains of oysters and other shell-fish in layers, as they live at present in the seas near our shores; of corals, in vast patches and beds, as they now occur in the waters of the Pacific; of shoals of fishes, of many different kinds, in immense abundance. Each of these beds of shells, of corals, and of fishes, must have required many years, perhaps many centuries, for the growth of the successive individuals and successive generations of which it consists: as long a time, perhaps, as the present inhabitants of the sea have lived therein: or many times longer, if there have been many such successive changes. And thus, while the present condition of the earth extends backwards to a period of vast but unknown antiquity; we have, offered to our notice, the evidence of a series of other periods, each of 100| which, so far as we can judge, may have been as long or

longer than that during which the dry land has had its present form.

13 But the most remarkable feature in the evidence is yet to come. We have spoken in general of the oysters, and corals, and fishes, which occur in the strata of our hills; as if they were creatures of the same kinds which we now designate by those names. But a more exact examination of these remains of organized beings, shows that this is not so The tribes of animals which are found petrified in our rocks. are almost all different, so far as our best natural historians can determine, from those which now live in our existing seas. They are different species; different genera. The creatures which we find thus imbedded in our mountains, are not only dead as individuals, but extinct as species. They belonged, not only to a terrestrial period, but to an animal creation, which is now past away. The earth is, it seems, a domicile which has outlasted more than one race of tenants.

14 It may seem rash and presumptuous in the natural historian to pronounce thus peremptorily that certain forms of life are nowhere to be found at present, even in the unfathomable and inaccessible depths of the ocean. But even if this were so, the proposition that the earth has changed its inhabitants, since the rocks were formed, of which our hills consist, does not depend for its proof on this assumption. For in the organic bodies which our strata contain, we find remains, not only of marine animals, but of animals which inhabit the fresh waters, and the land, and of plants. And the examination of such remains having been pursued with great zeal, and with all the aids which natural history can supply, the result has been, the proofs of a vast series of different tribes of animals and plants, which have successively occupied the earth and the seas; and of which the number, variety, multiplicity, and strangeness, exceed, by far, everything which could have been previously imagined. Thus

|101

Cuvier found, in the limestone strata on which Paris stands, animals of the most curious forms, combining in the most wonderful manner the qualities of different species of existing quadrupeds. In another series of strata, the Lias, which runs as a band across England from N. E. to S. W. we have the remains of lizards, or lacertine animals, different from those which now exist, of immense size and of extraordinary structure, some approaching to the form of fishes (*ichthyosaurus*); others, with the neck of a serpent; others with wings, like the fabled forms of dragons. Then beyond these, that is, anterior to them in the series of time, we have the immense collection of fossil plants, which occur in the Coal Strata; the shells and corals of the Mountain Limestone; the peculiar fishes, different altogether from existing fishes, of the Old Red Sandstone; and though, as we descend lower and lower, the traces of organic life appear to be more rare and more limited in kind, yet still we have, beneath these, in slates and in beds of limestone, many fossil remains, still differing from those which occur in the higher, and therefore, newer strata.

15 We have no intention of instituting any definite calculation with regard to the periods of time which this succession of forms of organic life may have occupied. This, indeed, the boldest geological speculators have not ventured to do. But the scientific discoveries, thus made, have a bearing upon the analogies of creation, quite as important as the discoveries of astronomy. And therefore we may state briefly some of the divisions of the series of terrestrial strata which have suggested themselves to geological inquirers. At the outset of such speculations, it was conceived that the lower rocks, composed of granite, slate, and the like, had existed before the earth was peopled with living things; and that these, being broken up into inclined positions, there were 102| deposited upon them, as the sediment of superincumbent

waters, strata more horizontal, containing organic remains. The former were then called *Primitive* or *Primary*, the latter, *Secondary* rocks. But it was soon found that this was too sweeping and peremptory a division. Rocks which had been classed as Primary, were found to contain traces of life: and hence, an intermediate class of *Transition* strata was spoken of. But this too was soon seen to be too narrow a scheme of arrangement, to take in the rapidly accumulating mass of facts, organic and others, which the geological record of the earth's history disclosed. It appeared that among the fossil-bearing strata there might be discerned a long series of Formations: the term *Formation* being used to imply a collection of successive strata, which, taking into account all the evidence, of materials, position, relations, and organic remains, appears to have been deposited during some one epoch or period; so as to form a natural group, chronologically and physiologically distinct from the others. In this way it appeared that, taking as the highest part of the Secondary series, the beds of chalk, which, marked by characteristic fossils, run through great tracts of Europe, with other beds, of sand and clay, which generally accompany these; there was, below this *Cretaceous Formation*, an *Oolitic Formation*, still more largely diffused, and still more abundant in its peculiar organic remains. Below this, we have, in England, the *New Red Sandstone Formation*, which, in other countries, is accompanied by beds abundant in fossils, as the *Muschelkalk* of Germany. Below this again we have the *Coal Formation*, and the *Mountain Limestone*, with their peculiar fossils. Below these, we have the Old Red Sandstone or Devonian System, with its peculiar fishes and other fossils. Beneath these, occur still numerous series of distinguishable strata; which have been arranged by Sir Roderick Murchison as the members of the *Silurian* formation; the researches by which it was established having been |103 carried on, in the first place, in South Wales, the ancient country of the Silures. Including the lower part of this formation, and descending still lower in order, is the *Cambrian* formation of Professor Sedgwick. And since the races of organic beings, as we thus descend through successive strata, seem to be fewer and fewer in their general types, till at last they disappear; these lower members of the geological series have been termed, according to their succession, *Palæozoic*, *Protozoic*, and *Hypozoic* or *Azoic*. The general impression on the minds of geologists has been, that, as we descend in this long staircase of natural steps, we are brought in view of a state of the earth in which life was scantily manifested, so as to appear to be near its earliest stages.

16 Each of these formations is of great thickness. Several of the members of each formation are hundreds, many of them thousands of feet thick. Taken altogether, they afford an astounding record of the time during which they must have been accumulating, and during which these successive groups of animals must have been brought into being, lived, and continued their kinds.

17 We must add, that over the Secondary strata there are found, in patches, generally of more limited extent, another, and of course, newer mass of strata, which have been termed *Tertiary Formations*. Of these, the strata, near and under Paris, lying in a hollow of the subjacent strata, and hence termed the *Paris Basin*, attracted prominent notice in the first place. And these were found to contain an immense quantity of remains of animals, which, being well preserved, and being subjected to a careful and scientific scrutiny by the great naturalist George Cuvier, had an eminent share in establishing in the minds of geologists the belief of the extinct character of fossil species, and of the possibility of reconstructing, from such remains, the animals, different from those which now live, which had formerly tenanted the earth.

104|

18 We have, in this enumeration, a series of groups of strata, each of which, speaking in a general way, has its own population of animals and plants, and is separated, by the peculiarities of these, from the groups below and above it. Each group may, in a general manner, be considered as a separate creation of animal and vegetable forms;—creatures which have lived and died, as the races now existing upon the earth live and die; and of which the living existence may, and according to all appearance, must, have occupied ages, and series of ages, such as have been occupied by the present living generations of the earth. This series of creations, or of successive periods of life, is, no doubt, a very striking and startling fact, very different from anything which the imagination of man, in previous stages of investigation of the earth's condition, had conceived; but still, is established by evidence so complete, drawn from an examination and knowledge of the structures of living things so exact and careful, as to leave no doubt whatever of the reality of the fact, on the minds of those who have attended to the evidence; founded, as it is, upon the analogies, offices, anatomy, and combinations of organic structures. The progress of human knowledge on this subject has been carried on and established by the same alternations of bold conjectures and felicitous confirmations of them,—of minute researches and large generalizations,—which have given reality and solidity to the other most certain portions of human knowledge. That the strata of the earth, as we descend from the highest to the lowest, are distinguished in general by characteristic or organic fossils, and that these forms of organization are different from those which now live on the earth, are truths as clearly and indisputably established in the minds of those who have the requisite knowledge of geology and natural history, as that the planets revolve round the sun, and the satellites round the planets. That these epochs of creation are something quite |105

different from anything which we now see taking place on the earth, no more disturbs the belief of those facts, which scientific explorers entertain, than the seemingly obvious difference between the nebulæ which are regarded as yet unformed planetary systems, and the solar system to which our earth belongs, disturbs the belief of astronomers, that such nebulæ, as well as our system, really exist. Indeed we may say, as we shall hereafter see, that the fact of our earth having passed through the series of periods of organic life which geologists recognize, is, hitherto, incomparably better established, than the fact that the nebulæ, or any of them, are passing through a series of changes, such as may lead to a system like ours; as some eminent astronomers in modern times have held. In this respect, the history of the world, and its place in the universe, are far more clearly learnt from geology than from astronomy.

19 But with regard to this series of Organic *Creations*, if, for the sake of brevity, we may call them so; we may naturally ask, in what manner, by what agencies, at what intervals, they succeeded each other on the earth? Now do the researches of geologists give us any information on these points, which may be brought to bear upon our present speculations? If we ask these questions, we receive, from different classes of geologists, different answers. A little while ago, most geologists held, probably the greater number still hold, that the transitions from one of these periods of organic life to another, were accompanied generally by seasons of violent disruption and mutation of the surface of the earth, exceeding anything which has taken place since the surface assumed its present general form; in the same proportion as the changes of its organic population go beyond any such changes which we can discern to be at present in operation. And there were found to be changes of other 106| kinds, which seemed to shew that these epochs of organic

transition had also been epochs of mechanical violence, upon a vast and wonderful scale. It appeared that, at some of these epochs at least, the strata previously deposited, as if in comparative tranquillity, had been broken, thrust up from below, or drawn or cast downwards; so that strata which must at first have been nearly level, were thrown into positions highly inclined, fractured, set on edge, contorted, even inverted. Over the broken edges of these strata, thus disturbed and fractured, were found vast accumulations of the fragments which such rude treatment might naturally produce; these fragmentary ruins being spread in beds comparatively level, over the bristling edges of the subjacent rocks, as if deposited in the fluid which had overwhelmed the previous structure; and with few or no traces of life appearing in this mass of ruins; while, in the strata which lay over them, and which appeared to have been the result of quieter times, new forms of organic life made their appearance in vast abundance. Such is, for example, the relation of the coal strata in a great part of England; broken into innumerable basins, ridges, vallies, strips and shreds, lying in all positions; and then filled into a sort of level, by the conglomerate of the magnesian limestone, and the superincumbent red sandstone and oolites. In other cases it appeared as if there were the means of tracing, in these dislocations, the agency of igneous stony matter, which had been injected from below, so as to form mountain-chains, or the cores of such; and in which the period of the convulsion could be traced, by the strata to which the disturbance extended; *those* strata being supposed to have been deposited before the eruption, which were thrust upwards by it into highly inclined positions; while those strata which, though near to these scenes of mechanical violence, were still comparatively horizontal, as they had been originally deposited, were naturally inferred to have been formed in the waters, after the |107 catastrophe had passed away. By such reasonings as these, M. Elie de Beaumont has conceived that he can ascertain the relative ages, (according to the vast and loose measurements of age which belong to this subject,) of the principal ranges of mountains of the earth's surface.

20 Such estimations of age can, indeed, as we have intimated, be only of the widest and loosest kind; yet they all concur in assigning very great and gigantic periods of time, as having been occupied by the events which have formed the earth's strata, and brought them into their present position. For not only must there have been long ages employed, as we have said, while the successive generations of each group of animals lived, and died, and were entombed in the abraded fragments of the then existing earth; but the other operations, which intervened between these apparently more tranquil processes, must also have occupied, it would seem, long ages at each interval. The dislocation, disruption, and contortion of the vast masses of previously existing mountains, by which their frame-work was broken up, and its ruins covered with beds of its own rubbish, many thousand feet thick, and gradually becoming less coarse and smoother, as the higher beds were deposited upon the lower; could hardly take place, it would seem, except in hundreds and thousands of years. And then again, all these processes of deposition, thus arranging loose masses of material into level beds, must have taken place at the bottom of deep oceans: and the beds of these oceans must have been elevated into the position of mountain ridges which they now occupy, by some mighty operation of nature, which must have been comparatively tranquil, since it has not much disturbed those more level beds; and which therefore must have been comparatively long continued. If we accept, as so many eminent geologists have done, this evidence of a vast series of successive periods of alternate violence and repose, we must

108|

assign to each such period a duration which cannot but be immense, compared with the periods of time with which we are commonly conversant. In the periods of comparative quiet, such as now exist on the earth's surface, and such as seem to be alone consistent with continued life and successive generation, deposits at the bottom of lakes and seas take place, it would seem, only at the rate of a few feet in a year, or perhaps, in a century. When, therefore, we find strata, bearing evidence of such a mode of deposit, and piled up to the amount of thousands and tens of thousands of feet, we are naturally led to regard them as the productions of myriads of years; and to add new myriads, as often as, in the prosecution of geological research, we are brought to new masses of strata of the like kind; and again, to interpolate new periods of the same order, to allow for the transition from one such group to another.

21 Nor is there anything which need startle us, in the necessity of assuming such vast intervals of time, when we have once brought ourselves to deal with the question of the antiquity of the earth upon scientific evidence alone. For if geology thus carries us far backwards through thousands, it may be, millions of years, astronomy does not offer the smallest argument to check this regressive supposition. On the contrary, all the most subtle and profound investigations of astronomers have led them to the conviction, that the motions of the earth may have gone on, as they now go on, for an indefinite period of past time. There is no tendency to derangement in the mechanism of the solar system, so far as science has explored it. Minute inequalities in the movements exist, too small to produce any perceptible effect on the condition of the earth's surface; and even these inequalities, after growing up through long cycles of ages, to an amount barely capable of being detected by astronomical scrutiny, reach a maximum; and, diminishing by the same

slow degrees by which they increased, correct themselves, and disappear. The solar system, and the earth as part of it, constitute, so far as we can discover, a Perpetual Motion.

22 There is therefore nothing, in what we know of the Cosmical conditions of our globe, to contradict the Terrestrial evidence for its vast antiquity, as the seat of organic life. If, for the sake of giving definiteness to our notions, we were to assume that the numbers which express the antiquity of these four Periods;—the Present organic condition of the earth; the Tertiary Period of geologists, which preceded that; the Secondary Period, which was anterior to that; and the Primary Period which preceded the Secondary; were on the same scale as the numbers which express these four magnitudes:—the magnitude of the Earth; that of the Solar System compared with the Earth; the distance of the nearest Fixed Stars compared with the solar system; and the distance of the most remote Nebulæ compared with the nearest fixed stars; there is, in the evidence which geological science offers, nothing to contradict such an assumption.

23 And as the infinite extent which we necessarily ascribe to space, allows us to find room, without any mental difficulty, for the vast distances which astronomy reveals, and even leaves us rather embarrassed with the infinite extent which lies beyond our farthest explorations; so the infinite duration which we, in like manner, necessarily ascribe to past time, makes it easy for us, so far as our powers of intellect are concerned, to go millions of millions of years backwards, in order to trace the beginning of the earth's existence,—the first step of terrestrial creation. It is as easy for the mind of man to reason respecting a system which is billions or trillions of miles in extent, and has endured through the like number of years, or centuries, as it is to reason about a system, (the earth, for instance,) which is

110|

forty million feet in extent, and has endured for a hundred thousand million of seconds, that is, a few thousand years.

24 This statement is amply sufficient for the argument which we have to found upon it; but before I proceed to do that, I will give another view which has recently been adopted by some geologists, of the mode in which the successive periods of creation, which geological research discloses to us, have passed into one another. According to this new view, we find no sufficient reason to believe that the history of the earth, as read by us in the organic and mechanical phenomena of its superficial parts, has consisted of such an alternation of periods of violence and of repose, as we have just attempted to describe. According to these theorists, strata have succeeded strata, one group of animals and plants has followed another, through a season of uniform change; with no greater paroxysm or catastrophe, it may be, than has occurred during the time that man has been an observer of the earth. It may be asked, how is this consistent with the phenomena which we have described;—with the vast masses of ruin, which mark the end of one period and the beginning of another, as is the case in passing from the coal measures of England to the superincumbent beds;—with the highly inclined strata of the central masses, and the level beds of the upper formations which have been described as marking the mountain ranges of Europe? To these questions, a reply is furnished, we are told, by a more extensive and careful examination of the strata. It may be, that in certain localities, in certain districts, the transition, from the mountain limestone and the coal, to the superjacent sandstones and oolites, is abrupt and seemingly violent; marked by *unconformable* positions of the upper upon the lower strata, by beds of conglomerate, by the absence of organic remains in certain of these beds. But if we follow these very strata into other parts of the world, or even |III

into other parts of this island, we find that this abruptness and incongruity between the lower and the higher strata disappears. Between the mountain-limestone and the red sandstone which lies over it, certain new beds are found, which fill up the incoherent interval; which offer the same evidence as the strata below and above them, of having been produced tranquilly; and which do not violently differ in position from either group. The appearance of incoherence in the series arose from the occurrence, in the region first examined, of a gap, which is here filled up,—a blank which is here supplied. Hence it is inferred, that whatever of violence and extreme disturbance is indicated by the dislocations and ruins there observed, was local and partial only; and that, at the very time when these fragmentary beds, void of organized beings, were forming in one place, there were, at the same time, going on, in another part of the earth's surface, not far removed, the processes of the life, death and imbedding of species, as tranquilly as at any other period. And the same assertion is made with regard to the more general fact, before described, of the stratigraphical constitution of mountain chains. It is asserted that the unconformable relation of the strata which compose the different parts of those chains, is a local occurrence only; and that the same strata, if followed into other regions, are found conformable to each other; or are reduced to a virtually continuous scheme, by the interpolation of other strata, which make a transition, in which no evidence of exceptional violence appears.

25 We shall not attempt, (it is not at all necessary for us to do so,) to decide between the doctrines of the two geological schools which thus stand in this opposition to each other. But it will be useful to our argument to state somewhat further the opinions of this latter school on one main point. We must explain the view which these geo-

112|

logists take of the mode of succession of one group of *organized* beings to another; by which, as we have said, the different successive strata are characterized. Such a phenomenon, it would at first seem, cannot be brought within the ordinary rules of the existing state of things. The species of plants and animals which inhabit the earth, do not change from age to age; they are the same in modern times, as they were in the most remote antiquity, of which we have any record. The dogs and horses, sheep and cattle, lions and wolves, eagles and swallows, corn and vines, oaks and cedars, which occupy the earth now, are not, we have the strongest reasons to believe, essentially different now from what they were in the earliest ages. At least, if one or two species have disappeared, no new species have come into existence. We cannot conceive a greater violation of the known laws of nature, than that such an event as the appearance of a new species should have occurred. Even those who hold the uniformity of the mechanical changes of the earth, and of the rate of change, from age to age, and from one geological period to another; must still, it would seem, allow that the zoological and phytological changes of which geology gives her testimony, are complete exceptions to what is now taking place. The formation of strata at the bottom of the ocean from the ruin of existing continents, may be going on at present. Even the elevation of the bed of the ocean in certain places, as a process imperceptibly slow, may be in action at this moment, as these theorists hold that it is. But still, even when the beds thus formed are elevated into mountain chains, if that should happen, in the course of myriads of years, (according to the supposition it cannot be effected in a less period,) the strata of such mountain chains will still contain only the species of such creatures as now inhabit the waters; and we shall have, even then, no succession of organic epochs, such as geology discovers in the existing mountains of the earth.

26 The answer which is made to this objection appears to me to involve a license of assumption on the part of the *uniformitarian* geologist, (as such theorists have been termed,) which goes quite beyond the bounds of natural philosophy: but I wish to state it; partly, in order to shew that the most ingenious men, stimulated by the exigencies of a theory, which requires some hypothesis concerning the succession of species, to make it coherent and complete, have still found it impossible to bring the creation of species of plants and animals within the domain of natural science; and partly, to shew how easily and readily geological theorists are led to assume periods of time, even of a higher order than those which I have ventured to suggest.

27 It must, however, be first stated, as a fact on which the assumption is founded which I have to notice, that the organic groups by which these successive strata are characterized, are not so distinct and separate, as it was convenient, for the sake of explanation, to describe them in the first instance. Although each body of strata is marked by predominant groups of genera and species, yet it is not true, that all the species of each formation disappear, when we proceed to the next. Some species and genera endure through several successive groups of strata; while others disappear, and new forms come into view, as we ascend. And thus, the change from one set of organic forms to another, as we advance in time, is made, not altogether by abrupt transitions, but in part continuously. The uniformitarian, in the case of organic, as in the case of mechanical change, obliterates or weakens the evidence of sudden and catastrophic leaps, by interposing intermediate steps, which involve, partly the phenomena of the preceding, and partly

those of the subsequent condition. As he allows no universal transition from one deposit to a succeeding discrepant and unconformable deposit, so he allows no abrupt and complete transition from one collection of organic beings,—one creation, as we may call it,—to another. If creation must needs be an act out of the region of natural science, he will have it to be at least an act not exercised at distant intervals, and on peculiar occasions; but constantly going on, and producing its effects, as much at one time in the geological history of the world, as at another.

28 And this he holds, not only with regard to the geological periods which have preceded the existing condition of the earth, but also with regard to the transition from those previous periods to that in which we live. The present population of the earth is not one in which all previous forms are extinct. The past population of the earth was not one in which there are found no creatures still living. On the contrary, he finds that there exists a vast mass of strata, superior to the secondary strata, which are characterized by extinct forms, and are yet inferior to those deposits which are now going on by the agency of obvious causes. These masses of strata contain a population of creatures, partly extinct species, and partly such species as are still living on our land and in our waters. The proportion in which the old and the new species occur in such strata, is various; and the strata are so numerous, so rich in organic remains, so different from each other, and have been so well explored, that they have been classified and named according to the proportion of new and of old species which they contain. Those which contain the largest proportion of species still living, have been termed *Pliocene,* as containing a *greater* number of *new* or recent species. Below these, are strata which are termed *Miocene,* implying a *smaller* number of *new* species. Below these again, are others which have |115

been termed *Eocene*, as containing few new species indeed, but yet enough to mark the *dawn*, the *Eos*, of the existing state of the organic world. These strata are, in many places, of very considerable thickness; and their number, their succession, and the great amount of extinct species which they contain, shows, in a manner which cannot be questioned, (if the evidence of geology is accepted at all,) in what a gradual manner, a portion at least, of the existing forms of organic life have taken the place of a different population previously existing on the surface of the globe.

29 And thus the uniformitarian is led to consider the facts which geology brings to light, as indicating a slow and almost imperceptible, but, upon the whole, constant series of changes, not only in the position of the earth's materials, but in its animal and vegetable population. Land becomes sea and sea becomes land; the beds of oceans are elevated into mountain regions, carrying with them the remains of their inhabitants; sheets of lava pour from volcanic vents and overwhelm the seats of life; and these, again, become fields of vegetation; or, it may be, descend to the depths of the sea, and are overgrown with groves of coral; lakes are filled with sediment, imbedding the remains of land animals, and form the museums of future zoologists; the deltas of mighty rivers become the centers of continents, and are excavated as coal-fields by men in remote ages. And yet all this time, so slow is the change, that man is unaware such changes are going on. He knows that the mountains of Scandinavia are rising out of the Baltic at the rate of a few feet in a century; he knows that the fertile slope of Etna has been growing for thousands of years by the addition of lava streams and parasitic volcanos; he knows that the delta of the Mississippi accumulates hundreds of miles of vegetable matter every generation; he knows that the shores

116| of Europe are yielding to the sea; but all these appear to him minute items, not worth summing; infinitesimal quantities, which he cannot integrate. And so, in truth, they are, for him. His ephemeral existence does not allow him to form a just conception, in any ordinary state of mind, of the effects of this constant agency of change, working through countless thousands of years. But Time, inexhausted and unremitting, sums the series, integrates the formula of change; and thus passes, with sure though noiseless progress, from one geological epoch to another.

30 And in the meanwhile, to complete the view thus taken by the uniformitarian of the geological history of the earth, by some constant but inscrutable law, creative agency is perpetually at work, to introduce, into this progressive system of things, new species of vegetable and animal life. Organic forms, ever and ever new ones, are brought into being, and left, visible footsteps, as it were, of the progress which Time has made;—marks placed between the rocky leaves of the book of creation; by which man, when his time comes, may turn back and read the past history of his habitation. But the point for us to remark is, the immeasurable, the inconceivable length of time, if any length of time could be inconceivable, which is required of our thoughts, by this new assumption of the constant production of new species, as a law of creation. We might feel ourselves well nigh overwhelmed, when, by looking at processes which we see producing only a few feet of height or breadth or depth during the life of man, we are called upon to imagine the construction of Alps and Andes,—when we have to imagine a world made a few inches in a century. But there, at least, we had *something* to start from: the element of change was small, but there *was* an element of change: we had to expand, but we had not to originate. But in conceiving that all the myriads of successive species, which we find in the earth's strata, have come into being by

|117

a law which is now operating, we have *nothing* to start from. We have seen, and know of, no such change; all sober and skilful naturalists reject it, as a fact not belonging to our time. We have here to build a theory without materials;—to sum a series of which every term, so far as we know, is nothing;—to introduce into our scientific reasonings an assumption contrary to all scientific knowledge.

31 This appears to me to be the real character of the assumption of the constant creation of new species. But, as I have said, it is not my business here, to pronounce upon the value or truth of this assumption. The only use which I wish to make of it is this:—If any persons, who have adopted the geological view which I have just been explaining, should feel any interest in the speculations here offered to their notice, they must needs be (as I have no doubt they will be) even more willing than other geologists, to grant to our argument a scale of time for geological succession, corresponding in magnitude to the scale of distances which astronomy teaches us, as those which measure the relation of the universe to the earth.

This being supposed to be granted, I am prepared to proceed with my argument.

CHAPTER VI.

The Argument from Geology.

1 I HAVE endeavoured to explain that, according to the discoveries of geologists, the masses of which the surface of the earth is composed, exhibit indisputable evidence that, at different successive periods, the land and the waters which occupy it, have been inhabited by successive races of plants and animals; which, when taken in large groups, according to the ascending or descending order of the strata, consist of species different from those above and below them. Many of these groups of species are of forms so different from any living things which now exist, as to give to the life of those ancient periods an aspect strangely diverse from that which life now displays, and to transfer us, in thought, to a creation remote in its predominant forms from that among which we live. I have shewn also, that the life and successive generations of these groups of species, and the events by which the rocks which contain these remains have been brought into their present situation and condition, must have occupied immense intervals of time;—intervals so large as to deserve to be compared, in their numerical expression, with the intervals of space which separate the planets and stars from each other. It has been seen, also, that the best geologists and natural historians have not been able to devise any hypothesis to account for the successive introduction of these new species into the earth's population; except the exercise of a series of acts of creation, by which they have been brought into being; either in groups at once, or in a perpetual succession of one or a few species, which the course of long intervals of time might accumulate into groups of |119

species. It is true, that some speculators have held that by the agency of natural causes, such as operate upon organic forms, one species might be transmuted into another; external conditions of climate, food, and the like, being supposed to conspire with internal impulses and tendencies, so as to produce this effect. This supposition is, however, on a more exact examination of the laws of animal life, found to be destitute of proof: and the doctrine of the successive creation of species remains firmly established among geologists. That the *extinction* of species, and of groups of species, may be accounted for by natural causes, is a proposition much more plausible, and to a certain extent, probable; for we have good reason to believe that, even within the time of human history, some few species have ceased to exist upon the earth. But whether the extinction of such vast groups of species as the ancient strata present to our notice, can be accounted for in this way, at least without assuming the occurrence of great catastrophes, which must, for a time, have destroyed all forms of life in the district in which they occurred, appears to be more doubtful. The decision of these questions, however, is not essential to our purpose. What is important is, that immense numbers of tribes of animals have tenanted the earth for countless ages, before the present state of things began to be.

2 The present state of things is that to which the existence and the history of MAN belong; and the remark which I now have to make is, that the existence and the history of Man are facts of an entirely different order from any which existed in any of the previous states of the earth; and that this history has occupied a series of years which, compared with geological periods, may be regarded as very brief and limited.

3 The remains of man are nowhere found in the strata which contain the records of former states of the earth.

120

Skeletons of vast varieties of creatures have been disinterred from their rocky tombs; but these cemeteries of nature supply no portion of a human skeleton. In earlier periods of natural science, when comparative anatomy was as yet very imperfectly understood, no doubt, many fossil bones were supposed to be human bones. The remains of giants and of antediluvians were frequent in museums. But a further knowledge of anatomy has made it appear that such bones all belong to animals, of one kind or another; often, to animals utterly different, in their form and skeleton, from man. Also some bones, really human, have been found petrified in situations in which petrification has gone on in recent times, and is still going on. Human skeletons, imbedded in rock by this process, have been found in the island of Guadaloupe, and elsewhere. But this phenomenon is easily distinguishable from the petrified bones of other animals, which are found in rocks belonging to really geological periods; and does not at all obliterate the distinction between the geological and the historical periods.

4 Indeed not bones only, but objects of art, produced by human workmanship, are found fossilized and petrified by the like processes; and these, of course, belong to the historical period. Human bones, and human works, are found in such deposits as morasses, sand-banks, lava-streams, mounds of volcanic ashes; and many of them may be of unknown, and, compared with the duration of a few generations, of very great antiquity: but such deposits are distinguishable, generally without difficulty, from the strata in which the geologist reads the records of former creations. It has been truly said, that the geologist is an *Antiquary*; for, like the antiquary, he traces a past condition of things in the remains and effects of it which still subsist; but it has also been truly said at the same time, that he is an antiquary *of a new Order*; for the remains which he studies are those

|121

which illustrate the history of the earth, not of man. The geologist's antiquity is not that of ornaments and arms, utensils and habiliments, walls and mounds; but of species and of genera, of seas and of mountains. It is true, that the geologist may have to study the works of man, in order to trace the effects of causes which produce the results which he investigates; as when he examines the pholad-pierced pillars of Puteoli, to prove the rise and the fall of the ground on which they stand; or notes the anchoring-rings in the wall of some Roman edifice, once a maritime fort, but now a ruin remote from the sea; or when he remarks the streets in the towns of Scania, which are now below the level of the Baltic*, and therefore shew that the land has sunk since these pavements were laid. But in studying such objects, the geologist considers the hand of man as only one among many agencies. Man is to him only one of the natural causes of change.

5 And if, with the illustrious author to whom we have just referred†, we liken the fossil remains, by which the geologist determines the age of his strata, to the Medals and Coins in which the antiquary finds the record of reigns and dynasties; we must still recollect that a *Coin* really discloses a vast body of characteristics of man, to which there is nothing approaching in the previous condition of the world. For how much does a Coin or Medal indicate? Property; exchange; government; a standard of value; the arts of mining, assaying, coining, drawing, and sculpture; language, writing, and reckoning; historical recollections, and the wish to be remembered by future ages. All this is involved in that small human work, a Coin. If the fossil remains of animals may (as has been said) be termed Medals struck by Nature to record the epochs of her history; Medals must be said to be, not merely, like fossil remains, records of material things; they are the records of thought, purpose, society,

122| * Lyell, II. 420. [6th Ed.] † Cuvier.

long continued, long improved, supplied with multiplied aids and helps; they are the permanent results, in a minute compass, of a vast progress, extending through all the ramifications of human life.

6 Not a coin merely, but any, the rudest work of human art, carries us far beyond the domain of mere animal life. There is no transition from man to animals. No doubt, there are races of men very degraded, barbarous, and brutish. No doubt there are kinds of animals which are very intelligent and sagacious; and some which are exceedingly disposed to and adapted to companionship with man. But by elevating the intelligence of the brute, we do not make it become the intelligence of the man. By making man barbarous, we do not make him cease to be man. Animals have their especial capacities, which may be carried very far, and may approach near to human sagacity, or may even go beyond it: but the capacity of man is of a different kind. It is a capacity, not for becoming sagacious, but for becoming rational; or rather, it is a capacity which he has in virtue of being rational. It is a capacity of progress. In animals, however sagacious, however well trained, the progress in skill and knowledge is limited, and very narrowly limited. The creature soon reaches a boundary, beyond which it cannot pass: and even if the acquired habits be transmitted by descent to another generation, (which happens in the case of dogs and several other animals,) still the race soon comes to a stand in its accomplishments. But in man, the possible progress from generation to generation, in intelligence and knowledge, and we may also say, in power, is indefinite: or if this be doubted; it is at least so vast, that compared with animals, his capacity is infinite. And this capacity extends to all races of men its characterizing efficacy: for we have good reason to believe that there is no race of human beings who may not, by a due course of culture, continued through gene-

|123

rations, be brought into a community of intelligence and power with the most intelligent and the most powerful races. This seems to be well established, for instance, with regard to the African negroes; so long regarded by most, by some probably regarded still, as a race inferior to Europeans. It has been found that they are abundantly capable of taking a share in the arts, literature, morality and religion of European peoples. And we cannot doubt that, in the same manner, the native Australians, or the Bushmen of the Cape of Good Hope, have human faculties and human capacities; however difficult it might be to unfold these, in one or two generations, into a form of intelligence and civilization in any considerable degree resembling our own.

7 It is not requisite for us, and it might lead to unnecessary difficulties, to fix upon any one attribute of man, as peculiarly characteristic, and distinguishing him from brutes. Yet it would not be too much to say that man is, in truth, universally and specifically characterized by the possession of *Language*. It will not be questioned that language, in its highest forms, is a wonderful vehicle and a striking evidence of the intelligence of man. His bodily organs can, by a few scarcely perceptible motions, shape the air into sounds which express the kinds, properties, actions and relations of things, under thousands of aspects, in forms infinitely more general and recondite than those in which they present themselves to his senses;—and he can, by means of these forms, aided by the use of his senses, explore the boundless regions of space, the far recesses of past time, the order of nature, the working of the Author of nature. This man does, by the exercise of his Reason, and by the use of Language, a necessary implement of his Reason for such purposes.

8 That Language, in such a stage, is a special character of man, will not be doubted. But it may be thought, there is little resemblance between Language in this exalted degree

124|

of perfection, and the seemingly senseless gibberish of the most barbarous tribes. Such an opinion, however, might easily be carried too far. All human language has in it the elements of indefinite intellectual activity, and the germs of indefinite development. Even the rudest kind of speech, used by savages, denotes objects by their kinds, their attributes, their relations, with a degree of generality derived from the intellect, not from the senses. The generality may be very limited; the relations which the human intellect is capable of apprehending may be imperfectly conveyed. But to denote kinds and attributes and actions and relations *at all*, is a beginning of generalization and abstraction;—or rather, is far more than a beginning. It is the work of a faculty which can generalize and abstract; and these mental processes once begun, the field of progress which is open to them is indefinite. Undoubtedly it may happen that weak and barbarous tribes are, for many generations, so hard pressed by circumstances, and their faculties so entirely absorbed in providing for the bare wants of the poorest life, that their thoughts may never travel to anything beyond these, and their language may not be extended so as to be applicable to any other purposes. But this is not the standard condition of mankind. It is not, by such cases, that man, or that human nature, is to be judged. The normal condition of man is one of an advance beyond the mere means of subsistence, to the arts of life, and the exercise of thought in a general form. To some extent, such an advance has taken place in almost every region of the earth and in every age.

9 Perhaps we may often have a tendency to think more meanly than they deserve, of so-called barbarous tribes, and of those whose intellectual habits differ much from our own. We may be prone to regard ourselves as standing at the summit of civilization; and all other nations and ages, as not only occupying inferior positions, but positions on a

|125

slope which descends till it sinks into the nature of brutes. And yet how little does an examination of the history of mankind justify this view! The different stages of civilization, and of intellectual culture, which have prevailed among them, have had no appearance of belonging to one single series, in which the cases differed only as higher or lower. On the contrary, there have been many very different kinds of civilization, accompanied by different forms of art and of thought; showing how universally the human mind tends to such habits, and how rich it is in the modes of manifesting its innate powers. How different have been the forms of civilization among the Chinese, the Indians, the Egyptians, the Babylonians, the Mexicans, the Peruvians! Yet in all, how much was displayed of sagacity and skill, of perseverance and progress, of mental activity and grasp, of thoughtfulness and power. Are we, in thinking of these manifestations of human capacity, to think of them as only a stage between us and brutes? or are we to think so, even of the stoical Red Indians of North America, or the energetic New Zealanders, and Caffres? And if not, why of the African Negroes, or the Australians, or the Bushmen? We may call their Language a jargon. Very probably it would, in its present form, be unable to express a great deal of what we are in the habit of putting into language. But can we refuse to believe that, with regard to matters with which they are familiar, and on occasions where they are interested, they would be to each other intelligible and clear? And if we suppose cases in which their affections and emotions are strongly excited, (and affections and emotions at least we cannot deny them,) can we not believe that they would be eloquent and impressive? Do we not know, in fact, that almost all nations which we call savage, are, on such occasions, eloquent in their own language? And since this is so, must not their language, after all, be a wonderful instru-

126|

ment as well as ours? Since it can convey one man's thoughts and emotions to many, clothed in the form which they assume in his mind; giving to things, it may be, an aspect quite different from that which they would have if presented to their own senses; guiding their conviction, warming their hearts, impelling their purposes;—can language, even in such cases, be otherwise than a wonderful produce of man's internal, of his mental, that is, of his peculiarly *human* faculties? And is not language, therefore, even in what we regard as its lowest forms, an endowment which completely separates man from animals which have no such faculty?—which cannot regard, or which cannot convey, the impressions of the individual in any such general and abstract form? Probably we should find, as those who have studied the languages of savages always have found, that every such language contains a number of curious and subtle practices,—*contrivances,* we cannot help calling them,—for marking the relations, bearings and connexions of words; contrivances quite different from those of the languages which we think of as more perfect; but yet, in the mouths of those who use such speech, answering their purpose with great precision. But without going into such details, the use of any *articulate* language is, as the oldest Greeks spoke of it, a special and complete distinction of man as man.

10 It would be an obscure and useless labour, to speculate upon the question whether animals have among themselves anything which can properly be called *Language.* That they have anything which can be termed Language, in the sense in which we here speak of it, as admitting of general expressions, abstractions, address to numbers, eloquence, is utterly at variance with any interpretation which we can put upon their proceedings. The broad distinction of Instinct and Reason, however obscure it may be, yet seems to be most simply described, by saying, that animals do not |127

apprehend their impressions under general forms, and that man does. Resemblance, and consequent association of impressions, may often shew like generalization; but yet it is different. There is, in man's mind, a germ of general thoughts, suggested by resemblances, which is evolved and fixed in language: and by the aid of such an addition to the impressions of sense, man has thousands of intellectual pathways from object to object, from effect to cause, from fact to inference. His impressions are projected on a sphere of thought of which the radii can be prolonged into the farthest regions of the universe. Animals, on the contrary, are shut up in their sphere of sensation,—passing from one impression to another by various associations, established by circumstances; but still, having access to no wider intellectual region, through which lie lines of transition purely abstract and mental. That they have their modes of communicating their impressions and associations, their affections and emotions, we know; but these modes of communication do not make a language; nor do they disturb the assignment of Language as a special character of man; nor the belief that man differs in his Kind, and we may say, using a larger phrase, in his Order, from all other creatures.

11 We may sometimes be led to assign much of the development of man's peculiar powers, to the influence of external circumstances. And that the development of those powers is so influenced, we cannot doubt; but their development only, not their existence. We have already said that savages, living a precarious and miserable life, occupied incessantly with providing for their mere bodily wants, are not likely to possess language, or any other characteristic of humanity, in any but a stunted and imperfect form. But, that manhood is debased and degraded under such adverse conditions, does not make man cease to be man. Even from such an abject race, if a child be taken and brought

128|

up among the comforts and means of development which civilized life supplies, he does not fail to shew that he possesses, perhaps in an eminent degree, the powers which specially belong to man. The evidences of human tendencies, human thoughts, human capacities, human affections and sympathies, appear conspicuously, in cases in which there has been no time for external circumstances to operate in any great degree, so as to unfold any difference between the man and the brute; or in which the influence of the most general of external agencies, the impressions of several of the senses, have been intercepted. Who that sees a lively child, looking with eager and curious eyes at every object, uttering cries that express every variety of elementary human emotion in the most vivacious manner, exchanging looks and gestures, and inarticulate sounds, with his nurse, can doubt that already he possesses the germs of human feeling, thought and knowledge? that already, before he can form or understand a single articulate word, he has within him the materials of an infinite exuberance of utterance, and an impulse to find the language into which such utterance is to be moulded by the law of his human nature? And perhaps it may have happened to others, as it has to me, to know a child who had been both deaf, dumb, and blind, from a very early age. Yet she, as years went on, disclosed a perpetually growing sympathy with the other children of the family in all their actions, with which of course she could only acquaint herself by the sense of touch. She sat, dressed, walked, as they did; even imitated them in holding a book in her hand when they read, and in kneeling when they prayed. No one could look at the change which came over her sightless countenance, when a known hand touched hers, and doubt that there was a human soul within the frame. The human soul seemed not only to be there, but to have been fully developed; though the means by which it could receive such communications as generally constitute human education, were thus cut off. And such modes of communication with her companions as had been taught her, or as she had herself invented, well bore out the belief, that her mind was the constant dwellingplace, not only of human affections, but of human thoughts. So plainly does it appear that human thought is not produced or occasioned by external circumstances only; but has a special and indestructible germ in human nature.

12 I have been endeavouring to illustrate the doctrine that man's nature is different from the nature of other animals; as subsidiary to the doctrine that the Human Epoch of the earth's history is different from all the preceding Epochs. But in truth, this subsidiary proposition is not by any means necessary to my main purpose. Even if barbarous and savage tribes, even if man under unfavourable circumstances, be little better than the brutes, still no one will doubt that the most civilized races of mankind, that man under the most favourable circumstances, is far, is, indeed, immeasurably elevated above the brutes. The history of man includes not only the history of Scythians and Barbarians, Australians and Negroes, but of ancient Greeks and of modern Europeans: and therefore there can be no doubt that the period of the Earth's history, which includes the history of man, is very different indeed from any period which preceded that. To illustrate the peculiarity, the elevation, the dignity, the wonderful endowments of man, we might refer to the achievements, the recorded thoughts and actions, of the most eminent among those nations;—to their arts, their poetry, their eloquence; their philosophers, their mathematicians, their astronomers; to the acts of virtue and devotion, of patriotism, generosity, obedience, truthfulness, love, which took place among them;—to their piety, their reverence for the deity, their resignation to his will, their hope

of immortality. Such characteristic traits of man as man, (which all examples of intelligence, virtue, and religion, are,) might serve to show that man is, in a sense quite different from other creatures, 'fearfully and wonderfully made;' but I need not go into such details. It is sufficient for my purpose to sum up the result in the expressions which I have already used; that man is an intellectual, moral, religious, and spiritual being.

13 But the existence of man upon the earth being thus an event of an order quite different from any previous part of the earth's history, the question occurs, how long has this state of things endured? What period has elapsed since this creature, with these high powers and faculties, was placed upon the earth? How far must we go backward in time, to find the beginning of his wonderful history?—so utterly wonderful compared with anything which had previously occurred. For as to that point, we cannot feel any doubt. The wildest imagination cannot suggest that corals and madrepores, oysters and sepias, fishes and lizards, may have been rational and moral creatures; nor even those creatures which come nearer to human organization; megatheriums and mastodons, extinct deer and elephants. Undoubtedly the earth, till the existence of man, was a world of mere brute creatures. How long then has it been otherwise? How long has it been the habitation of a rational, reflective, progressive race? Can we by any evidence, geological or other, approximate to the beginning of the Human History?

14 This is a large and curious question, and one on which a precise answer may not be within our reach. But an answer not precise, an approximation, as we have suggested, may suffice for our purpose. If we can determine, in some measure, the order and scale of the period during which man has occupied the earth, the determination may serve to support the analogy which we wish to establish.

15 The geological evidence with regard to the existence of man is altogether negative. Previous to the deposits and changes which we can trace as belonging obviously to the present state of the earth's surface, and the operation of causes now existing, there is no vestige of the existence of man, or of his works. As was long ago observed *, we do not find, among the shells and bones which are so abundant in the older strata, any weapons, medals, implements, structures, which speak to us of the hand of man, the workman. If we look forwards ten or twenty thousand years, and suppose the existing works of man to have been, by that time, ruined and covered up by masses of rubbish, inundations, morasses, lava-streams, earthquakes; still, when the future inhabitant of the earth digs into and explores these coverings, he will discover innumerable monuments that man existed so long ago. The materials of many of his works, and the traces of his own mind, which he stamps upon them, are as indestructible as the shells and bones which give language to the oldest work. Indeed in many cases the oldest fossil remains are the results of objects of seemingly the most frail and perishable material;—of the most delicate and tender animal and vegetable tissues and filaments. That no such remains of textures and forms, moulded by the hand of man, are anywhere found among these, must be accepted as indisputable evidence that man did not exist, so as to be contemporary with the plants and animals thus commemorated. According to geological evidence, the race of man is a novelty upon the earth;—something which has succeeded to all the great geological changes.

16 And in this, almost all geologists are agreed. Even those who hold that, in other ways, the course of change has been uniform;—that even the introduction of man, as a new species of animal, is only an event of the same kind as

* By Bishop Berkeley. See Lyell, III. 346.

myriads of like events which have occurred in the history of the earth;—still allow that the introduction of man, as a moral being, is an event entirely different from any which had taken place before; and that this event is, geologically speaking, recent. The changes of which we have spoken, as studied by the geologist in connexion with the works of man, the destruction of buildings on sea-coasts by the incursions of the ocean, the removal of the shore many miles away from ancient harbours, the overwhelming of cities by earthquakes or volcanic eruptions; however great when compared with the changes which take place in one or two generations; are minute and infinitesimal, when put in comparison with the changes by which ranges of mountains, and continents have been brought into being, one after another, each of them filled with the remains of different organic creations.

17 Further than this, geology does not go on this question. She has no chronometer which can tell us when the first buildings were erected, when man first dwelt in cities, first used implements or arms; still less, language and reflection. Geology is compelled to give over the question to History. The external evidences of the antiquity of the species fail us, and we must have recourse to the internal. Nature can tell us so little of the age of man, that we must inquire what he can tell us himself.

18 What man can tell us of his own age—what history can say of the beginning of history,—is necessarily very obscure and imperfect. We know how difficult it is to trace to its origin the History of any single Nation: how much more, the History of All Nations! We know that all such particular histories carry us back to periods of the migrations of tribes, confused mixtures of populations, perplexed and contradictory genealogies of races; and as we follow these further and further backwards, they become more and

more obscure and uncertain; at least in the histories which remain to us of most nations. Still, the obscurity is not such as to lead us to the conviction that research is useless and unprofitable. It is an obscurity such as naturally arises from the lapse of time, and the complexity of the subject. The aspect of the world, however far we go back, is still historical and human; historical and human, in as high a degree, as it is at the present day. Men, as described in the records of the oldest times, are of the same nature, act with the same views, are governed by the same motives, as at present. At all points, we see thought, purpose, law, religion, progress. If we do not find a beginning, we find at least evidence that, in approaching the beginning, the condition of man does not, in any way, cease to be that of an intellectual, moral, and religious creature.

19 There are, indeed, some histories which speak to us of the beginning of man's existence upon earth; and one such history in particular, which comes to us recommended by indisputable evidence of its own great antiquity, by numerous and striking confirmations from other histories, and from facts still current, and by its connexion with that religious view of man's condition, which appears to thoughtful men to be absolutely requisite to give a meaning and purpose to man's faculties and endowments. I speak, of course, of the Hebrew Scriptures. This history professes to inform us how man was placed upon the earth; and how, from one center, the human family spread itself in various branches into all parts of the world. This genealogy of the human race is accompanied by a chronology, from which it results that the antiquity of the human race does not exceed a few thousand years. Even if we accept this history as true and authoritative, it would not be wise to be rigidly tenacious of the chronology, as to its minute exactness. For, 134| in the first place, of three different forms in which this

history appears, the chronology is different in all the three: I mean the Hebrew, the Samaritan, and the Septuagint versions of the Old Testament. And even if this were not so, since this chronology is put in the form of genealogies, of which many of the steps may very probably have a meaning different from the simple succession of generations in a family, (as some of them certainly have,) it would be unwise to consider ourselves bound to the exact number of years stated, in any of the three versions, or even in all. It makes no difference to our argument, nor to any purpose in which we can suppose this narrative to have a bearing, whether we accept six thousand or ten thousand years, or even a longer period, as the interval which has now elapsed since the creation of man took place, and the peopling of the earth began.

20 And, in our speculations at least, it will be well for us to take into account the view which is given us of the antiquity of the human race, by other histories as well as by this. A satisfactory result of such an investigation would be attained if, looking at all these histories, weighing their value, interpreting their expressions fairly, discovering their sources of error, and of misrepresentation, we should find them all converge to one point; all give a consistent and harmonious view of the earliest stages of man's history; of the times and places in which he first appeared as man. If all nations of men are branches of the same family, it cannot but interest us, to find all the family traditions tending upwards towards the same quarter; indicating a divergence from the same point; exhibiting a recollection of the original domicile, or of the same original family circle.

21 To a certain extent at least, this appears to be the result of the historical investigations which have been pursued, relative to this subject. A certain group of nations is brought before us by these researches which, a few thousands

|135

of years ago, were possessed of arts, and manners, and habits, and belief, which make them conspicuous, and which we can easily believe to have been contemporaneous successors of a common, though, it may be even then, remote stock. Such are the Jews, Egyptians, Chaldeans, and Assyrians. The histories of these nations are connected with and confirm each other. Their languages, or most of them, have certain affinities, which glossologists, on independent grounds, have regarded as affinities implying an original connexion. Their chronologies, though in many respects discrepant, are not incapable of being reduced into an harmony by very probable suppositions. Here we have a very early view of the condition of a portion of the earth as the habitation of man, and perhaps a suggestion of a condition earlier still.

22 It is true, that there are other nations also, which claim an antiquity for their civilization equal to or greater than that which we can ascribe to these. Such are the Indians and the Chinese. But while we do not question that these nations were at a remote period in possession of arts, knowledge, and regular polity, in a very eminent degree, we are not at all called upon to assent to the immense numbers, tens of thousands and hundreds of thousands of years, by which such nations, in their histories, express their antiquity. For, in the first place, such numbers are easily devised and transferred to the obscure early stages of tradition, when the art of numeration is once become familiar. These vast intervals, applied to series of blank genealogies, or idle fables, gratify the popular appetite for numerical wonders, but have little claim on critical conviction.

23 And in the next place, we discover that not numeration only, but a more recondite art, had a great share in the fabrication of these gigantic numbers of years. Some of the nations of whom we have thus spoken, the Indians, for example, had, at an early period, possessed themselves

of a large share of astronomical knowledge. They had observed and examined the motions of the Sun, the Moon, the Planets, and the Stars, till they had discovered Cycles, in which, after long and seemingly irregular wanderings in the skies, the heavenly bodies came round again to known and regular positions. They had thus detected the order that reigns in the seeming disorder; and had, by this means, enabled themselves to know beforehand when certain astronomical events would occur; certain configurations of the Planets, for instance, and eclipses; and knowing how such events would occur in future, they were also able to calculate how the like events had occurred in the past. They could thus determine what eclipses and what planetary configurations had occurred, in thousands and tens of thousands of years of past time; and could, if they were disposed to falsify their early histories, and to confirm the falsification by astronomical evidence, do so with a very near approximation to astronomical truth. Such astronomical confirmation of their assertions, so incapable in any common apprehension of being derived from any other source than actual observation of the fact, naturally produced a great effect upon common minds; and still more, on those who examined the astronomical fact, enough only to see that it was, approximately, at least, true. But in recent times the fallacy of this evidence has been shown, and the fabrication detected. For though the astronomical rules which they had devised were approximately true, they were true approximately only. The more exact researches of modern European astronomy discovered that their cycles, though nearly exact, were not quite so. There was in them, an error which made the cycle, at every revolution of its period, when it was applied to past ages, more and more wrong; so that the astronomical events which they asserted to have happened, as they had calculated that they would have happened, the

|137

better informed astronomer of our day, knows would not have happened exactly so, but in a manner differing more and more from their statement, as the event was more and more remote. And thus the fact which they asserted to have been observed, had not really happened; and the confirmation, which it had been supposed to lend to their history, disappeared. And thus, there is not, in the asserted antiquity of Indian civilization and Indian astronomy, anything which has a well-founded claim to disturb our belief that the nations of the more western regions of Asia had a civilization as ancient as theirs. And considerations of nearly the same kind may be applied to the very remote astronomical facts which are recorded as having been observed in the history of some others of the ancient nations above mentioned.

24 Still less need we be disturbed by the long series of dynasties, each occupying a large period of years, which the Egyptians are said to have inserted in their early history, so as to carry their origin beyond the earliest times which I have mentioned. If they spoke of the Greek nations as children compared with their own long continued age, as Plato says they did, a few thousands of years of previous existence would well entitle them to do so. So far as such a period goes, their monuments and their hieroglyphical inscriptions give a reality to their pretensions, which we may very willingly grant. And even the history of the Jews supposes that the Egyptians had attained a high point in arts, government, knowledge, when Abraham, the father of the Jewish nation, was still leading the life of a nomad. But this supposition is not inconsistent with the account which the Jewish scriptures give, of the origin of nations; especially if, as we have said, we abstain from any rigid and narrow interpretation of the chronology of those scriptures; as on every ground, it is prudent to do.

25 It appears then not unreasonable to believe, that a very few thousands, or even a few hundreds of years before the time of Abraham, the nations of central and western Asia offer to us the oldest aspect of the life of man upon the earth; and that in reasoning concerning the antiquity of the human race, we may suppose that at that period, he was in the earliest stages of his existence. Although, in truth, if we were to accept the antiquity claimed by the Egyptians, the Indians, or the Chinese, the nature of our argument would not be materially altered; for ten thousand, or even twenty thousand years, bears a very small proportion to the periods of time which geology requires for the revolutions which she describes; and, as I have said, we have geological evidence also, to show how brief the human period has been, when compared with the period which preceded the existence of man. And if this be so; if such peoples as those who have left to us the monuments of Egypt and of Assyria, the pyramids and ancient Thebes, the walls of Nineveh and Babylon, were the first nations which lived as nations; or if they were separated from such only by the interval by which the Germans of to-day are separated from the Germans of Tacitus; we may well repeat our remark, that the history of man in the earliest times, is as truly a history of a wonderful, intellectual, social, political, spiritual creature, as it is at present. We see, in the monuments of those periods, evidences so great and so full of skill, that even now, they amaze us, of arts, government, property, thought, the love of beauty, the recognition of deity; evidences of memory, foresight, power. If London or Berlin were now destroyed, overwhelmed, and, four thousand years hence, disinterred, these cities would not afford stronger testimony of those attributes, as existing in modern Europeans, than we have of such qualities in the ancient Babylonians and Egyptians. |139

The history of man, as that of a creature pre-eminent in the creation, is equally such, however far back we carry our researches.

26 Nor is there anything to disturb this view, in the fact of the existence of the uncultured and barbarous tribes which occupy, and always have occupied, a large portion of the earth's surface. For, in the first place, there is not, in the aspect of the fact, or in the information which history gives us, any reason to believe that such tribes exhibit a form of human existence, which, in the natural order of progress, is earlier than the forms of civilized life, of which we have spoken. The opinion that the most savage kind of human life, least acquainted with arts, and least provided with resources, is the state of nature out of which civilized life has everywhere gradually emerged, is an opinion which, though at one time popular, is unsupported by proof, and contrary to probability*. Savage tribes do not so grow into civilization; their condition is, far more probably, a condition of civilization degraded and lost, than of civilization incipient and prospective. Add to this, that if we were to assume that this were otherwise; if man thus originally and naturally savage, did also naturally tend to become civilized; this *tendency* is an endowment no less wonderful, than those endowments which civilization exhibits. The capacity is as extraordinary as the developed result; for the capacity involves the result. If savage man be the germ of the most highly civilized man, he differs from all other animal germs, as man differs from brute. And add to this again, that in the tribes which we call savage, and whose condition most differs, in external circumstances, from ours, there are, after

* A recent popular writer, who has asserted the self-civilizing tendency of man, has not been able, it would seem, to adduce any example of the operation of this tendency, except a single tribe of North American Indians, in whom it operated for a short time, and to a small extent.

all, a vast mass of human attributes: thought, purpose, language, family relations; generally property, law, government, contract, arts, and knowledge, to no small extent: and in almost every case, religion. Even uncivilized man is an intellectual, moral, social, religious creature; nor is there, in his condition, any reason why he may not be a spiritual creature, in the highest sense in which the most civilized man can be so.

27 Here then we are brought to the view which, it would seem, offers a complete reply to the difficulty, which astronomical discoveries appeared to place in the way of religion:—the difficulty of the opinion that man, occupying this speck of earth, which is but as an atom in the Universe, surrounded by millions of other globes, larger, and, to appearance, nobler than that which he inhabits, should be the object of the peculiar care and guardianship, of the favour and government, of the Creator of All, in the way in which Religion teaches us that He is. For we find that man, (the human race, from its first origin till now,) has occupied but an atom of time, as he has occupied but an atom of space:—that as he is surrounded by myriads of globes which may, like this, be the habitations of living things, so he has been preceded, on this earth, by myriads of generations of living things, not possibly or probably only, but certainly; and yet that, comparing his history with theirs, he has been, certainly has been fitted to be, the object of the care and guardianship, of the favour and government, of the Master and Governor of All, in a manner entirely different from anything which it is possible to believe with regard to the countless generations of brute creatures which had gone before him. If we will doubt or overlook the difference between man and brutes, the difficulty of ascribing to man peculiar privileges, is made as great by the revelations of geology, as of astronomy. The scale of man's insignificance |141

is, as we have said, of the same order in reference to time, as to space. There is nothing which at all goes beyond the magnitude which observation and reasoning suggest for geological periods, in supposing that the tertiary strata occupied, in their deposition and elevation, a period as much greater than the period of human history, as the solar system is larger than the earth:—that the secondary strata were as much longer than these in their formation, as the nearest fixed star is more distant than the sun:—that the still earlier masses, call them primary, or protozoic, or what we will, did, in their production, extend through a period of time as vast, compared with the secondary period, as the most distant nebula is remoter than the nearest star. If the earth, as the habitation of man, is a speck in the midst of an infinity of space, the earth, as the habitation of man, is also a speck at the end of an infinity of time. If we are as nothing in the surrounding universe, we are as nothing in the elapsed eternity; or rather, in the elapsed organic antiquity, during which the earth has existed and been the abode of life. If man is but one small family in the midst of innumerable possible households, he is also but one small family, the successor of innumerable tribes of animals, not possible only, but actual. If the planets *may* be the seats of life, we know that the seas which have given birth to our mountains *were* the seats of life. If the stars may have hundreds of systems of tenanted planets rolling round them, we know that the secondary group of rocks does contain hundreds of tenanted beds, witnessing of as many systems of organic creation. If the nebulæ may be planetary systems in the course of formation, we know that the primary and transition rocks either shew us the earth in the course of formation, as the future seat of life, or exhibit such life as already begun.

28 How far that which astronomy thus asserts as possible, is probable:—what is the value of these possibilities of

142

life in distant regions of the universe, we shall hereafter consider. But in what geology asserts, the case is clear. It is no possibility, but a certainty. No one will now doubt that shells and skeletons, trunks and leaves, prove animal and vegetable life to have existed. Even, therefore, if Astronomy could demonstrate all that her most fanciful disciples assume, Geology would still have a complete right to claim an equal hearing;—to insist upon having her analogies regarded. She would have a right to answer the questions of Astronomy, when she says, How can we believe this? and to have her answers accepted.

29 Astronomy claims a sort of dignity over other sciences, from her *antiquity*, her *certainty*, and the *vastness* of her discoveries. But the antiquity of astronomy as a science had no share in such speculations as we are discussing; and if it had had, new truths are better than old conjectures; new discoveries must rectify old errors; new answers must remove old difficulties. The vigorous youth of Geology makes her fearless of the age of Astronomy. And as to the certainty of Astronomy, it has just as little to do with these speculations. The certainty stops, just when these speculations begin. There may, indeed, be some danger of delusion on this subject. Men have been so long accustomed to look upon astronomical science as the mother of certainty, that they may confound astronomical discoveries with cosmological conjectures; though these be slightly and illogically connected with those. And then, as to the vastness of astronomical discoveries,—granting that character, inasmuch as it is to a certain degree, a matter of measurement,—we must observe, that the discoveries of geology are no less vast: they extend through time, as those of astronomy do through space. They carry us through millions of years, that is, of the earth's revolutions, as those of astronomy do through millions of the earth's diameters, or of diameters of the

earth's orbit. Geology fills the regions of duration with events, as astronomy fills the regions of the universe with objects. She carries us backwards by the relation of cause and effect, as astronomy carries us upwards by the relations of geometry. As astronomy steps on from point to point of the universe by a chain of triangles, so geology steps from epoch to epoch of the earth's history by a chain of mechanical and organical laws. If the one depends on the axioms of geometry, the other depends on the axioms of causation.

30 So far then, Geology has no need to regard Astronomy as her superior; and least of all, when they apply themselves together to speculations like these. But in truth, in such speculations, Geology has an immeasurable superiority. She has the command of an implement, in addition to all that Astronomy can use; and one, for the purpose of such speculations, adapted far beyond any astronomical element of discovery. She has, for one of her studies,—one of her means of dealing with her problems,—the knowledge of Life, animal and vegetable. Vital organization is a subject of attention which has, in modern times, been forced upon her. It is now one of the main parts of her discipline. The geologist must study the traces of life in every form; must learn to decypher its faintest indications and its fullest developement. On the question, then, whether there be in this or that quarter, evidence of life, he can speak with the confidence derived from familar knowledge; while the astronomer, to whom such studies are utterly foreign, because he has no facts which bear upon them, can offer, on such questions, only the loosest and most arbitrary conjectures; which, as we have had to remark, have been rebuked by eminent men, as being altogether inconsistent with the acknowledged maxims of his science.

31 When, therefore, Geology tells us that the earth, 144| which has been the seat of human life for a few thousand

years only, has been the seat of animal life for myriads, it may be, millions of years, she has a right to offer this, as an answer to any difficulty which Astronomy, or the readers of astronomical books, may suggest, derived from the considerations that the Earth, the seat of human life, is but one globe of a few thousand miles in diameter, among millions of other globes, at distances millions of times as great.

32 Let the difficulty be put in any way the objector pleases. Is it that it is unworthy of the greatness and majesty of God, according to our conceptions of Him, to bestow such peculiar care on so small a part of His creation? But we know, from geology, that He has bestowed upon this small part of His creation, mankind, this special care;—He has made their period, though only a moment in the ages of animal life, the only period of intelligence, morality, religion. If then, to suppose that He has done this, is contrary to our conceptions of His greatness and majesty, it is plain that our conceptions are erroneous; they have taken a wrong direction. God has not judged, as to what is worthy of Him, as we have judged. He has found it worthy of Him to bestow upon man His special care, though he occupies so small a portion of time; and why not, then, although he occupies so small a portion of space?

33 Or is the objection this; that if we suppose the earth only to be occupied by inhabitants, all the other globes of the universe are wasted;—turned to no purpose? Is waste of this kind considered as unsuited to the character of the Creator? But here again, we have the like waste, in the occupation of the earth. All its previous ages, its seas and its continents, have been wasted upon mere brute life; often, so far as we can see, for myriads of years, upon the lowest, the least conscious forms of life; upon shell-fish, corals, sponges. Why then should not the seas and conti-

nents of other planets be occupied at present with a life no higher than this, or with no life at all? Will it be said that, so far as material objects are occupied by life, they are not wasted; but that they are wasted, if they are entirely barren and blank of life? This is a very arbitrary saying. Why should the life of a sponge, or a coral, or an oyster, be regarded as a good employment of a spot of land and water, so as to save it from being wasted? No doubt, if the coral or the oyster be there, there is a reason why it is so, consistently with the attributes of God. But then, on the same ground, we may say that if it be not there, there is a reason why it is not so. Such a mode of regarding the parts of the universe can never give us reasons why they should or should not be inhabited, when we have no other grounds for knowing whether they are. If it be a sufficient employment of a spot of rock or water that it is the seat of organization —of organic powers; why may it not be a sufficient employment of the same spot that it is the seat of attraction, of cohesion, of crystalline powers? All the planets, all parts of the universe, we have good reason to believe, are pervaded by attraction, by forces of aggregation and atomic relation, by light and heat. Why may not these be sufficient to prevent the space being wasted, in the eyes of the Creator? as, during a great part of the earth's past history, and over large portions of its present mass, they are actually held by Him sufficient; for they are all that occupy those portions. This notion, then, of the improbability of there being, in the universe, so vast an amount of waste spaces, or waste bodies, as is implied in the opinion that the earth alone is the seat of life, or of intelligence, is confuted by the fact, that there are vast spaces, waste districts, and especially waste times, to an extent as great as such a notion deems improbable. The avoidance of such waste, according to our notions of waste, is no part of the economy of creation, so far as we can discern that economy, in its most certain exemplifications.

34 Or will the objection be made in this way; that such a peculiar dignity and importance given to the earth is contrary to the analogy of creation;—that since there are so many globes, similar to the earth,—like her, revolving round the sun, like her, revolving on their axes, several of them, like her, accompanied by satellites; it is reasonable to suppose that their destination and office is the same as hers;—that since there are so many stars, each like the sun, a source of light, and probably, of heat, it is reasonable to suppose that, like the sun, they are the centers of systems of planets, to which their light and heat are imparted, to uphold life:—is it thought that such a resemblance is a strong ground for believing that the planets of our system, and of other systems, are inhabited as the earth is? If such an astronomical analogy be insisted on, we must again have recourse to geology, to see what such analogy is worth. And then, we are led to reflect, that if we were to follow such analogies, we should be led to suppose that all the successive periods of the earth's history were occupied with life of the same order; that as the earth, in its present condition, is the seat of an intelligent population, so must it have been, in all former conditions. The earth, in its former conditions, was able and fitted to support life; even the life of creatures closely resembling man in their bodily structure. Even of monkeys, fossil remains have been found. But yet, in those former conditions, it did not support human life. Even those geologists who have dwelt most on the discovery of fossil monkeys, and other animals nearest to man, have not dreamt that there existed, before man, a race of rational, intelligent, and progressive creatures. As we have seen, geology and history alike refute such a

|147

fancy. The notion, then, that one period of time in the history of the earth must resemble another, in the character of its population, because it resembles it in physical circumstances, is negatived by the facts which we discover in the history of the earth. And so, the notion that one part of the universe must resemble another in its population, because it resembles it in physical circumstances, is negatived as a law of creation. Analogy, further examined, affords no support to such a notion. The analogy of time, the events of which we know, corrects all such guesses founded on a supposed analogy of space, the furniture of which, so far as this point is concerned, we have no sufficient means of examining.

35 But in truth, we may go further. Not only does the analogy of creation not point to any such entire resemblance of similar parts, as is thus assumed, but it points in the opposite direction. Not entire resemblance, but universal difference is what we discover; not the repetition of exactly similar cases, but a series of cases perpetually dissimilar, presents itself; not constancy, but change, perhaps advance; not one permanent and pervading scheme, but preparation and completion of successive schemes; not uniformity and a fixed type of existences, but progression and a climax. This may be said to be the case in the geological aspect of the world; for, without occupying ourselves with the question, how far the monuments of animal life, which we find preserved in the earth's strata, exhibit a gradual progression from ruder and more imperfect forms to the types of the present terrestrial population; from sponges and mollusks, to fish and lizards, from cold-blooded to warm-blooded animals, and so on, till we come to the most perfect vertebrates;—a doctrine which many eminent geologists have held, and still hold;—without discussing this question, or assuming that the fact is so; this at least cannot be denied or doubted, that man is incomparably the most perfect and

highly endowed creature which ever has existed on the earth. How far previous periods of animal existence were a necessary preparation of the earth, as the habitation of man, or a gradual progression towards the existence of man, we need not now inquire. But this at least we may say; that man, now that he is here, forms a climax to all that has preceded; a term incomparably exceeding in value all the previous parts of the series; a complex and ornate capital to the subjacent column; a personage of vastly greater dignity and importance than all the preceding line of the procession. The analogy of nature, in this case at least, appears to be, that there should be inferior, as well as superior provinces, in the universe; and that the inferior may occupy an immensely larger portion of time than the superior; why not then of space? The intelligent part of creation is thrust into the compass of a few years, in the course of myriads of ages; why not then into the compass of a few miles, in the expanse of systems? The earth was brute and inert, compared with its present condition, dark and chaotic, so far as the light of reason and intelligence are concerned, for countless centuries before man was created. Why then may not other parts of creation be still in this brute and inert and chaotic state, while the earth is under the influence of a higher exercise of creative power? If the earth was, for ages, a turbid abyss of lava and of mud, why may not Mars or Saturn be so still? If the germs of life were, gradually, and at long intervals, inserted in the terrestrial slime, why may they not be just inserted, or not yet inserted, in Jupiter? Or why should we assume that the condition of those planets resembles ours, even so far as such suppositions imply? Why may they not, some or all of them, be barren masses of stone and metal, slag and scoriæ, dust and cinders? That some of them are composed of such materials, we have better reason to believe, than we have to believe anything |149 else respecting their physical constitution, as we shall hereafter endeavour to shew. If then, the earth be the sole inhabited spot in the work of creation, the oasis in the desert of our system, there is nothing in this contrary to the analogy of creation. But if, in some way which perhaps we cannot discover, the earth obtained, for accompaniments, mere chaotic and barren masses, as conditions of coming into its present state; as it may have required, for accompaniments, the brute and imperfect races of former animals, as conditions of coming into its present state, as the habitation of man; the analogy is against, and not in favour of, the belief that they too (the other masses, the planets, &c.) are habitations. I may hereafter dwell more fully on such speculations: but the possibility that the planets are such rude masses, is quite as tenable, on astronomical grounds, as the possibility that the planets resemble the earth, in matters of which astronomy can tell us nothing. We say, therefore, that the example of geology refutes the argument drawn from the supposed analogy of one part of the universe with another; and suggests a strong suspicion that the force of analogy, better known, may tend in the opposite direction.

36 When such possibilities are presented to the reader, he may naturally ask, if we are thus to regard man as the climax of creation, in space, as in time, can we point out any characters belonging to him, which may tend to make it conceivable that the Creator should thus distinguish him, and care for him:—should prepare his habitation, if it be so, by ages of chaotic and rudimentary life, and by accompanying orbs of brute and barren matter. If Man be, thus, the head, the crowned head of the creation, is he worthy to be thus elevated? Has he any qualities which make it conceivable that, with such an array of preparation and accompaniment, he should be placed upon the earth, his throne?

Or rather, if he be thus the chosen subject of God's care, has he any qualities, which make it conceivable that he should be thus selected; taken under such guardianship; admitted to such a dispensation; graced with such favour. The question with which we began again recurs: What is man that God should be thus mindful of him? After the views which have been presented to us, does any answer now occur to us?

37 The answer which we have to give, is that which we have already repeatedly stated. Man is an intellectual, moral, religious, and spiritual creature. If we consider these attributes, we shall see that they are such as give him a special relation to God, and as we conceive, and must conceive, God to be; and may therefore be, in God, the occasion of special guardianship, special regard, a special dispensation towards man.

38 As an intellectual creature, he has not only an intelligence which he can apply to practical uses, to minister to the needs of animal and social life; but also an intellect by which he can speculate about the relations of things, in their most general form; for instance, the properties of space and time, the relations of finite and infinite. He can discover truths, to which all things, existing in space and time, must conform. These are conditions of existence to which the creation conforms, that is, to which the Creator conforms; and man, capable of seeing that such conditions are true and necessary, is capable, so far, of understanding some of the conditions of the Creator's workmanship. In this way, the mind of man has some community with the mind of God: and however remote and imperfect this community may be, it must be real. Since, then, man has thus, in his intellect, an element of community with God, it is so far conceivable that he should be, in a special manner, the object of God's care and favour. The human mind, with |151

its wonderful and perhaps illimitable powers, is something of which we can believe God to be 'mindful.'

39 Again: man is a moral creature. He recognizes, he cannot help recognizing, a distinction of right and wrong in his actions: and in his internal movements which lead to action. This distinction he recognizes as the reason, the highest and ultimate reason, for doing or for not doing. And this law of his own reason, he is, by reflection, led to recognize as a Law of the Supreme Reason; of the Supreme Mind which has made him what he is. The Moral Law, he owns and feels as God's Law. By the obligation which he feels to obey this Law, he feels himself God's subject; placed under his government; compelled to expect his judgment, his rewards, and punishments. By being a moral creature, then, he is, in a special manner, the subject of God; and not only we can believe that, in this capacity, God cares for him; but we cannot believe that he *does not* care for him. He cares for him, so as to approve of what he does right, and to condemn what he does wrong. And he has given him, in his own breast, an assurance that he will do this; and thus, God cares for man, in a peculiar and special manner. As a moral creature, we have no difficulty in conceiving that God may think him worthy of his regard and government.

40 The developement of man's moral nature, as we have just described it, leads to, and involves the developement of his religious nature. By looking within himself, and seeing the Moral Law, he learns to look upwards to God, the Author of the Law, and the Awarder of the rewards and penalties which follow moral good and evil. But the belief of such a dispensation carries us, or makes us long to be carried, beyond the manifestations of this dispensation, as they appear in the ordinary course of human life. By 152| thinking on such things, man is led to ascribe a wider range

to the Moral Government of God: —to believe in methods of reward and punishment, which do not appear in the natural course of events: to accept events, out of the order of nature, which announce that God has provided such methods: to accept them, when duly authenticated, as messages from God: and thus, when God provides the means, to allow himself to be placed in intercourse with God. Since man is capable of this; since, as a religious creature, this is his tendency, his need, the craving of his heart, without which, when his religious nature is fully unfolded, he can feel no comfort nor satisfaction; we cannot be surprised that God should deem him a proper object of a special fatherly care; a fit subject for a special dispensation of his purposes, as to the consequences of human actions. Man being this, we can believe that God is not only 'mindful of him,' but 'visits him.'

41 As we have said, the soul of man, regarded as the subject of God's religious government, is especially termed his *Spirit:* the course of human being which results from the intercourse with God, which God permits, is a *spiritual* existence. Man is capable, in no small degree, of such an existence, of such an intercourse with God; and, as we are authorized to term it, of such a life with God, and in God, even while he continues in his present human existence. I say *authorized,* because such expressions are used, though reverently, by the most religious men; who are, at any rate, authority as to their own sentiments; which are the basis of our reasoning. Whatever then, may be the imperfection, in this life, of such a union with God, yet since man can, when sufficiently assisted and favoured by God, enter upon such a union, we cannot but think it most credible and most natural, that he should be the object of God's special care and regard, even of his love and presence.

42 That men are, only in a comparatively small num- |153

ber of cases, intellectual, moral, religious, and spiritual, in the degree which I have described, does not, by any means, deprive our argument of its force. The capacity of man is, that he may become this; and such a capacity may well make him a special object in the eyes of Him under whose guidance, and by whose aid, such a developement and elevation of his nature is open to him. However imperfect and degraded, however unintellectual, immoral, irreligious, and unspiritual, a great part of mankind may be, still they all have the germs of such an elevation of their nature: and a large portion of them make, we cannot doubt, no small progress in this career of advancement to a spiritual condition. And with such capacities, and such practical exercise of those capacities, we can have no difficulty in believing, if the evidence directs us to believe, that that part of the creation in which man has his present appointed place, is the special field of God's care and love; by whatever wastes of space, and multitudes of material bodies, it may be surrounded; by whatever races it may have been previously occupied, of brutes that perish, and that, compared with man, can hardly be said to have lived.

CHAPTER VII.

The Nebulæ.

---

1 I HAVE attempted to show that, even if we suppose the other bodies of the universe to resemble the Earth, so far as to seem, by their materials, forms, and motions, no less fitted than she is to be the abodes of life; yet that, knowing what we do of man, we can believe that the Earth is tenanted by a race who are the *special* objects of God's care. Even if the tendency of the analogies of creation were, to incline us to suppose that the other planets are as well suited as our globe, to have inhabitants, still it would require a great amount of evidence, to make us believe that they have such inhabitants as we are; while yet such evidence is altogether wanting. Even if we knew that the stars were the centers of revolving systems, we should have an immense difficulty in believing that an Earth, with such a population as ours, revolves about any of them. If astronomy made a plurality of worlds probable, we have strong reasonings, drawn from other subjects, to think that the other worlds are not like ours.

2 The admirers of astronomical triumphs may perhaps be disposed to say, that when so much has been discovered, we may be allowed to complete the scheme by the exercise of fancy. I have attempted to show that we are not in such a state of ignorance, when we look at other relations of the earth and of man, as to allow us to do this. But now we may go a little onwards in our argument; and may ask, whether Astronomy really does what is here claimed for her:—whether she carries us so securely to the bounds of

the visible universe, that our Fancy may take up the task, and people the space thus explored:—whether the bodies which Astronomy has examined, be really as fitted as our Earth, to sustain a population of living things:—whether the most distant objects in the universe do really seem to be systems, or the beginnings of systems:—whether Astronomy herself may not incline in favour of the condition of man, as being the sole creature of his kind?

3 In making this inquiry, it will of course be understood, that I do so with the highest admiration for the vast discoveries which Astronomy has really made; and for the marvellous skill and invention of the great men who have, in all ages of the world, and not least, in our time, been the authors of such discoveries. From the time when Galileo first discovered the system of Jupiter's satellites, to the last scrutiny of the structure of a nebula by Lord Rosse's gigantic telescope, the history of the telescopic exploration of the sky, has been a history of genius felicitously employed in revealing wonders. In this history, the noble labours of the first and the second Herschel, relative to the distribution of the fixed stars, the forms and classes of nebulæ, and the phenomena of double stars, especially bear upon our present speculations; to which we may add, the examination of the aspect of each planet, by various observers, as Schroeter, and of the moon by others, from Huyghens to Mädler and Beer. The achievements which are most likely to occur to the reader's mind are those of the Earl of Rosse; as being the latest addition to our knowledge, and the result of the greatest instrumental powers. By the energy and ingenuity of that eminent person, an eye is directed to the heavens, having a pupil of six feet diameter, with the most complete optical structure, and the power of ranging about for its objects over a great extent of sky: 156| and thus the quantity of light which the eye receives from

any point of the heavens is augmented, it may be, fifty thousand times. The rising Moon is seen from the Observatory in Ireland with the same increase of size and light, as if her solid globe, two thousand miles in diameter, retaining all its illumination, really rested upon the summits of the Alps, to be gazed at by the naked eye. An object which appears to the naked eye a single star, may, by this telescope, so far as its power of seeing is concerned, be resolved into fifty thousand stars, each of the same brightness as the obvious star. What seems to the unassisted vision a nebula, a patch of diluted light, in which no distinct luminous point can be detected, may, by such an instrument, be discriminated or *resolved* into a number of bright dots; as the stippled shades of an engraving are resolved into dots by the application of a powerful magnifying glass. Similar results of the application of great telescopic power had of course been attained long previously: but, as the nature of scientific research is, each step adds something to our means of knowledge; and the last addition assumes, includes, and augments the knowledge which we possessed before. The discussions in which we are engaged, belong to the very boundary region of science;—to the frontier where knowledge, at least astronomical knowledge, ends, and ignorance begins. Such discoveries, therefore, as those made by Lord Rosse's telescope, require our especial notice here.

4 We may begin, at what appear to us the outskirts of creation, the Nebulæ. At one time it was conceived by astronomers in general, that these patches of diffused light, which are seen by them in such profusion in the sky, are not luminous bodies of regular forms and definite boundaries, apparently solid, as the stars are supposed to be; but really, as even to good telescopes many of them seem, masses of luminous cloud or vapour, loosely held together, as clouds |157

and vapours are, and not capable by any powers of vision of being resolved into distinct visible elements. This opinion was for a time so confidently entertained, that there was founded upon it an hypothesis, that these were gaseous masses, out of which suns and systems might afterwards be formed, by the concentration of these luminous vapours into a solid central sun, more intensely luminous; while detached portions of the mass, flying off, and cooling down so as to be no longer self-luminous, might revolve round the central body, as planets and satellites. This is the *Nebular Hypothesis*, suggested by the elder Herschel, and adopted by the great mathematician Laplace.

5 But the result of the optical scrutiny of the nebulæ by more modern observers, especially by Lord Rosse in Ireland, and Mr Bond in America, has been, that many celestial objects which were regarded before as truly nebulous, have been resolved into stars; and this resolution has been extended to so many cases of nebulæ, of such various kinds, as to have produced a strong suspicion in the minds of astronomers that *all* the nebulæ, however different in their appearance, may really be resolved into stars, if they be attacked with optical powers sufficiently great.

6 If this were to be assumed as done, and if each of the separate points, into which the nebulæ are thus resolved, were conceived to be a star, which looks so small only because it is so distant, and which really is as likely to have a system of planets revolving about it, as is a star of the first magnitude:—we should then have a view of the immensity of the visible universe, such as I presented to the reader in the beginning of this essay. All the distant nebulæ appear as nebulæ, only because they are so distant: if truly seen, they are groups of stars, of which each may be as important as our sun, being, like it, the center of a planetary system. And thus, a patch of the heavens, one hundredth

158|

or one thousandth part of the visible breadth of our sun, may contain in it more life, not only than exists in the solar system, but in as many such systems as the unassisted eye can see stars in the heavens, on the clearest winter night.

7 This is a stupendous view of the greatness of the creation; and, to many persons, its very majesty, derived from magnitude and number, will make it so striking and acceptable, that, once apprehended, they will feel as if there were a kind of irreverence in disturbing it. But if this view be really not tenable when more closely examined, it is, after all, not wise to connect our feelings of religious reverence with it, so that they shall suffer a shock when we are obliged to reject it. I may add, that we may entertain an undoubting trust that any view of the creation which is found to be true, will also be found to supply material for reverential contemplation. I venture to hope that we may, by further examination, be led to a reverence of a deeper and more solemn character than a mere wonder at the immensity of space and number.

8 But whatever the result may be, let us consider the evidence for this view. It assumes that all the Nebulæ are resolvable into stars, and that they appear as nebulæ only because they are more distant than the region in which they can appear as stars. Are there any facts, any phenomena in the heavens, which may help us to determine whether this is a probable opinion?

9 It is most satisfactory for us, when we can, in such inquiries, know the thoughts which have suggested themselves to the minds of those who have examined the phenomena with the most complete knowledge, the greatest care, and the best advantages; and have speculated upon these phenomena in a way both profound and unprejudiced. Some remarks of Sir John Herschel, recommended by these precious characters, seem to me to bear strongly upon the ques- |159

tion which I have just had to ask:—Do all the nebulæ owe their nebulous appearance to their being too distant to be seen as groups of distinct stars, though they really are such groups?

10 Herschel, in the visit which he made to the Cape of Good Hope, for the purpose of erecting to his father the most splendid monument that son ever erected,—the completed survey of the vault of heaven,—had full opportunity of studying a certain pair of remarkable bright spaces of the skies, filled with a cloudy light, which lie near the Southern pole; and which, having been unavoidably noticed by the first Antarctic voyagers, are called the *Magellanic Clouds.* When the larger of these two clouds is examined through powerful telescopes, it presents, we are told, a constitution of astonishing complexity: 'large patches and tracts of nebulosity in every stage of resolution, from light, irresolvable with 18 inches of reflecting aperture, up to perfectly separated stars like the Milky Way, and clustering groups sufficiently insulated and condensed to come under the designation of irregular, and in some cases pretty rich clusters. But besides these, there are also nebulæ in abundance, both regular and irregular; globular clusters in every stage of condensation, and objects of a nebulous character quite peculiar, and which have no analogies in any other region of the heavens*.' He goes on to say, that these nebulæ and clusters are far more crowded in this space than they are in any other, even the most crowded parts, of the nebulous heavens. This *Nubecula Major*, as it is termed, is of a round or oval form, and its diameter is about six degrees, so that it is about twelve times the apparent diameter of the moon. The *Nubecula Minor* is a smaller patch of the same kind. If we suppose the space occupied by the various objects which the nubecula major includes, to be, in a general

* Herschel, *Outl. of Astr.* Art. 893.

way, spherical, its nearest and most remote parts must (as its angular size proves) differ in their distance from us by little more than a tenth part of our distance from its center. That the two nubeculæ are thus approximately spherical spaces, is in the highest degree probable; not only from the peculiarity of their contents, which suggests the notion of a peculiar group of objects, collected into a limited space; but from the barrenness, as to such objects, of the sky in the neighbourhood of these Magellanic Clouds. To suppose (the only other possible supposition,) that they are two columns of space, with their ends turned towards us, and their lengths hundreds and thousands of times their breadths, would be too fantastical a proceeding to be tolerated; and would, after all, not explain the facts without further altogether arbitrary assumptions.

11 It appears, then, that, in these groups, there are stars of various magnitudes, clusters of various forms, nebulæ regular and irregular, nebulous tracts and patches of peculiar character; and all so disposed, that the most distant of them, whichever these may be, are not more than one-tenth more distant than the nearest. If the nearest star in this space be at nine times the distance of Sirius, the farthest nebulæ, contained in the same space, will not be at more than ten times the distance of Sirius. Of course, the doctrine that nebulæ are seen as nebulæ, merely because they are so distant, requires us to assume all nebulæ to be hundreds and thousands of times more distant than the smallest stars. If stars of the eighth magnitude (which are hardly visible to the naked eye) be eight times as remote as Sirius, a nebula containing a thousand stars, which is invisible to the naked eye, must be more than eight thousand times as remote as Sirius. And thus if in the whole galaxy, we reckon only the stars as far as the eighth magnitude, and suppose all the stars of the galaxy to form a nebula, which is visible to the spec- |161

tators in a distant nebula, only as their nebula is visible to us; we must place them at eight thousand times two hundred thousand times the distance of the Sun; and, even so, we are obviously vastly understating the calculation. These are the gigantic estimates with which some astronomical speculators have been in the habit of overwhelming the minds of their listeners; and these views have given a kind of majesty to the aspect of the nebulæ; and have led some persons to speak of the discovery of every new streak of nebulous light in the starry heavens, as a discovery of new worlds, and still new worlds. But the Magellanic Clouds show us very clearly that all these calculations are entirely baseless. In those regions of space, there coexist, in a limited compass, and in indiscriminate position, stars, clusters of stars, nebulæ, regular and irregular, and nebulous streaks and patches. These, then, are different kinds of things in themselves, not merely different to us. There are such things as nebulæ side by side with stars, and with clusters of stars. Nebulous matter resolvable occurs close to nebulous matter irresolvable. The last and widest step by which the dimensions of the universe have been expanded in the notions of eager speculators, is checked by a completer knowledge and a sager spirit of speculation. Whatever inference we may draw from the resolvability of some of the nebulæ, we may not draw this inference;—that they are more distant, and contain a larger array of systems and of worlds, in proportion as they are difficult to resolve.

12 But indeed, if we consider this process, of the resolution of nebulæ into luminous points, on its own ground, without looking to such facts as I have just adduced, it will be difficult, or impossible, to assign any reason why it should lead to such inferences as have been drawn from it. Let us look at this matter more clearly. An astronomer, armed with a powerful telescope, *resolves* a nebula, discerns that a

162|

luminous cloud is composed of shining dots:—but what are these dots? Into *what* does he resolve the nebula? Into *Stars*, it is commonly said. Let us not wrangle about words. By all means let these dots be Stars, if we know about what we are speaking: if a *Star* merely mean a luminous dot in the sky. But that these stars shall resemble, in their nature, stars of the first magnitude, and that such stars shall resemble our Sun, are surely very bold structures of assumption to build on such a basis. Some nebulæ are resolvable; are resolvable into distinct points; certainly a very curious, probably an important discovery. We may hereafter learn that *all* nebulæ are resolvable into distinct points: that would be a still more curious discovery. But what would it amount to? What would be the simple way of expressing it, without hypothesis, and without assumption? Plainly this: that the substance of all nebulæ is not continuous, but discrete;—separable, and separate into distinct luminous elements;—nebulæ are, it would then seem, as it were, of a curdled or granulated texture: they have run into *lumps* of light, or have been formed originally of such lumps. Highly curious. But what are these lumps? How large are they? At what distances? Of what structure? Of what use? It would seem that he must be a bold man who undertakes to answer these questions. Certainly he must appear to ordinary thinkers to be *very* bold, who, in reply, says, gravely and confidently, as if he had unquestionable authority for his teaching:—'These lumps, O man, are Suns; they are distant from each other as far as the Dog-star is from us; each has its system of Planets, which revolve around it; and each of these Planets is the seat of an animal and vegetable creation. Among these Planets, some, we do not yet know how many, are occupied by rational and responsible creatures, like Man; and the only matter which perplexes us, holding this belief on astronomical grounds, is, that we do |163 not quite see how to put our theology into its due place and form in our system.'

13 In discussing such matters as these, where our knowledge and our ignorance are so curiously blended together, and where it is so difficult to make men feel that so much ignorance can lie so close to so much knowledge;—to make them believe that they have been allowed to discover so much, and yet are not allowed to discover more:—we may be permitted to illustrate our meaning, by supposing a case of blended knowledge and ignorance, of real and imaginary discovery. Suppose that there were carried from a scientific to a more ignorant nation, excellent maps of the world, finely engraved; the mountain-ranges shaded in the most delicate manner, and the sheet crowded with information of all kinds, in writing large, small, and microscopic. Suppose also, that when these maps had been studied with the naked eye, so as to establish a profound respect for the knowledge and skill of the author of them, some of those who perused them should be furnished with good microscopes, so as to carry their examination further than before. They might then find that, in several parts, what before appeared to be merely crooked lines, was really writing, stating, it may be, the amount of population of a province, or the date of foundation of a town. To exhaust all the information thus contained in the maps, might be a work of considerable time and labour. But suppose that, when this was done, a body of resolute microscopists should insist that the information which the map contained was not exhausted: that they should continue peering perseveringly at the lines which formed the shading of the mountains, maintaining that these lines also were writing, if only it might be deciphered; and should go on increasing, with immense labour and ingenuity, the powers of their microscopes, in order to discover the legend contained in these unmeaning lines. We

164|

should, perhaps, have here an image of the employment of those astronomers, who now go on looking in nebulæ for worlds. And we may notice in passing, that several of the arguments which are used by such astronomers, might be used, and would be used, by our microscopists:—how improbable it was that a person so full of knowledge, and so able to convey it, as the author of the maps was known to be, should not have a design and purpose in every line that he drew: what a waste of space it would be to leave any part of the sheet blank of information: and the like. To which the reply is to us obvious; that the design of shading the mountains was design enough; and that the information conveyed was all that was necessary or convenient. Nor does this illustration at all tend to show that such astronomical scrutiny, directed intelligently, with a right selection of the points examined, may not be highly interesting and important. If the microscopists had examined the map with a view to determine the best way in which mountains can be indicated by shading, they would have employed themselves upon a question which has been the subject of multiplied and instructive discussion in our own day.

14 But to return to the subject of Nebulæ, we may further say, with the most complete confidence, that whether or not nebulous matter be generally resolvable into shining dots, it cannot possibly be true that its being, or not being so resolvable by our telescopes, depends merely upon its smaller or greater distance from the observer. For, in the first place, that there is matter, to the best assisted eye not distinguishable from nebulous matter, which is not so resolvable, is proved by several facts. The tails of Comets often resemble nebulæ; so much so that there are several known nebulæ, which are, by the less experienced explorers of the sky, perpetually mistaken for comets, till they are proved not to be so, by their having no cometary motion. Such is |165

the nebula in Andromeda, which is visible to the naked eye*. But the tails and nebulous appendages of comets, though they alter their appearance very greatly, according to the power of the telescope with which they are examined, have never been resolved into stars, or any kind of dots; and seem, by all investigations, to be sheets or cylinders or cones of luminous vapour, changing their form as they approach to or recede from the sun, and perhaps by the influence of other causes. Yet some of them approach very near the earth; all of them come within the limits of our system. Here, then, we have (probably, at least,) nebulous matter, which when brought close to the eye, compared with the stellar nebulæ, still appears as nebulous.

15 Again, as another phenomenon, bearing upon the same question, we have the Zodiacal Light. This is a faint cone of light† which, at certain seasons, may be seen extending from the horizon obliquely upwards, and following the course of the ecliptic, or rather, of the sun's equator. It appears to be a lens-shaped envelope of the sun, extending beyond the orbits of Mercury and Venus, and nearly attaining that of the earth; and in Sir John Herschel's view, may be regarded as placing the sun in the list of nebulous stars. No one has ever thought that this nebulous appearance was resolvable into luminous points; but if it were, probably not even the most sanguine of speculators on the multitude of suns would call these points *suns*.

16 But indeed the nebulæ themselves, and especially the most remote of the nebulæ, or at least those which most especially require the most powerful telescopes, offer far more decisive proofs that their resolvability or non-resolvability,—their apparent constitution as diffused and vaporous masses,—does not depend upon their distance. A remarkable fact in

* Herschel. *Outl. of Ast.* Art. 874, and Plate 11, Fig. 3.

† Ibid. Art. 897.

the irregular, and in some of the regular nebulæ* is, that they consist of long patches and streaks, which stretch out in various directions, and of which the form† and extent vary according to the visual power which is applied to them. Many of the nebulæ, and especially of the fainter ones, entirely change their form with the optical power of the instrument by which they are scrutinized; so that, as seen in the mightier telescopes of modern times, the astronomer scarcely recognizes the figures in which the earlier observers have recorded what they saw in the same place. Parts which, before, were separate, are connected by thin bridges of light which are now detected; and where the nebulous space appeared to be bounded, it sends off long tails of faint light into the surrounding space. Now no one can suppose that these newly seen portions of the nebula are immensely further off than the other parts. However little we know of the nature of the object, we must suppose it to be one connected object, with all its parts, as to sense, at the same distance from us. Whether therefore it be resolvable or no, there must be some other reason, besides the difference of distance, why the brighter parts were seen, while the fainter parts were not. The obvious reason is, that the latter were not seen because they were thin films which required more light to see them. We are led, irresistibly as it seems, to regard the whole mass of such a nebula, as an aggregation of vaporous rolls and streaks, assuming such forms as thin volumes of smoke or vapour often assume in our atmosphere, and assuming, like them, different shapes according to the quantity of light which comes to us from them. If, as soon as one of these new filaments or webs of a nebula comes into view, we should say, Here we have a new array of suns and of worlds, we should judge as fantastically, as any one who should combine the like imaginations with the vary-

* Hersch. 874. † Ibid. 881—8.

ing cloud-work of a summer-sky. To suppose that all the varied streaks by which the patch of nebulous light shades off into the surrounding darkness, and which change their form and extent with every additional polish which we can give to a reflecting or refracting surface, disclose, with every new streak, new worlds, is a wanton indulgence of fancy, to which astronomy gives us no countenance*.

17 Undoubtedly all true astronomers, taught caution and temperance of thought by the discipline of their magnificent science, abstain from founding such assumptions upon their discoveries. They know how necessary it is to be upon their guard against the tricks which fancy plays with the senses; and if they see appearances of which they cannot interpret the meaning, they are content that they should have no meaning for them, till the due explanation comes. We have innumerable examples of this wise and cautious temper, in all periods of astronomy. One has occurred lately. Several careful astronomers, observing the stars by day, had been surprized to see globes of light glide across the field of view of their telescopes, often in rapid succession and in great numbers. They did not, as may be supposed, rush to the assumption that these globes were celestial bodies of a new kind, before unseen; and that from the peculiarity of their appearance and movement, they were probably inhabited by beings of a peculiar kind. They proceeded very differently; they altered the focus of their telescopes, looked with other glasses, made various changes and trials, and finally discovered that these globes of light were the winged seeds of certain plants which were wafted through the air; and which,

* At the recent meeting of the British Association (Sept. 1853), drawings were exhibited of the same nebulæ, as seen through Lord Rosse's large telescope, and through a telescope of three feet aperture. With the smaller telescopic power, all the characteristic features were lost. The spiral structure (see next Article but one,) has been almost entirely brought to light by the large telescope.

illuminated by the sun, were made globular by being at distances unsuited to the focus of the telescope*.

18 But perhaps something more may be founded on the ramified and straggling form which belongs to many of the nebulæ. Under the powers of Lord Rosse's telescope, a considerable number of them assume a shape consisting of several spiral films diverging from one center, and growing broader and fainter as they diverge, so as to resemble a curled feather, or whirlpool of light†. This form, though generally deformed by irregularities, more or less, is traceable in so many of the nebulæ, that we cannot easily divest ourselves of the persuasion that there is some general reason for such a form;—that something, in the mechanical causes which have produced the nebulæ, has tended to give them this shape. Now when this thought has occurred to us, since mathematicians have written a great deal concerning the mechanics of the universe, it is natural to ask, whether any of the problems which they have solved give a result like that thus presented to our eyes. Do such spirals as we here see, occur in any of the diagrams which illustrate the possible motions of celestial bodies? And to this, a person acquainted with mathematical literature might reply, that in the second Book of Newton's *Principia*, in the part which has especial reference to the Vortices of Descartes, such spirals appear upon the page. They represent the path which a body would describe if, acted upon by a central force, it had to move in a medium of which the resistance was considerable;—considerable, that is, in comparison with the other forces which act; as for example, the forces which deflect the motion

* See Monthly Notices of the Royal Astronomical Society, Dec. 13, 1850.

† The frontispiece to this volume represents two of these Spiral Nebulæ; those denominated 51 Messier, and 99 Messier, as given by Lord Rosse in the *Phil. Trans. for* 1850. The former of these two has a lateral focus, besides the principal focus or pole.

from a straight line. Indeed that in such a case a body would describe a spiral, of which the general form would be more or less oval, is evident on a little consideration. And in this way, for instance, Encke's comet, which, if the resistance to its motion were insensible, would go on describing an ellipse about the sun, always returning upon the same path after every revolution; does really describe a path which, at each revolution, falls a little within the preceding revolution, and thus gradually converges to the center. And if we suppose the comet to consist of a luminous mass, or a string of masses, which should occupy a considerable arc of such an orbit, the orbit would be marked by a track of light, as an oval spiral. Or if such a comet were to separate into two portions, as we have, with our own eyes, recently seen Biela's comet do; or into a greater number; then these portions would be distributed along such a spiral. And if we suppose a large mass of cometic matter thus to move in a highly resisting medium, and to consist of patches of different densities, then some would move faster and some more slowly; but all, in spirals such as have been spoken of; and the general aspect produced would be, that of the spiral nebulæ which I have endeavoured to describe. The luminous matter would be more diffused in the outer and more condensed in the central parts, because to the center of attraction all the spirals converge.

19 This would be so, we say, if the luminous matter moved in a greatly resisting medium. But what is the measure of *great* resistance? It is, as we have already said, that the resistance which opposes the motion shall bear a considerable proportion to the force which deflects the motion. But what is that force? Upon the theory of the universal gravitation of matter, on which theory we here proceed, the force which deflects the motions of the parts of each system into curves, is the mutual attraction of the parts of the system; leaving out of the account the action of other systems,

170|

as comparatively insignificant and insensible. The condition, then, for the production of such spiral figures as I have spoken of, amounts really to this; that the mutual attraction of the parts of the luminous matter is slight; or, in other words, that the matter itself is very thin and rare. In that case, indeed, we can easily see that such a result would follow. A cloud of dust, or of smoke, which was thin and light, would make but a little way through the air, and would soon fall downwards: while a metal bullet shot horizontally with the same velocity, might fly for miles. Just so, a loose and vaporous mass of cometic matter would be pulled rapidly inwards by the attraction to the center: and supposing it also drawn into a long train, by the different density of its different parts, it would trace, in lines of light, a circular or elliptical spiral converging to the center of attraction, and resembling one of the branches of the spiral nebulæ. And if several such cometic masses thus travelled towards the center, they would exhibit the wheel-like figure with bent spokes, which is seen in the spiral nebulæ. And such a figure would all the more resemble some of those nebulæ, as seen through Lord Rosse's telescope, if the spirals were accompanied by exterior branches of thinner and fainter light, which nebulous matter of smaller density might naturally form. Perhaps too, such matter, when thin, may be supposed to cool down more rapidly from its state of incandescence; and thus to become less luminous. If this were so, a greater optical power would of course be required, to make the diverging branches visible at all.

20 There is one additional remark, which we may make, as to the resemblance of cometary* and nebular matter.

* I am aware that some astronomers do not consider it as proved that cometary matter is entirely self-luminous. Arago found that the light of a Comet contained a portion of polarized light, thus proving that it had been reflected, (*Cosmos*, I. p. 111, and III. p. 566). But I think the opinion that the greater part of the light is self-luminous, like

|171

That cometary matter is of very small density, we have many reasons to believe:—its transparency, which allows us to see stars through it undimmed;—the absence of any mechanical effect, weight, inertia, impulse, or attraction, in the nearest appulses of comets to planets and satellites:—and the fact that, in the recent remarkable event in cometic history, the separation of Biela's comet into two, the two parts did not appear to exert any perceptible attraction on each other, any more than two volumes of dust or of smoke would do on earth. Luminous cometary matter, then, is very light, that is, has very little weight or inertia. And luminous nebulous matter is also very light in this sense: if our account of the cause of spiral nebulæ has in it any truth. But yet, if we suppose the nebulæ to be governed by the law of universal gravitation, the attractive force of the luminous matter upon itself must be sufficient to bend the spirals into their forms. How are we to reconcile this; that the matter is so loose that it falls to the center in rapid spirals, and yet that it attracts so strongly that there is a center, and an energetic central force to curve the spirals thither? To this, the reply which we must make is, that the size of the nebular space is such, that though its rarity is extreme, its whole mass is considerable. One part does not perceptibly attract another, but the whole does perceptibly attract every part. This indeed need the less surprize us, since it is exactly the case with our earth. One stone does not visibly attract another. It is much indeed for man, if he can make perceptible the attraction of a mountain upon a plumb-line; or of a stratum of rock a thousand feet thick upon the going of a pendulum; or of large masses of metal upon a delicate balance. By such experiments, men of science have endeavoured to measure that

the nebulæ, generally prevails. Any other supposition is scarcely consistent with the rapid changes of brightness which occur in a comet during its motion to and from the Sun.

minute thing, the attraction of one portion of terrestrial matter upon another; and thus, to weigh the whole mass of the earth. And equally great, at least, may be the disproportion between the mutual attraction of two parts of a nebulous system, and the total central attraction; and thus, though the former be insensible, the latter may be important.

21 It has been shewn by Newton, that if any mass of matter be distributed in a uniform sphere, or in uniform concentric spherical shells, the total attraction on a point without the sphere, will be the same as if the whole mass were collected in that single point, the center. Now, proceeding upon the supposition of such a distribution of the matter in a nebula (which is a reasonable average supposition,) we may say, that if our sun were expanded into a nebula reaching to the extreme bounds of the known solar system, namely, to the newly-discovered planet Neptune, or even hundreds of times further; the attraction on an external point would remain the same as it is, while the attraction on points within the sphere of diffusion would be less than it is; according to some law, depending upon the degree of condensation of the nebular matter towards the center; but still, in the outer regions of the nebula, not differing much from the present solar attraction. If we could discover a mass of luminous matter, descending in a spiral course towards the center of such a nebula, that is, towards the sun, we should have a sort of element of the spiral nebulæ which have now attracted so much of the attention of astronomers. But, by an extraordinary coincidence, recent discoveries have presented to us such an element. Encke's comet, of which we have just spoken, appears to be describing such a spiral curve towards the sun. It is found that its period is, at every revolution, shorter and shorter; the amplitude of its sweep, at every return within the limits of our observation, narrower and narrower; so that in the

|173

course of revolutions and ages, however numerous, still, not such as to shake the evidence of the fact, it will fall into the sun.

22 Here then we are irresistibly driven to calculate what degree of resemblance there is, between the comet of Encke, and the luminous elements of the spiral nebulæ, which have recently been found to exist in other regions of the universe. Can we compare its density with theirs? Can we learn whether the luminous matter in such nebulæ is more diffused or less diffused, than that of the comet of Encke? Can we compare the mechanical power of getting through space, as we may call it, that is, the ratio of the inertia to the resistance, in the one case, and in the other? If we can, the comparison cannot fail, it would seem, to be very curious and instructive. In this comparison, as in most others to which cosmical relations conduct us, we must expect that the numbers to which we are led, will be of very considerable amount. It is not equality in the density of the two luminous masses which we are to expect to find; if we can mark their proportions by thousands of times, we shall have made no small progress in such speculations.

23 The comet of Encke describes a spiral, gradually converging to the sun; but at what rate converging? In how many revolutions will it reach the sun? Of how many folds will its spire consist, before it attains the end of its course? The answer is:—Of very many. The retardation of Encke's Comet is very small; so small, that it has tasked the highest powers of modern calculation to detect it. Still however, it is there; detected, and generally acknowledged, and confirmed by every revolution of the comet, which brings it under our notice; that is, commonly, about every three years. And having this fact, we must make what we can of it, in reasoning on the condition of the universe. No 174| accuracy of calculation is necessary for our purpose: it is

enough, if we bring into view the kind of scale of numbers to which calculation would lead us.

24 Encke's Comet revolves round the sun in 1211 days. The period diminishes at present, by about $\frac{1}{9}$ of a day every revolution. This amount of diminution will change, as the orbit narrows; but for our purpose, it will be enough to consider it unchangeable. The orbit therefore will cease to exist in a number of periods expressed by 9 times 1211; that is, in something more than 10,000 revolutions; and of course sooner than this, in consequence of its coming in contact with the body of the sun. In 30,000 years then, it may be, this comet will complete its spiral, and be absorbed by the central mass. This long time, this long series of ten thousand revolutions, are long, because the resistance is so small, compared with the inertia of the moving mass. However thin, and rare, and unsubstantial the comet may be, the medium which resists it is much more so.

25 But this spiral, converging to its pole so slowly that it reaches it only after 10,000 circuits, is very different indeed from the spirals which we see in the nebulæ of which we have spoken. In the most conspicuous of those, there are only at most three or four circular or oval sweeps, in each spiral, or even the spiral reaches the center before it has completed a single revolution round it. Now what are we to infer from this? How is it, that the comet has a spiral of so many revolutions, and the nebulæ of so few? What difference of the mechanical conditions is indicated by this striking difference of form? Why, while the Comet thus lingers long in the outer space, and approaches the sun by almost imperceptible degrees, does the Nebular Element rush, as it were, headlong to its center, and show itself unable to circulate even for a few revolutions?

26 Regarding the question as a mechanical problem, the answer must be this:—It is so, because the nebula is so |175

much more rare than the matter of the comet, or the resisting medium so much more dense; or combining the two suppositions, because in the case of the comet, the luminous matter has *much* more inertia, more mechanical reality and substance, than the medium through which it moves; but in the nebula, very *little* more.

27 The numbers of revolutions of the spiral, in the two cases, may not exactly represent the difference of the proportions; but, as I have said, they may serve to shew the scale of them; and thus we may say, that if Encke's comet, approaching the center by 10,000 revolutions, is 100,000 times as dense as the surrounding medium, the elements of the nebula, which reach the center in a single revolution, are only *ten* times as dense as the medium through which they have to move*.

28 Nor does this result (that the bright element of the nebulæ is so few times denser than the medium in which it moves) offer anything which need surprize us: for, in truth, in a diffused nebula, since we suppose that its parts have mechanical properties, the nebula itself is a resisting medium. The rarer parts, which may very naturally have cooled down in consequence of their rarity, and so, become non-luminous, will resist the motions of the more dense and still-luminous portions. If we recur to the supposition, which we lately made, that the Sun were expanded into a nebulous sphere, reaching the orbit of Neptune, the diffused matter would offer a far greater resistance to the motions of comets than they now experience. In that case, Encke's comet might be brought to the center after a few revolutions; and if, while it were thus descending, it were to be

* We assume here that the number of revolutions to the center is greater in proportion as the relative density of the resisting medium is less; which is by no means mechanically true; but the calculation may serve, as we have said, to show the scale of the numbers involved.

drawn out into a string of luminous masses, as Biela's comet has begun to be, these comets, and any others, would form separate luminous spiral tracks in the solar system; and would convert it into a spiral nebula of many branches, like those which are now the most recent objects of astronomical wonder.

29 It seems allowable to regard it as one of those coincidences, in the epochs of related yet seeming unconnected discoveries, which have so often occurred in the history of science; that we should, nearly at the same time, have had brought to our notice, the prevalence of spiral nebulæ, and the circumstances, in Biela's and in Encke's comets, which seem to explain them: the one by showing the origin of luminous broken lines, one part drifting on faster than another, according to its different density, as is usual in incoherent masses*; and the other by showing the origin of the spiral form of those lines, arising from the motion being in a resisting medium.

30 But though I have made suppositions by which our Solar System might become a spiral nebula, undoubtedly it is at present something very different; and the leading points of difference are very important for us to consider. And the main point is, that which has already been cursorily noticed: that instead of consisting of matter all nearly of the same density, and a great deal of it luminous, our Solar System consists of kinds of matter immensely different in density, and of large and regular portions which are not luminous. Instead of a diffused nebula with vaporous comets trailing spiral tracks through a medium little rarer than themselves; we have a

* Humboldt, whom nothing relative to the history of science escapes, quotes from Seneca a passage in which mention is made of a Comet which divided into two parts: and from the Chinese Annals, a notice of three 'coupled Comets,' which in the year 896 appeared, and described their paths together. *Cosmos*, III. p. 570, and the notes.

central sun, and the dark globes of the solid planets rolling round him, in a medium so rare, that in thousands of revolutions not a vestige of retardation can be discovered by the most subtle and persevering researches of astronomers. In the solar system, the luminous matter is collected into the body of the sun; the non-luminous matter, into the planets. And the comets and the resisting medium, which offer a small exception to this account, bear a proportion to the rest which the power of numbers scarce suffices to express.

31 Thus with regard to the density of matter in the solar system; we have supposed, as a mode of expression, that the density of a comet, Encke's Comet for instance, is 100,000 times that of the resisting medium. Probably this is greatly understated; and probably also we greatly understate the matter, when we suppose that the tail of a comet is 100,000 times rarer than the matter of the sun*. And thus the resisting medium would be, at a very low calculation, 10,000 millions of times more rare than the substance of the sun.

32 And thus we are not, I think, going too far, when we say, that our Solar System, compared with spiral nebulous systems, is a system completed and finished, while they are mere confused, indiscriminate, incoherent masses. In the Nebulæ, we have loose matter of a thin and vaporous constitution, differing as more or less rare, more or less luminous, in a small degree; diffused over enormous spaces, in straggling and irregular forms; moving in devious and brief curves, with no vestige of order or system, or even of separation of different kinds of bodies. In the Solar System, we have the luminous separated from the non-luminous, the hot

* Laplace has proved that the masses of comets are very small. He reckons their mean mass as very much less than 1-100000th of the Earth's mass. And hence, considering their great size, we see how rare they must be. See *Expos. du Syst. du Monde.*

178|

from the cold, the dense from the rare; and all, luminous and non-luminous, formed into globes, impressed with regular and orderly motions, which continue the same for innumerable revolutions and cycles*. The spiral nebulæ, compared with the solar system, cannot be considered as other than a kind of chaos; and not even a chaos, in the sense of a state preceding an orderly and stable system; for there is no indication, in those objects, of any tendency towards such a system. If we were to say that they appear mere shapeless masses, flung off in the work of creating solar systems, we might perhaps disturb those who are resolved to find everywhere worlds like ours; but it seems difficult to suggest any other reason for not saying so.

33 The same may be said of the other very irregular nebulæ, which spread out patches and paths of various degrees of brightness; and shoot out, into surrounding space, faint branches which are of different form and extent, according to the optical power with which they are seen. These irregular forms are incapable of being permanent according to the laws of mechanics. They are not figures of equilibrium; and therefore, must change by the attraction of the matter upon itself. But if the tenuity of the matter is extreme, and the resistance of the medium in which it floats considerable, this tendency to change and to condensation may be almost nullified; and the bright specks may long keep their straggling forms, as the most fantastically shaped clouds of a summer-sky often do. It is true, it may be said that the reason why we see no change in the form of such nebulæ, is that our observations have not endured long enough; all visible changes in the stars requiring an immense time, according to the gigantic scale of celestial mechanism.

* Humboldt repeatedly expresses his conviction that our Solar system contains a greater variety of forms than other systems. (*Cosmos*, III. 373 and 587.)

But even this hypothesis (it is no more,) tends to establish the extreme tenuity of the nebulæ: for more solid systems, like our solar system, require, for the preservation of their form, motions which are perceptible, and indeed conspicuous, in the course of a month; namely, the motions of the planets. All therefore concurs to prove the extreme tenuity of the substance of irregular nebulæ.

34 Nebulæ which assume a regular, for instance, a circular or oval shape, with whatever variation of luminous density from the inner to the outer parts, may have a form of equilibrium, if their parts have a proper gyratory motion. Still, we see no reason for supposing that these differ so much from irregular nebulæ, as to be denser bodies, kept in their forms by rapid motions. We are rather led to believe that, though perhaps denser than the spiral nebulæ, they are still of extremely thin and vaporous character. It would seem very unlikely that these vast clouds of luminous vapor should be as dense as the tail of a comet; since a portion of luminous matter so small as such a tail is, must have cooled down from its most luminous condition; and must require to be more dense than nebular matter in order to be visible at all by its own light.

35 Thus we appear to have good reason to believe that nebulæ are vast masses of incoherent or gaseous matter, of immense tenuity, diffused in forms more or less irregular, but all of them destitute of any regular system of solid moving bodies. We seem therefore to have made it certain that *these* celestial objects at least are not inhabited. No speculators have been bold enough to place inhabitants in a comet; except, indeed, some persons who have imagined that such a habitation, carrying its inmates alternately into the close vicinity of the sun's surface, and far beyond the orbit of Uranus, and thus exposing them to the fierce

180| extremes of heat and cold, might be the seat of penal

inflictions on those who had deserved punishment by acts done in their life on one of the planets. But even to give coherence to this wild imagination, we must further suppose that the tenants of such prison-houses, though still sensible to human suffering from extreme heat and cold, have bodies of the same vaporous and unsubstantial character as the vehicle in which they are thus carried about the system: for no frame of solid structure could be sustained by the incoherent and varying volume of a comet. And probably, to people the nebulæ with such thin and fiery forms, is a mode of providing them with population, that the most ardent advocates of the plurality of worlds are not prepared to adopt.

36 So far then as the Nebulæ are concerned, the improbability of their being inhabited, appears to mount to the highest point that can be conceived. We may, by the indulgence of fancy, people the summer-clouds, or the beams of the aurora borealis, with living beings, of the same kind of substance as those bright appearances themselves; and in doing so, we are not making any bolder assumption than we are, when we stock the Nebulæ with inhabitants, and call them in that sense, 'distant worlds.'

CHAPTER VIII.

The Fixed Stars.

---

1 WE appear, in the last chapter, to have cleared away the supposed inhabitants of the outskirts of creation, so far as the Nebulæ are the outskirts of creation. We must now approach a little nearer, in appearance at least, to our own system. We must consider the Fixed Stars; and examine any evidence which we may be able to discover, as to the probability of their containing, in themselves or in accompanying bodies, as planets, inhabitants of any kind. Any special evidence which we can discern on this subject, either way, is indeed slight. On the one side we have the asserted analogy of the parts of the universe; of which point we have spoken, and may have more to say hereafter. Each Fixed Star is conceived to be of the nature of our Sun; and therefore, like him, the center of a planetary system. On the other side, it is extremely difficult to find any special facts relative to the nature of the fixed stars, which may enable us in any degree to judge how far they really are of a like nature with the Sun, and how far this resemblance goes. We may, however, notice a few features in the starry heavens, with which, in the absence of any stronger grounds, we may be allowed to connect our speculations on such questions. The assiduous scrutiny of the stars which has been pursued by the most eminent astronomers, and the reflections which their researches have suggested to them, may have a new interest, when discussed under this point of view.

2 Next after the Nebulæ, the cases which may most 182| naturally engage our attention, are Clusters of stars. The

cases, indeed, in which these clusters are the closest, and the stars the smallest, and in which, therefore, it is only by the aid of a good telescope that they are resolved into stars, do not differ from the resolvable nebulæ, except in the degree of optical power which is required to resolve them. We may, therefore, it would seem, apply to such clusters, what we have said of resolvable nebulæ: that when they are thus, by the application of telescopic power, resolved into bright points, it seems to be a very bold assumption to assume, without further proof, that these bright points are suns, distant from each other as far as we are from the nearest stars. The boldness of such an assumption appears to be felt by our wisest astronomers*. That several of the clusters which are visible, some of them appearing as if the component stars were gathered together in a nearly spherical form, are systems bound together by some special force, or some common origin, we may regard, with those astronomers, as in the highest degree probable. With respect to the stability of the form of such a system, a curious remark has been made by Sir John Herschel†, that if we suppose a globular space filled with equal stars, uniformly dispersed through it, the particular stars might go on for ever, describing ellipses about the center of the globe, in all directions, and of all sizes; and all completing their revolutions in the same time. This follows, because, as Newton has shown, in such a case, the compound force which tends to the center of the sphere would be everywhere proportional to the distance from the center; and under the action of such a force, ellipses about the center would be described, all the periods being of the same amount. This kind of symmetrical and simple systematic motion, presented by Newton as a mere exemplification of the results of his mechanical principles, is perhaps realized, approximately at least, in some of the globular

* Herschel, 866. † Ibid. 866.

clusters. The motions will be swift or slow, according to the total mass of the groups. If, for instance, our Sun were thus broken into fragments, so as to fill the sphere girdled by the earth's orbit, all the fragments would revolve round the center in a year. Now there is no symptom, in any cluster, of its parts moving nearly so fast as this; and therefore we have, it would seem, evidence that the groups are much less dense than would be the space so filled with fragments of the sun. The slowness of the motions, in this case, as in the nebulæ, is evidence of the weakness of the forces, and therefore, of the rarity of the mass: and till we have some gyratory motion discovered in these groups, we have nothing to limit our supposition of the extreme tenuity of their total substance.

3 Let us then go on to the cases in which we have proof of such gyratory motions in the stars; for such are not wanting. Fifty years ago, Herschel the father, had already ascertained that there are certain pairs of stars, very near each other (so near, indeed, that to the unassisted eye they are seen as single stars only,) and which revolve about each other. These Binary Sidereal Systems have since been examined with immense diligence and profound skill by Herschel the son, and others; and the number of such binary systems has been found, by such observers, to be very considerable. The periods of their revolutions are of various lengths, from 30 or 40 years to several hundreds of years. Some of those pairs which have the shortest periods, have already, since the nature of their movements was discovered, performed more than a complete revolution*; thus leaving no room for doubting that their motions are really of this gyratory kind. Not only the fact, but the law of this orbital motion, has been investigated: and the investigations, which naturally were commenced on the hypothesis that these distant bodies were governed by that Law of universal Gravitation, which prevails

* Herschel, 846.

throughout the solar system, and so completely explains the minutest features of its motions, have ended in establishing the reality of that Law, for several Binary Systems, with as complete evidence as that which carries its operations to the orbits of Uranus and Neptune.

4 Being able thus to discern, in distant regions of the universe, bodies revolving about each other, we have the means of determining, as we do in our own solar system, the masses of the bodies so revolving. But for this purpose, we must know their distance from each other; which is, to our vision, exceedingly small, requiring, as we have said, high magnifying powers to make it visible at all. And again, to know what linear distance this small visible distance represents, we must know the distance of the stars from us, which is, for every star, as we know, immensely great; and for most, we are destitute of all means of determining how great it is. There are, however, some of these binary systems, in which astronomers conceive that they have sufficiently ascertained the value of both these elements, (the distance of the two stars from each other, and from us,) to enable them to proceed with the calculation of which I have spoken; the determination of the masses of the revolving bodies. In the case of the star *Alpha Centauri*, the first star in the constellation of the Centaur, the period is reckoned to be 77 years; and as, by the same calculator, the apparent semi-axis of the orbit described is stated at 15 seconds of space, while the annual parallax of each star is about one second, it is evident that the orbit must have a radius about 15 times the radius of the earth's orbit; that is, an orbit greater than that of Saturn, and approaching to that of Uranus. In the solar system, a revolution in such an orbit would occupy a time greater than that of Saturn, which is 30 years, and less than that of Uranus, which is about 80 years: it would, in fact, be about 58 years. And since, in the binary star, the period is greater

than this, namely 77 years, the attraction which holds together its two elements must be less than that which holds together the Sun and a planet at the same distance; and therefore the masses of the two stars together are considerably less than the mass of our sun.

5 A like conclusion is derived from another of these conspicuous double stars, namely, the one termed by astronomers 61 *Cygni*; of which the annual parallax has lately been ascertained to be one third of a second of space, while the distance of the two stars is 15 seconds. Here therefore we have an orbit 45 times the size of the Earth's orbit; larger than that of the new-discovered planet Neptune, whose orbit is 30 times as large as the earth's, and his period nearly 165 years. The period of 61 Cygni is however, it appears, probably not short of 500 years; and hence it is calculated that the sum of the masses of the two stars which make up this pair is about one third of the mass of our Sun*.

6 These results give some countenance to the opinion, that the quantity of luminous matter, in other systems, does not differ very considerably from the mass of our Sun. It differs in these cases as 1 to 3, or thereabouts. In what degree of condensation, however, the matter of these binary systems is, compared with that of our solar system, we have no means whatever of knowing. Each of the two stars may have its luminous matter diffused through a globe as large as the earth's orbit; and in that case, would probably not be more dense than the tail of a comet†. It is observed by astronomers, that in the pairs of binary stars which we have mentioned, the two stars of each pair are of different colours; the stars being of a high yellow, approaching to orange colour‡,

* Herschel, 848.

† That these systems have not condensed to *one* center, appears to imply a less complete degree of condensation than exists in those systems which have done so.

186|

‡ Herschel, 850.

but the smaller individual being in each case of a deeper tint. This might suggest to us the conjecture that the smaller mass had cooled further below the point of high luminosity than the larger; but that both these degrees of light belong to a condition still progressive, and probably still gaseous. Without attaching any great value to such conjectures, they appear to be at least as well authorized as the supposition that each of these stars, thus different, is nevertheless precisely in the condition of our sun.

7 But, even granting that each of the individuals of this pair were a sun like ours, in the nature of its material and its state of condensation, is it probable that it resembles our Sun also in having planets revolving about it? A system of planets revolving around or among a pair of suns, which are, at the same time, revolving about one another, is so complex a scheme, so impossible to arrange in a stable manner, that the assumption of the existence of such schemes, without a vestige of evidence, can hardly require confutation. No doubt, if we were really required to provide such a binary system of suns with attendant planets, this would be best done by putting the planets so near to one sun, that they should not be sensibly affected by the other; and this is accordingly what has been proposed*. For, as has been well said of the supposed planets, in making this proposal, 'Unless closely nestled under the protecting wing of their immediate superior, the sweep of the other sun in his perihelion passage round their own, might carry them off, or whirl them into orbits utterly inconsistent with the existence of their inhabitants.' To assume the existence of the inhabitants, in spite of such dangers, and to provide against the dangers by placing them so close to one sun as to be out of the reach of the other, though the whole distance of the two may not, and as we have seen, in some cases does not, exceed

* Herschel, 847.

the dimensions of our solar system, is showing them all the favour which is possible. But in making this provision, it is overlooked that it may not be possible to keep them in permanent orbits so near to the selected center: their sun may be a vast sphere of luminous vapour; and the planets, plunged into this atmosphere, may, instead of describing regular orbits, plough their way in spiral paths through the nebulous abyss to its central nucleus.

8 Clustered stars, then, and double stars, appear to give us but little promise of inhabitants. We must next turn our attention to the single stars, as the most hopeful cases. Indeed, it is certain that no one would have thought of regarding the individual stars of clusters, or of pairs, as the centers of planetary systems, if the view of insulated stars, as the centers of such systems, had not already become familiar, and, we may say, established. What, then, is the probability of that view? Is there good evidence that the Fixed Stars, or some of them, really have planets revolving round them? What is the kind of proof which we have of this?

9 To this we must reply, that the only proof that the fixed stars are the centers of planetary systems, resides in the assumption that those stars are *like the Sun*;—resemble him in their qualities and nature, and therefore, it is inferred, must have the same offices, and the same appendages. They are, as the Sun is, independent sources of light, and thence, probably, of heat; and therefore they must have attendant planets, to which they can impart their light and heat; and these planets must have inhabitants, who live under and enjoy those influences. This is, probably, the kind of reasoning on which those rely, who regard the fixed stars as so many worlds, or centers of families of worlds.

10 Every thing in this argument, therefore, depends upon this: that the Stars are *like the Sun:* and we must

188|

consider, what evidence we have of the exactness of this likeness.

11 The Stars are like the Sun in this, that they shine with an independent light, not with a borrowed light, as the planets shine. In this, however, the stars resemble, not only the Sun, but the nebulous patches in the sky, and the tails of comets; for these also, in all probability, shine with an original light. Probably it will hardly be urged that we see, by the very appearance of the stars, that they are of the nature of the Sun: for the appearance of luminaries in the sky is so far from enabling us to discriminate the nature of their light, that to a common eye, a planet and a fixed star appear alike as stars. There is no obvious distinction between the original light of the stars and the reflected light of the planets. The stars, then, being like the sun in being luminous, does it follow that they are, like the sun, definite dense masses*? Or are they, or many of them, luminous masses in a far more diffused state; visually contracted to points, by the immense distance from us at which they are?

12 We have seen that some of those stars, which we have the best means of examining, are, in mass, one third, or less, of our Sun. If such a mass, at the distance of the fixed stars, were diffused through a sphere equal in radius to the earth's orbit, it would still appear to us as a point; as is evident by this, that the fixed stars, for the most part, have no discoverable annual parallax; that is, the earth's orbit appears to them a point. If one of the fixed stars, Sirius, for instance, be in this diffused condition, such a circumstance will not, mechanically speaking, prevent his having planets revolving round him; for, as we have said, the attraction of his whole mass, in whatever state of spherical diffusion, will be the same as if it were collected at the center. But such a state of diffusion will make him so un-

* The density of the sun is about as great as the density of water.

like our Sun, as much to break the force of the presumption that he must have planets because our Sun has. If the luminous matter of the stars gradually cools, grows dark, and solidifies, such diffusion would imply that the time of solidification is not yet begun; and therefore that the solid planets which accompany the luminous central body are not yet brought into being. If there be any truth in this hypothetical account of the changes, through which the matter of the stars successively passes; and if, by such changes, planetary systems are formed; how many of the fixed stars may never yet have reached the planetary state! how many, for want of some necessary mechanical condition, may never give rise to permanent orbits at all!

13 And that the matter of the stars does go through changes, we have evidence, in many such changes which have actually been observed *; and perhaps in the different colours of different stars; which may, not improbably, arise from their being at different stages of their progress. That planetary systems, once formed, go through mighty changes, we have evidence in the view which geology gives us of the history of this earth: and in that view, we see also, how unique, and how far elevated in its purpose, the last period of this history may be, compared with the preceding periods; and, up to the present time at least, how comparatively brief in its duration. If, therefore, stellar globes can become planetary systems in the progress of ages, it will not be at all inconsistent with what we know of the order of nature, that only a few, or even that only one, should have yet reached that condition. All the others, but the one, may be systems yet unformed, or fragments struck off in the forming of the one. If any one is not satisfied with this account of the degree of resemblance between the fixed stars and the sun, but would make the likeness greater than this; we have only to say,

* Herschel, 827—832.

that the proof that it is so lies upon him. Such a resemblance as we have supposed, is all that the facts suggest. That the stars are independent luminaries, we see: but whether they are as dense as the sun, or globes a hundred or a thousand times as rare, we have no means whatever of knowing. And, to assume that besides these luminous bodies which we see, there are dark bodies which we do not see, revolving round the others in permanent orbits, which require special mechanical conditions; and to suppose this, in order that we may build upon this assumption a still larger one, that of living inhabitants of these dark bodies; is a hypothetical procedure, which it seems strange that we should have to combat, at the present stage of the history of science, and in dealing with those whose minds have been disciplined by the previous events in the progress of astronomy.

14 Let us consider, however, further, how far astronomy authorizes us to regard the Fixed Stars as being, like our Sun, the centers of systems of Planets. Those who hold this, consider them as having a permanent condition of brightness, as our Sun has had for an indefinite period, so far as we have any knowledge on the subject. Yet, as we have said, no small number of the stars undergo changes of brightness; and some of them undergo such changes, in a manner which is not discernibly periodical; and which must therefore be regarded as progressive. This phenomenon countenances the opinion of such a progress from one material condition to another; which, we have seen, is suggested by the analogy of the probable formation of our own solar system. The very star which is so often taken as the probable center of a system, Sirius, has, in the course of the last 2000 years, changed its light from red to white. Ptolemy notes it as a red star: in Tycho's time it was already, as it is now, a white one*. The star *Eta Argus* changes

* *Cosmos*, III. 169, 205, and 641.

both its degree of light and its colour; ranging, in seemingly irregular intervals of time, from the fourth to the first magnitude*, and from yellow to red. Several other examples of the like kind have been observed. Mr Hind† gives an example in which he has, quite recently, observed in two years a star change its colour from very red to bluish. These variable unperiodical stars are probably very numerous. Also, some stars, observed of old, are now become invisible. 'The lost Pleiad,' by the loss of which the cluster, called the Seven Stars, offers now only six to the naked eye, is an example of a change of this kind already noted in ancient times. There are several others, of which the extinction is recognized by astronomers as proved‡. In other cases, new stars have appeared, and have then seemed to die away and vanish. The appearance of a new star in the time of the Greek astronomer Hipparchus induced him to construct his famous Catalogue of the Stars. Others are recorded to have appeared in the middle ages. The first which was observed by modern astronomers was the celebrated star seen by Tycho Brahe in 1572. It appeared suddenly in the constellation Cassiopeia, was fixed in its place like the neighbouring stars, had no nebula or tail, exceeded in splendour all other stars, being as bright as Venus when she is nearest the earth. It soon began to diminish in brightness, and passing through various diminishing degrees of magnitude, vanished altogether after seventeen months. This star also passed through various colours; being first white, then yellow, then red. In like manner, in 1604, a new star of great magnitude blazed forth in the constellation Serpentarius; and was seen by Kepler. And this also, like that of 1572, after a few months, declined and vanished.

* *Cosmos*, III. 172 and 252.

† *Astron. Soc. Notices*, Dec. 13, 1850.

‡ See Grant's *Hist. of Physical Astronomy*, p. 538.

15 These appearances led Tycho to frame an hypothesis like that which Sir William Herschel afterwards proposed, that the stars are formed by the condensation of luminous nebulous matter. Nor is it easy to think of such phenomena (of which several others have been observed, though none so conspicuous as these,) without regarding them as showing that the matter of the fixed stars, occasionally at least, passes through changes of consistence as great as would be the condensation and extinction of a luminous vapour. And if such changes have been but few within the recorded period of man's observation of the stars, we must recollect how small that period is, compared with the period during which the stars have existed. The stars themselves give us testimony of their having been in being for millions of years. For according to the best estimates we can form of their distances, the time which light would employ in reaching us from the most remote of them, would be millions of years: and, therefore, we now see those remote stars by means of the light emitted from them millions of years ago. And if, in the 2000 years during which such observations are recorded, only 200 stars have undergone such changes in a degree visible to the earth's inhabitants; in a million of years, change going on at the same rate, 100,000 stars would exhibit visible progressive change, showing that they had not yet reached a permanent condition. And how much of change may go on in any star without its being in any degree perceptible to the most exact astronomical scrutiny!

16 The tendency of these considerations is, to lead us to think that the fixed stars are not generally in that permanent condition in which our sun is; and which appears to be alone consistent with the existence of a system such as the solar system*. These views, therefore, fall in with

* I am aware of certain speculations, and especially of some recent ones, tending to show that even our Sun is wasting away by the

that which we have been led to by this consideration of the Nebulæ: that the Solar System is in a more complete and advanced state, as a system, than many at least of the stellar systems can be; it may be, than any other.

17 It has been alleged, as a proof of the likeness of the Fixed Stars to our Sun, that like him, they revolve upon their axes*. This has been supposed to be proved with regard to many of them, by their having periodical recurrences of fainter and brighter lustre; as if they were revolving orbs, with one side darkened by spots. Such facts are not very numerous or definite in the heavens. *Omicron*† in the constellation *Cetus*, is the longest known of them: and is held to revolve in 831 days. From the curious phenomena now spoken of, it has been called *Mira Ceti*‡. *Algol*, the second star (*Beta*) of *Perseus*, called also *Caput Medusæ*, is another, with a period of 2 days 21 hours: and in this case, the obscuration of the light, and the restoration of it, are so sudden, that from the time when it was first remarked, (by Goodricke, in 1782,) it suggested the hypothesis of an opaque body revolving round the star. The star *Delta* in the constellation *Cepheus*, is another, with a period of 5 days 9 hours. The star *Beta* in the *Lyre*, has a period of 6 days 10 hours, or perhaps 12 days 21 hours, one revolution having been taken for two. Another such star is *Eta Aquilæ*, with a period of 7 days 4 hours. These five are all the periodical stars of which astronomers can speak with precision§. But about thirty more are supposed to be sub-

emission of light and heat: but these opinions, even if established, do not much affect our argument, one way or the other.

* Chalmers's *Astron. Disc.* p. 39. † Hersch. 820.

‡ The periodical character of this star was discovered by David Fabricius, a parish priest in East Friesland, the father of John Fabricius, who discovered the solar spots. (*Cosmos*, III. 234.)

§ Hersch. 825. In Humboldt's *Cosmos*, III. 243, Argelander, who has most carefully observed and studied these periodical stars, ject to such change, though their periods, epochs, and phases of brightness, cannot at present be given exactly.

18 That these periodical changes in certain of the fixed stars are a curious and interesting astronomical fact, is indisputable. Nothing can be more probable also, than that it indicates, in the stellar masses, a revolution on their axes; which cannot surprise us, seeing that revolution upon an axis is, so far as we know, a universal law of all the large compact masses of matter which exist in the universe; and may be conceived to be a result derived from their origin, and a condition of any permanent or nearly permanent figure. But this can prove little or nothing as to their being like the sun, in any way which implies their having inhabitants, in themselves or in accompanying planets. The rotation of our Sun is not, in any intelligible way, connected with its having near it the inhabited Earth.

19 If we were to suppose some of the stars to be centers of planetary systems, we can hardly suppose it likely that these alone rotate, and that the others stand still. Probably all the stars rotate, more or less regularly, according as they are permanent or variable in form; but the most regular may still have no planets: and if they have, those planets may be as blank of inhabitants as our moon will be proved to be.

20 The revolution of Algol seems to approach the nearest to a fact in favour of a star being the center of a revolving system: and from the first, as we have said, the periodical change, and the sudden darkening and brightening of this luminary, suggested the supposition of an opaque body revolving about it. But this body cannot be a planet. The planets which revolve about our Sun are not, any of them, nor all of them together, large enough to produce a perceptible obscuration of his light, to a spectator outside the system. But

has given a catalogue containing 24, with the most recent determinations of their periods.

in Algol, the phenomena are very different from this*. The star is usually visible as a star of the second magnitude; but during each period of 2 days 21 hours, (or 69 hours,) it suffers a kind of eclipse, which reduces it to a star of the fourth magnitude. During this eclipse, the star diminishes in splendour for $3\frac{1}{2}$ hours; is at its lowest brightness for a quarter of an hour; and then, in $3\frac{1}{2}$ hours more, is restored to its original splendour. According to these numbers, if the obscuration be produced by a dark body revolving round a central luminary, and describing a circular orbit, as the regular recurrence of the obscuration implies, the space of the orbit during which the eclipsing body is interposed must be about one ninth of the circumference; for the obscuration occupies $7\frac{1}{4}$ hours out of 69. And therefore the space during which the eclipsing body obscures the central one must be about one *sixth* of the *diameter* of its orbit. But in order that the revolving body may, through this space, obscure the central one, the latter must extend over this space, namely, one sixth of the diameter of the orbit. But we may remark that there is no proof, in the phenomena, that the darkening body is detached from the bright mass. The effect would be the same if the dark mass were a part of the revolving star itself. It may be that the star has not yet assumed a spherical form, but is an oblong nebular mass with one part (perhaps from being thinner in texture,) cooled down and become opaque. And the amount of obscuration, reducing the star from the second to

* Hersch. 821. Humboldt (*Cosmos*, III. 238 and 246,) gives the period as 68 hours 49 minutes, and says that it is 7 or 8 hours in its less bright state. If we could suppose the times of the waning, and of the greatest eclipse, given by Herschel, to be exactly determined, as $3\frac{1}{2}$ and $\frac{1}{4}$, that is, in the proportion of 14 to 1, the darkening body must have its effective breadth $\frac{14}{15}$ of that of the star. But this is on the supposition that the orbit of the darkening body has the spectator's eye in its plane: if this be not so, the darkening body may be much larger.

the fourth magnitude, implies that the obscuring mass is large (perhaps one half the diameter, or much more) compared with the luminous mass. If this be a probable hypothesis to account for the phenomena, they are much more against than for the supposition of the star being the center of seats of habitation. And even if we have a planet nearly as large as its sun, revolving at the distance of only six of the sun's radii, how unlike is this to the solar system!

21 In fact, all these periodical stars, in so far as they are periodical, are proved, not to be like, but to be *unlike* our sun. It is true that the sun has spots, by means of which his rotation has been determined by astronomers. But these spots, besides being so small that they produce no perceptible alteration in his brightness, and are never, or very rarely, visible to the naked eye, are not permanent. A star with a permanent dark side would be very unlike our sun. The largest known of these stars, *Mira*, as the old astronomers called it, becomes invisible to the naked eye for 5 months during a period of 11 months. It must therefore have nearly one half its surface quite dark. This is very unlike the condition of the sun; and is a condition, it would seem, very little fitted to make this star the center of a planetary system like ours.

22 But there are other remarkable phenomena respecting these periodical stars, which have a bearing on our subject. Their periods are not quite regular, but are subject to certain variations. Thus it has been supposed that the period of Mira is subject to a cyclical fluctuation, embracing 88 of its periods; that is, about 80 years. But this notion of a cycle of so long a duration, requires confirmation; the fact of fluctuation in the period is alone certain. In like manner, Algol's periods are not quite uniform. All these facts agree with our suggestion, that the periodical stars are bodies of luminous matter which have not yet assumed a permanent form; and |197

which therefore, as they revolve about their axes, and turn to us their darker and their brighter parts, do so at intervals, and in an order, somewhat variable. And this suggestion appears to be remarkably confirmed, by a result which recent observations have disclosed relative to this star, Algol; namely, that its periods become shorter and shorter. For if the luminous matter, which is thus revolving, be gradually gathering into a more condensed form;—becoming less rare, or more compact; as, for instance, it would do, if it were collecting itself from an irregular, or elongated, into a more spherical form; such a shortening of the period of revolution would take place; for a mass which contracts while it is revolving, accelerates its rate of revolution, by mechanical principles. And thus we do appear to have, in this observed acceleration of the periods of Algol, an evidence that that luminous mass has not yet reached its final and permanent condition.

23 It is true, it has been conjectured, by high authority *, that this accelerated rapidity of the periods of Algol will not continue; but will gradually relax, and then be changed to an increase; like many other cyclical combinations in astronomy. But this conjecture seems to have little to support it. The cases in which an acceleration of motion is retarded, checked, and restored, all belong to our Solar System; and to assume that Algol, like the solar system, has assumed a permanent and balanced condition, is to take for granted precisely the point in question. We know of no such cycles among the fixed stars, at least with any certainty: for the cycle proposed for Mira must be considered as greatly needing confirmation; considering how long is the cycle, and how recent the suggestion of its existence.

* Hersch. *Outl. Ast.* 821. Another explanation of the variable period of Algol, is that the star is moving towards us, and therefore the light occupies less and less time to reach us.

24 And even in the solar system, we have accelerated motions, in which no mathematician or astronomer looks for a check or regress of the acceleration. No one expects that Encke's comet will cease to be accelerated, and to revolve in periods continually shorter; though all the other motions hitherto observed in the system are cyclical. In the case of a fixed star, we have much less reason to look for such a cycle, than we have in Encke's comet. But further: with regard to the existence of such a cycle of faster and slower motion in the case of Algol, the most recent observed facts are strongly against it; for it has been observed by Argelander, that not only there is a diminution of the period, but that this diminution proceeds with accelerated rapidity; a course of events which, in no instance, in the whole of the cosmical movements, ends in a regression, retardation, and restoration of the former rate. We are led to believe, therefore, that this remarkable luminary will go on revolving faster and faster, till its extreme point of condensation is attained. And in the mean time, we have very strong reasons to believe that this mutable body is not, like the sun, a permanent center of a permanent system; and that any argument drawn from its supposed likeness to the sun, in favour of the supposition that the regions which are near it are the seats of habitation, is quite baseless.

25 There are other phenomena of the Fixed Stars, and other conjectures of astronomers respecting them, which I need not notice, as they do not appear to have any bearing upon our subject. Such are the 'proper motions' of the stars, and the explanation which has been suggested of some of them; that they arise from the stars revolving round other stars which are dark, and therefore invisible. Such again is the attempt to show that the Sun, carrying with it the whole Solar System, is in motion; and the further attempt to show the direction of this motion: and again, the hypothesis that

|199

the Sun itself revolves round some distant body in space. These minute inquiries and bold conjectures, as to the movements of the masses of matter which occupy the universe, do not throw any light on the question whether any part besides the earth is inhabited; any more than the investigation of the movements of the ocean, and of their laws, could prove or disprove the existence of marine plants and animals. They do not on that account cease to be important and interesting subjects of speculation; but they do not belong to our subject.

26 In Fontenelle's *Dialogues on the Plurality of Worlds*, a work which may be considered as having given this subject a place in popular literature, he illustrates his argument by a comparison, which it may be worth while to look at for a moment. The speaker who asserts that the moon, the planets, and the stars, are the seats of habitation, describes the person who denies this, as resembling a citizen of Paris, who, seeing from the towers of Notre Dame the town of Saint Denis, (it being supposed that no communication between the two places had ever occurred,) denies that it is inhabited, because he cannot see the inhabitants. Of course the conclusion is easy, if we may thus take for granted that what he sees is a town. But we may modify this image, so as to represent our argument more fairly. Let it be supposed that we inhabit an island, from which innumerable other islands are visible; but the art of navigation being quite unknown, we are ignorant whether any of them are inhabited. In some of these islands, are seen masses more or less resembling churches; and some of our neighbours assert that these are churches; that churches must be surrounded by houses; and that houses must have inhabitants. Others hold that the seeming churches are only peculiar forms of rocks. In this state of the debate, everything depends upon the degree of resemblance to churches which the forms exhibit. But suppose

200|

that telescopes are invented, and employed with diligence upon the questionable shapes. In a long course of careful and skilful examination, no house is seen, and the rocks do not at all become more like churches, rather the contrary. So far, it would seem, the probability of inhabitants in the islands is lessened. But there are other reasons brought into view. Our island is a long extinct volcano, with a tranquil and fertile soil; but the other islands are apparently somewhat different. Some of them are active volcanoes, the volcanic operations covering, so far as we can discern, the whole island; others undergo changes, such as weather or earthquakes may produce; but in none of them can we discover such changes as show the hand of man. For these islands, it would seem the probability of inhabitants is further lessened. And so long as we have no better materials than these for forming a judgment, it would, surely, be accounted rash, to assert that the islands in general are inhabited; and unreasonable, to blame those who deny or doubt it. Nor would such blame be justified by adducing theological or *a priori* arguments; as, that the analogy of island with island makes the assumption allowable; or that it is inconsistent with the plan of the Creator of islands to leave them uninhabited. For we know that many islands are, or were long, uninhabited. And if ours were an island occupied by a numerous, well-governed, moral, and religious race, of which the history was known, and of which the relation to the Creator was connected with its history; the assumption of a history, more or less similar to ours, for the inhabitants of the other islands, whose existence was utterly unproved, would, probably, be generally deemed a fitter field for the romance-writer than for the philosopher. It could not, at best, rise above the region of vague conjecture.

27 Fontenelle, in the agreeable book just referred to, says, very truly, that the formula by which his view is urged |201

on adversaries is, *Pourquoy non?* which he holds to be a powerful figure of logic. It is, however, a figure which has this peculiarity, that it may, in most cases, be used with equal force on either side. When we are asked Why the Moon, Mercury, Saturn, the system of Sirius, should *not* be inhabited by intelligent beings; we may ask, Why the Earth in the ages previous to man might not be so inhabited? The answer would be, that we have proof *how* it *was* inhabited. And as to the fact in the other case, I shall shortly attempt to give proof that the Moon is certainly not, and Mercury and Saturn probably not inhabited. With regard to the Fixed Stars, it is more difficult to reason; because we have the means of knowing so little of their structure. But in this case also, we might easily ask on our side, *Pourquoy non?* Why should not the Solar System be the chief and most complete system in the universe, and the Earth the principal planet in that System? So far as we yet know, the Sun is the largest Sun among the stars; and we shall attempt to show, that the Earth is the largest solid opaque globe in the solar system. Some System must be the largest and most finished of all; why not ours? Some planet must be the largest planet; why not the Earth?

28 It should be recollected that there must be some system which is the most complete of all systems, some planet which is the largest of all planets. And if that largest planet, in the most complete system, be, after being for ages tenanted by irrational creatures, at last, and alone of all, occupied by a rational race, that race must necessarily have the power of asking such questions as these: Why they should be alone rational? Why their planet should be alone thus favoured? If the case be ours, we may hope to be then able to answer these questions, when we can explain the most certain fact which they involve; Why the Earth was occupied so long by irrational creatures, before

202|

the rational race was placed upon it? The mere power of asking such questions can prove or disprove nothing; for it is a power which must equally subsist, whether the human inhabitants of the earth be or be not the only rational population which the universe contains. If there be a race thus favoured by the Creator, they must, at that stage of their knowledge in which man now is, be able to doubt, as man does, of the extent and greatness of the privilege which they enjoy.

29 The argument that the Fixed Stars are like the Sun, and therefore the centers of inhabited systems as the Sun is, is sometimes called an argument from Analogy; and this word *Analogy* is urged, as giving great force to the reasoning. But it must be recollected, that precisely the point in question is, whether there *is* an analogy. The stars, it is said, are like the Sun. In what respects? We know of none, except in being self-luminous: and this they have in common with the nebulæ, which, as we have seen, are not centres of inhabited systems. Nor does this quality of being self-luminous at all determine the degree of condensation of a star. Sirius may be less than a hundredth or a thousandth of the density of the Sun. But the Stars, it may be further urged, are like the Sun in turning on their axes. To this we reply, that we know this only of those stars in which, the very phenomenon which proves their revolution, proves also that they are unlike the Sun, in having one side darker than the other. Add to which, their revolution is not connected with the existence of planets, still less of inhabitants of planets, in any intelligible manner. The resemblance, therefore, so far as it bears upon the question, is confined to one single point, in the highest degree ambiguous and inconclusive: and any argument drawn from this one point of resemblance, has little claim to be termed an argument from analogy*.

* Humboldt, very justly, regards the force of analogy as tending in the opposite direction. 'After all,' he asks, (*Cosmos*, III. 373,) 'is the assumption of satellites to the Fixed Stars so absolutely neces-

|203

30 On a subject on which we know so little, it is difficult to present any view which deserves to be regarded as an analogy. We see, among the stars, nebulæ more or less condensed, which are possibly, in some cases, stages of a connected progress towards a definite star; and it may be, to a star with planets in permanent orbits. We see, in our planet, evidence of successive stages of a connected series of brute animals, preceded perhaps by various stages of lifeless chaos. If the histories of the Sun, and of all the stars, are governed by a common analogy, the nebulous condensation, and the stages of animal life, may be parts of the same continued series of events; and different stars may be at different points of that series. But even on this supposition, but a few of the stars may be the seats of conscious life, and none, of intelligence. For among the stars which have condensed to a permanent form, how many may have failed in throwing off a permanent planet! How many may be in some stage of lifeless chaos! We must needs suppose a vast number of stages between a nebular chaos and the lowest forms of conscious life. Perhaps as many as there are fixed stars; and far more than there are of stars which become fertile of life: so that no two systems may be at the same stage of the planetary progress. And if this be so,—our system being so complicated, that we must suppose it peculiarly developed, having the largest Sun that we know of, and our Earth being (as we shall hereafter attempt to prove,) the largest solid planet that we know of,—this Earth may be the sole seat of the highest stage of planetary developement.

31 The assumption that there is anything of the nature of a regular law or order of progress from nebular matter to conscious life,—a law which extends to all the stars, or to

sary? If we were to begin from the outer planets, Jupiter, &c., analogy might seem to require that all planets have satellites. But yet this is not true for Mars, Venus, Mercury.' To which we may further 204| add the *twenty-three* Planetoids. In this case there is a much greater number of bodies which have not satellites, than which have them.

many of them,—is in the highest degree precarious and unsupported: but since it is sometimes employed in such speculations as we are pursuing, we may make a remark or two connected with it. If we suppose, on the planets of other systems, a progress in some degree analogous to that which geology shows to have occurred on the Earth, there may be, in those planets, creatures in some way analogous to our vegetables and animals; but analogy also requires that they should differ far more from the terrestrial vegetables and animals of any epoch, than those of one epoch do from those of another; since they belong to a different stellar system, and probably exist under very different conditions from any that ever prevailed on the Earth. We are forbidden, therefore, by analogy, to suppose that on any other planet there was such an anatomical progression towards the form of man, as we can discern, (according to some eminent physiologists,) among the tribes which have occupied the Earth. Are we to conceive that the creatures on the planets of other systems are, like the most perfect terrestrial animals, symmetrical as to right and left, vertebrate, with fore limbs and hind limbs, heads, organs of sense in their heads, and the like? Every one can see how rash and fanciful it would be to make such suppositions. Those who have, in the play of their invention, imagined inhabitants of other planets, have tried to avoid this servile imitation of terrestrial forms. Here is Sir Humphry Davy's account of the inhabitants of Saturn. 'I saw moving on the surface below me, immense masses, the forms of which I find it impossible to describe. They had systems for locomotion similar to that of the morse or sea-horse, but I saw with great surprize that they moved from place to place by six extremely thin membranes, which they used as wings. I saw numerous convolutions of tubes, more analogous to the trunk of the elephant, than to anything else I can imagine, |205 occupying what I supposed to be the upper parts of the body *.' The attendant Genius informs the narrator, that though these creatures look like zoophytes, they have a sphere of sensibility and intellectual enjoyment far superior to that of the inhabitants of the Earth. If we were to reason upon a work of fancy like this, we might say, that it was just as easy to ascribe superior sensibility and intelligence to zoophyte-formed creatures upon the Earth, as in Saturn. Even fancy cannot aid us in giving consistent form to the inhabitants of other planets.

32 But even if we could assent to the opinion, as probable, that there may occur, on some other planet, progressions of organized forms analogous in some way to that series of animal forms which has appeared upon the earth, we should still have no ground to assume that this series must terminate in a rational and intelligent creature like man. For the introduction of reason and intelligence upon the Earth is no part nor consequence of the series of animal forms. It is a fact of an entirely new kind. The transition from brute to man does not come within the analogy of the transition from brute to brute. The thread of analogy, even if it could lead us so far, would break here. We may conceive analogues to other animals, but we could have no analogue to man, except man. Man is not merely a higher kind of animal; he is a creature of a superior order, participating in the attributes of a higher nature; as we have already said, and as we hope hereafter further to show. Even, therefore, if we were to assume the general analogy of the Stars and of the Sun, and were to join to that, the information which geology gives us of the history of our own planet; though we might, on this precarious path, be led to think of other planets as peopled with unimagined monsters; we should still find a chasm in our reasoning, if we tried, in this way, to find intelligent and rational creatures in planets which
206| may revolve round Sirius or Arcturus.

* *Consolations in Travel.* Dial. 1.

33 The reasonable view of the matter appears to be this. The assumption that the Fixed Stars are of exactly the same nature as the Sun, was, at first, when their vast distance and probable great size were newly ascertained, a bold guess; to be confirmed or refuted by subsequent observations and discoveries. Any appearances, tending in any degree to confirm this guess, would have deserved the most considerate attention. But there has not been a vestige of any such confirmatory fact. No planet, nor anything which can fairly be regarded as indicating the existence of a planet, revolving about a star, has anywhere been discerned. The discovery of nebulæ, of binary systems, of clusters of stars, of periodical stars, of varying and accelerated periods of such stars, all seem to point the other way. And if all these facts be held to be but small in amount, as to the information which they convey, about the larger, and perhaps nearer stars; still they leave the original assumption a mere guess, unsupported by all that three centuries of most diligent, and in other respects successful research, has been able to bring to light. That Copernicus, that Galileo, that Kepler, should believe the stars to be Suns, in every sense of the term, was a natural result of the expansion of thought which their great discoveries produced, in them and their contemporaries. Nor are we yet called upon to withdraw from them our sympathy; or entitled to contradict their conjecture. But all the knowledge that the succeeding times have given us; the extreme tenuity of much of the luminous matter in the skies; the existence of gyratory motion among the stars, quite different from planetary systems; the absence of any observed motions at all resembling such systems; the appearance of changes in stars, quite inconsistent with such permanent systems; the disclosure of the history of our own planet, as one in which changes have constantly been going on; the certainty that by far the greater part of the dura- |207

tion of its existence, it has been tenanted by creatures entirely different from those which give an interest, and thence, a persuasiveness, to the belief of inhabitants in worlds appended to each star; the impossibility, which appears, on the gravest consideration, of transferring to other worlds such interests as belong to our own race in this world; all these considerations should, it would seem, have prevented that old and arbitrary conjecture from growing up, among a generation professing philosophical caution, and scientific discipline, into a settled belief.

34 Some of the moral and theological views which tend to encourage and uphold this belief, may be taken under our more special consideration hereafter: but here, where we are reasoning principally upon astronomical grounds, we may conclude what we have to remark about the Fixed Stars, as the centers of inhabited systems of worlds, by saying; that it will be time enough to speculate about the inhabitants of the planets which belong to such systems, when we have ascertained that there are such planets, or one such planet. When that is done, we can then apply to them any reasons which may exist, for believing that all, or many planets, are the seats of habitation of living things. What reasons of this kind can be adduced, and what is their force with regard to our own solar system, we must now proceed to discuss*.

* What is said in Art. 15, that in consequence of the time employed in the transmission of visual impressions, our seeing a star is evidence, not that it exists now, but that it existed, it may be, many thousands of years ago; may seem, to some readers, to throw doubts upon reasonings which we have employed. It may be said that a star which was a mere chaos, when the light, by which we see it, set out from it, may, in the thousands of years which have since elapsed, have grown into an orderly world. To which bare possibility, we may oppose another supposition at least equally possible:—that the distant stars were sparks or fragments struck off in the formation of the Solar System, which are really long since extinct; and survive in appearance, only by the light which they at first emitted.

CHAPTER IX.

The Planets.

1 WHEN it was discovered, by Copernicus and Galileo, that Mercury, Venus, Mars, Jupiter, Saturn, which had hitherto been regarded only as 'wandering fires, that move in mystic dance,' were really, in many circumstances, bodies resembling the Earth;—that they and the Earth alike, were opaque globes, revolving about the Sun in orbits nearly circular, revolving also about their own axes, and some of them accompanied by their Satellites, as the Earth is by the Moon;—it was inevitable that the conjecture should arise, that they too had inhabitants, as the Earth has. Each of these bodies was seemingly coherent and solid; furnished with an arrangement for producing day and night, summer and winter; and might therefore, it was naturally conceived, have inhabitants moving upon its solid surface, and reckoning their lives and their employments by days, and months, and years. This was an unavoidable guess. It was far less bold and sweeping than the guess that there are inhabitants in the region of the Fixed Stars, but still, like that, it was, for the time at least, only a guess; and like that, it must depend upon future explorations of these bodies and their conditions, whether the guess was confirmed or discredited. The conjecture could not, by any moderately cautious man, be regarded as so overwhelmingly probable, that it had no need of further proof. Its final acceptance or rejection must depend on the subsequent progress of astronomy, and of science in general.

2 We have to consider then how far subsequent discoveries have given additional value to this conjecture. And, as, in the first place, important among such discoveries, we

must note the addition of several new planets to our system. It was found, by the elder Herschel, (in 1781,) that, far beyond Saturn, there was another planet, which, for a time, was called by the name of its sagacious discoverer; but more recently, in order to conform the nomenclature of the planets to the mythology with which they had been so long connected, has been termed *Uranus.* This was a vast extension of the limits of the solar system. The Earth is, as we have already said, nearly a hundred millions of miles from the Sun. Jupiter is at more than five times, and Saturn nearly at ten times this distance: but Uranus, it was found, describes an orbit of which the radius is above nineteen times as great as that of the Earth. But this did not terminate the extension of the solar system which the progress of astronomy revealed. In 1846, a new planet, still more remote, was discovered: its existence having been divined, before it was seen, by two mathematicians, Mr Adams, of Cambridge, and M. Leverrier, of Paris, from the effects of its force upon Uranus. This new planet was termed Neptune: its distance from the Sun is about thirty times the Earth's distance. Besides these discoveries of large planets, a great number of small planets were detected in the region of the solar system which lies between the orbits of Mars and Jupiter. This series of discoveries began on the first day of 1801, when Ceres was detected by Piazzi at Palermo; and has gone on up to the present time, when twenty-three of these small bodies have been brought to light; and probably the group is not yet exhausted.

3 Now if we have to discuss the probability that all these bodies are inhabited, we may begin with the outermost of them at present known, namely Neptune. How far is it likely that this globe is occupied by living creatures which enjoy, like the creatures on the Earth, the light and heat of 210| the Sun, about which the planet revolves? It is plain, in

the first place, that this light and heat must be very feeble. Since Neptune is thirty times as far from the sun as the earth is, the diameter of the sun as seen from Neptune will only be one thirtieth as large as it is, seen from the earth. It will, in fact, be reduced to a mere star. It will be about the diameter under which Jupiter appears when he is nearest to us. Of course its brightness will be much greater than that of Jupiter; nearly as much indeed, as the sun is brighter than the moon, both being nearly of the same size: but still, with our full-moonlight reduced to the amount of illumination which we receive from a *full Jupiter*, and our sun-light reduced in nearly the same proportion, we should have but a dark, and also a cold world. In fact, the light and the heat which reach Neptune, so far as they depend on the distance of the sun, will each be about nine hundred times smaller than they are on the earth. Now are we to conceive animals, with their vital powers unfolded, and their vital enjoyments cherished, by this amount of light and heat? Of course, we cannot say, with certainty, that any feebleness of light and heat are inconsistent with the existence of animal life: and if we had good reason to believe that Neptune is inhabited by animals, we might try to conceive in what manner their vital scheme is accommodated to this scanty supply of heat and light. If it were certain that they were there, we might inquire how they could live there, and what manner of creatures they could be. If there were any general ground for assuming inhabitants, we might consider what modifications of life their particular conditions would require.

4 But is there any such general ground! Such a ground we should have, if we could venture to assume that *all* the bodies of the Solar System are inhabited;—if we could proceed upon such a principle, we might reject or postpone the difficulties of particular cases.

|211

5 But is such an assumption true? Is such a principle well founded? The best chance which we have of learning whether it is so, is to endeavour to ascertain the fact, in the body which is nearest to us; and thus, the best placed for our closer scrutiny. This is, of course, the Moon; and with regard to the Moon, we have, again, this advantage in beginning the inquiry with her:—that she, at least, is in circumstances, as to light and heat, so far as the Sun's distance affects them, which we know to be quite consistent with animal and vegetable life. For her distance from the Sun is not appreciably different from that of the Earth; her revolutions round the earth do not make nearly so great a difference, in her distance from the sun, as does the earth's different distances from the sun in summer and in winter: the fact also being, that the earth is considerably nearer to the sun in the winter of this our northern hemisphere, than in the summer. The moon's distance from the sun then, adapts her for habitation: is she inhabited?

6 The answer to this question, so far as we can answer it, may involve something more than those mere astronomical conditions, her distance from the sun, and the nature of her motions. But still, if we are compelled to answer it in the negative;—if it appear, by strong evidence, that the Moon is not inhabited; then is there an end of the general principle, that *all* the bodies of the solar system are inhabited, and that we must begin our speculations about each, with this assumption. If the Moon be not inhabited, then, it would seem, the belief that each special body in the system is inhabited, must depend upon reasons specially belonging to that body; and cannot be taken for granted without such reasons. Of the two bodies of the solar system which alone we can examine closely, so as to know anything about them, the Earth and the Moon, if the one be inhabited, and the other blank of inhabitants, we have then no right to assume

212|

at once, that any other body in the solar system belongs to the former of these classes rather than to the latter. If, even under terrestrial conditions of light and heat, we have a total absence of the phenomenon of life, known to us only as a terrestrial phenomenon; we are surely not entitled to assume that when these conditions fail, we have still the phenomenon, life. We are not entitled to *assume* it; however it may be capable of being afterwards proved, in any special case, by special reasons; a question afterwards to be discussed.

7 Is then the Moon inhabited? From the moon's proximity to us, (she is distant only thirty diameters of the earth, less than ten times the earth's circumference; a railroad carriage at its ordinary rate of travelling would reach her in a month,) she can be examined by the astronomer with peculiar advantages. The present powers of the telescope enable him to examine her mountains as distinctly as he could the Alps at a few hundred miles distance, with the naked eye; with the additional advantage, that her mountains are much more brilliantly illuminated by the Sun, and much more favourably placed for examination, than the Alps are. He can map and model the inequalities of her surface, as faithfully and exactly as he can those of the surface of Switzerland. He can trace the streams that seem to have flowed from eruptive orifices over her plains, as he can the streams of lava from the craters of Etna or Hecla.

8 Now this minute examination of the Moon's surface being possible, and having been made, by many careful and skilful astronomers, what is the conviction which has been conveyed to their minds, with regard to the fact of her being the seat of vegetable or animal life? Without exception, it would seem, they have all been led to the belief, that the Moon is not inhabited: that she is, so far as life and organization are concerned, waste and barren, like the streams of

|213

lava or of volcanic ashes on the earth, before any vestige of vegetation has been impressed upon them; or like the sands of Africa, where no blade of grass finds root. It is held, by such observers, that they can discern and examine portions of the moon's surface as small as a square mile*: yet in their examination, they have never perceived any alteration, such as the cycle of vegetable changes through the revolutions of seasons would produce. Sir William Herschel did not doubt that if a change had taken place on the visible part of the Moon, as great as the growth or the destruction of a great city, as great, for instance as the destruction of London by the great fire of 1666, it would have been perceptible to his powers of observation. Yet nothing of the kind has ever been observed. If there were lunar astronomers, as well provided as terrestrial ones are, with artificial helps of vision, they would undoubtedly be able to perceive the differences which the progress of generations brings about on the surface of our globe: the clearing of the forests of Germany or North America; the embankment of Holland; the change of the modes of culture which alter the colour of the ground in Europe; the establishment of great nests of manufactures which shroud portions of the land in smoke, as those which have their centers at Birmingham or at Manchester. However obscurely they might discern the nature of those changes, they would still see that change was going on. And so should we, if the like changes were going on upon the face of the Moon. Yet no such changes have ever been noticed. Nor even have such changes been remarked, as might occur in a mere brute mass without life;—the formation of new streams of lava, new craters, new crevices, new elevations. The Moon ex-

* More recently, at the meeting of the British Association in Sept. 1853, Professor Phillips has declared, that astronomers can discern the shape of a spot on the Moon's surface, which is a few hundred feet in breadth.

hibits strong evidences, which strike all telescopic observers, of an action resembling, in many respects, volcanic action, by which its present surface has been formed*. But, if it have been produced by such internal fires, the fires seem to be extinguished; the volcanoes to be burnt out. It is a mere cinder; a collection of sheets of rigid slag, and inactive craters. And if the Moon and the Earth were both, at first, in a condition in which igneous eruptions from their interior produced the ridges and cones which roughen their surfaces; the Earth has had this state succeeded by a series of states of life in innumerable forms, till at last it has become the dwellingplace of man; while the Moon, smaller in dimensions, has at an earlier period completely cooled down, as to its exterior at least, without ever being judged fit or worthy by its Creator of being the seat of life: and remains, hung in the sky, as an object on which man may gaze, and perhaps, from which he may learn something of the constitution of the universe: and among other lessons, this; that he must not take for granted, that all the other globes of the solar system are tenanted, like that on which he has his appointed place.

9 It is true, that in coming to this conclusion, the astronomers of whom I speak, have been governed by other reasons, besides those which I have mentioned, the absence of any changes, either rapid or slow, discoverable in the Moon's face. They have seen reason to believe that water and air, elements so essential to terrestrial life, do not exist in the Moon. The dark spaces on her disk, which were called *seas* by those who first depicted them, have an appearance inconsistent with their being oceans of water. They are not level and smooth, as water would be; nor

* A person visiting the Eifel, a region of extinct volcanoes, west of the Rhine, can hardly fail to be struck with the resemblance of the craters there, to those seen in the moon through a telescope.

uniform in their colour, but marked with permanent streaks and shades, implying a rigid form. And the absence of an atmosphere of transparent vapour and air, surrounding the moon, as our atmosphere surrounds the earth, is still more clearly proved, by the absence of all the optical effects of such an atmosphere, when stars pass behind the moon's disk, and by the phenomena which are seen in solar eclipses, when her solid mass is masked by the Sun*. This absence of moisture and air in the Moon, of course, entirely confirms our previous conclusion, of the absence of vegetable and animal life; and leaves us, as we have said, to examine the question for the other bodies, on their special grounds, without any previous presumption that such life exists. Undoubtedly the aspect of the case will be different in one feature, when we see reason to believe that other bodies have an atmosphere; and if there be in any planet sufficient light and heat, and clouds and winds, and a due adjustment of the power of gravity, and the strength of the materials of which organized frames consist, there may be, so far as we can judge, life of some kind or other. But yet, even in those cases, we should be led to judge also, by analogy, that the life which they sustain is more different from the terrestrial life of the present period of the earth, than that is from the terrestrial life of any former geological period, in proportion as the conditions of light and heat, and attraction and density, are more different on any other planet, than they can have been on the earth, at any period of its history.

10 Let us then consider the state of these elements of

* Bessel has discussed and refuted (it was hardly necessary) the conjecture of some persons (he describes them as 'the feeling hearts who would find sympathy even in the Moon'), that there may be in the Moon's vallies air enough to support life, though it does not rise above the hills.—*Populäre Vorlesungen*, p. 78.

being in the other planets. I have mentioned, among them, the force of gravity, and the density of materials; because these are important elements in the question. It may seem strange, that we are able, not only to measure the planets, but to weigh them; yet so it is. The wonderful discovery of universal gravitation, so firmly established, as the law which embraces every particle of matter in the solar system, enables us to do this, with the most perfect confidence. The revolutions of the satellites round their primary planets, give us a measure of the force by which the planets retain them in their orbits; and in this way, a measure of the quantity of matter of which each planet consists. And other effects of the same universal law, enable us to measure, though less easily and less exactly, the masses, even of those planets, which have no satellites. And thus we can, as it were, put the Earth, and Jupiter or Saturn, in the balance against each other; and tell the proportionate number of pounds which they would weigh, if so poised. And again, by another kind of experiment, we can, as we have said, weigh the earth against a known mountain; or even against a small sphere of lead duly adjusted for the purpose. And this has been done: and the results are extremely curious; and very important in our speculations relative to the constitution of the universe.

11 And in the first place, we may remark that the Earth is really much less heavy than we should expect, from what we know of the materials of which it consists. For, measuring the density, or specific gravity, of materials, (that is their comparative weight in the same bulk,) by their proportion to water, which is the usual way, the density of iron is 8, that of lead 11, that of gold 19: the ordinary rocks at the Earth's surface have a density of 3 or 4. Moreover, all the substances with which we are acquainted, contract into a smaller space, and have their density increased, by being |217

subjected to pressure. Air does this, in an obvious manner; and hence it is, that the lower parts of our atmosphere are denser than the upper parts; being pressed by a greater superincumbent weight, the weight of the superior parts of the atmosphere itself. Air is thus obviously and eminently elastic. But all substances, though less obviously and eminently, are still, really, and in some degree, elastic. They all contract by compression. Water for instance, if pressed by a column of water 100000 feet high, would be reduced to a bulk $\frac{1}{10}$ less than before. In the same manner iron, compressed by a column of iron 90000 feet high, loses $\frac{1}{10}$ of its bulk, and of course gains so much in density. And the like takes place, in different amounts, with all material whatever. This is the rate at which compression produces its effect of increasing the density, in bodies which are in the condition of those which lie around us. But if this law were to go on at the same rate, when the compression is greatly increased, the density of bodies deep down towards the center of the Earth must be immense. The Earth's radius is above 20 million feet. At a million feet depth we should have matter subjected to the pressure of a column of a million feet of superincumbent matter, heavier than water; and hence we should have a compression of water 10 times as great as we have mentioned; and, therefore, the bulk of the water would be reduced almost to nothing, its density increased almost indefinitely: and the same would be the case with other materials, as metals and stones. If, therefore, this law of compression were to hold for these great pressures, all materials whatever, contained in the depths of the Earth's mass, must be immensely denser, and immensely specifically heavier, than they are at the surface. And thus, the Earth consisting of these far denser materials towards the center, but, nearer the surface, of lighter materials, such as rock, and metals, in their ordinary

218| state, must, we should expect, be, on the whole, much heavier than if it consisted of the heaviest ordinary materials; heavier than iron, or than lead; hundreds of times perhaps heavier than stone.

12 This, however, is not found to be so. The expectation of the great density of the Earth, which we might have derived from the known laws of condensation of terrestrial substances, is not confirmed. The mass of the Earth being weighed, by means of such processes as we have already referred to, is found to be only five times heavier than so much water: less heavy than if it were made of iron: less than twice as heavy as if it were made of ordinary rock. This, of course, shows us that the condensation of the interior parts of the Earth's mass, is by no means so great as we should have expected it to be, from what we know of the laws of condensation here; and from considering the enormous pressure of superincumbent materials to which those interior parts are subjected. The laws of condensation, it would seem, do not go on operating for these enormous pressures, by the same progression as for smaller pressure. If a mass of a material is compressed into $\frac{9}{10}$ its bulk by the weight of a column of 100000 feet high, it does not follow that it will be again compressed into $\frac{9}{10}$ of its condensed bulk, by another column of 100000 feet high. The compression and condensation reach, or tend to, a limit; and probably, before they have gone very far. It may be possible to compress a piece of iron by one-thousandth part, even by such forces as we can use; and yet it may not be possible to compress the same piece of iron into one half its bulk, even by the weight of the whole Earth, if made to bear upon it. This appears to be probable: and this will explain, how it is, that the materials of the Earth are not so violently condensed as we should have supposed; and thus, why the Earth is so light.

13 We must avoid drawing inferences too boldly, on a

|219

subject where our means of knowledge are so obscure as they are with regard to the interior of the Earth; but yet, perhaps, we may be allowed to say, that the result which we have just stated, that the Earth is so light, suggests to us the belief that the interior consists of the same materials as the exterior, slightly condensed by pressure*. We find no encouragement to believe that there is a nucleus within, of some material, different from what we have on the outside; some metal, for instance, heavier than lead. If the earth were of granite, or of lava, to the center, it would, so far as we can judge, have much the same weight which it now has. Such a central mass, covered with the various layers of stone, which form the upper crust of the Earth, would naturally make this globe of at least the weight which it really has. And therefore, if we were to learn that a planet was much lighter than this, as to its materials,—much less dense, taking the whole mass together,—we should be compelled to infer that it was, throughout, or nearly so, formed of less compact matter than metal and stone; or else, that it had internal cavities, or some other complex structure, which it would be absurd to assume, without positive reasons.

14 Now having decided these views from an examination of the Earth, let us apply them to other planets, as bearing upon the question of their being inhabited; and in the first place, to Jupiter. We can, as we have said, easily compare the mass of Jupiter and of the Earth; for both of them have Satellites. It is ascertained, by this means, that the mass or weight of Jupiter is about 333 times the weight

* The doctrine that the interior nucleus of the Earth is fluid, whether accepted or rejected, does not materially affect this argument. It appears, that in some cases, at least, the melting of substances is prevented, by their being subjected to extreme pressure; but the density, the element from which we reason, is measured by methods quite independent of such questions.

of the earth: but as his diameter is also 11 times that of the earth, his bulk is 1331 times that of the earth: (the *cube* of 11 is 1331); and, therefore, the density of Jupiter is to that of the earth, only as 333 to 1331, or about 1 to 4. Thus the density of Jupiter, taken as a whole, is about a quarter of the earth's density; less than that of any of the stones which form the crust of the earth; and not much greater than the density of water. Indeed, it is tolerably certain, that the density of Jupiter is not greater than it would be, if his entire globe were composed of water; making allowance for the compression which the interior parts would suffer by the pressure of those parts superincumbent. We might, therefore, offer it as a conjecture not quite arbitrary, that Jupiter is a mere sphere of water.

15 But is there any thing further in the appearance of Jupiter, which may serve to contradict, or to confirm, this conjecture? There is one circumstance in Jupiter's form, which is, to say the least, perfectly consistent with the supposition, that he is a fluid mass; namely, that he is not an exact sphere, but oblate, like an orange. Such a form is produced, in a fluid sphere, by a rotation upon its axis. It is produced, even in a sphere which is (at present at least,) partly solid and partly fluid; and the oblateness of the earth is accounted for in this way. But Jupiter, who, while he is much larger than the earth, revolves much more rapidly, is much more oblate than the earth. His polar and equatoreal diameters are in the proportion of 13 to 14. Now it is a remarkable circumstance, that this is the amount of oblateness, which, on mechanical principles, would result from his time of revolution, if he were entirely fluid, and of the same density throughout *. So far, then, we have some

* Herschel, 512. Bessel, however, holds that the oblateness of Jupiter proves that his interior is somewhat denser than his exterior. *Pop. Vorles.* p. 91.

confirmation at least, of his being composed entirely of some fluid which in its density agrees with water.

16 But there are other circumstances in the appearances of Jupiter, which still further confirm this conjecture of his watery constitution. His belts,—certain bands of darker and lighter colour, which run parallel to his equator, and which, in some degree, change their form, and breadth, and place, from time to time,—have been conjectured, by almost all astronomers, to arise from lines of cloud, alternating with tracts comparatively clear, and having their direction determined by currents analogous to our trade-winds, but of a much more steady and decided character, in consequence of the greater rotatory velocity*. Now vapours, supplying the materials of such masses of cloud, would naturally be raised from such a watery sphere as we have supposed, by the action of the Sun; would form such lines; and would change their form from slight causes of irregularity, as the belts are seen to do. The existence of these lines of cloud does of itself show that there is much water on Jupiter's surface, and is quite consistent with our conjecture, that his whole mass is water †.

* Herschel, 513.

† A difficulty may be raised, founded on what we may suppose to be the fact, as to the extreme cold of those regions of the Solar System. It may be supposed that water under such a temperature could exist in no other form than ice. And that the cold must there be intense, according to our notion, there is strong reason to believe. Even in the outer regions of our atmosphere, the cold is probably very many degrees below freezing, and in the blank and airless void beyond, it may be colder still. It has been calculated by physical philosophers, on grounds which seem to be solid, that the cold of the space beyond our atmosphere is 100° below zero. The space near to Jupiter, if an absolute vacuum, in which there is no matter to receive and retain heat emitted from the Sun, may, perhaps, be no colder than it is nearer the Sun. And as to the effect the great cold would produce on Jupiter's watery material, we may remark, that if there be a free surface, there will be vapour produced by the Sun's heat; and

17 Perhaps some persons may be disposed to doubt whether, if Jupiter be, as we suppose, merely or principally a mass of water and of vapour, we are entitled to extend to him the law of universal gravitation, which is the basis of our speculations. But this doubt may be easily dismissed. We know that the waters of the earth are affected by gravitation; not only towards the earth, as shown by their weight, but towards those distant bodies, the Sun and the Moon; for this gravitation produces the tides of the ocean. And our atmosphere also has weight, as we know; and probably has also solar and lunar tides, though these are masked by many other causes of diurnal change. We have, then, the same reason for supposing that air and water, in other parts of the system, are governed by universal gravitation, and exercise themselves the attractive force of gravitation, which we have for making the like suppositions with regard to the most solid bodies. Whatever argument proves universal gravitation, proves it for all matter alike; and Newton, in the course of his magnificent generalization of the law, took care to demonstrate, by experiment, as well as by reasoning, that it might be so generalized.

18. As bearing upon the question of life in Jupiter, there is another point which requires to be considered; the force of gravity at his surface. Though, equal bulk for equal bulk, he is lighter than the earth, yet his bulk is so great that, as we have seen, he is altogether much heavier than the earth. This, his greater mass, makes bodies, at equal

if there be air, there will be clouds. We may add, that so far as we have reason to believe, below the freezing point, no accession of cold produces any material change in ice. Even in the expeditions of our Arctic navigators, a cold of 40° below zero was experienced, and ice was still but ice, and there were vapours and clouds as in our climate. It is quite an arbitrary assumption, to suppose that any cold which may exist in Jupiter would prevent the state of things which we suppose.

distances from the centers, ponderate proportionally more to him than they would do to the earth. And though his surface is 11 times further from his center than the earth's is, and therefore the gravity at the surface is thereby diminished, yet, even after this deduction, gravity at the surface of Jupiter is nearly two and a half times that on the earth*. And thus a man transferred to the surface of Jupiter would feel a stone, carried in his hands, and would feel his own limbs also, (for his muscular power would not be altered by the transfer,) become $2\frac{1}{2}$ times as heavy, as difficult to raise, as they were before. Under such circumstances animals of large dimensions would be oppressed with their own weight. In the smaller creatures on the earth, as in insects, the muscular power bears a great proportion to the weight, and they might continue to run and to leap, even if gravity were tripled or quadrupled. But an elephant could not trot with two or three elephants placed upon his back. A lion or tiger could not spring, with twice or thrice his own weight hung about his neck. Such an increase of gravity would be inconsistent then, with the present constitution and life of the larger terrestrial animals; and if we are to suppose planets inhabited, in which gravity is much more energetic than it is upon the earth, we must suppose classes of animals which are adapted to such a different mechanical condition.

19 Taking into account then, these circumstances in Jupiter's state; his (probably) bottomless waters; his light, if any, solid materials; the strong hand with which gravity presses down such materials as there are; the small amount of light and heat which reaches him, at 5 times the earth's distance from the sun; what kind of inhabitants shall we be led to assign to him? Can they have skeletons, where no

* Herschel, 508.

substance so dense as bone is found, at least in large masses? It would seem not probable*. And it would seem they must be dwellers in the waters, for against the existence there of solid land, we have much evidence. They must, with so little of light and heat, have a low degree of vitality. They must then, it would seem, be cartilaginous and glutinous masses; peopling the waters with minute forms: perhaps also with larger monsters; for the weight of a bulky creature, floating in the fluid, would be much more easily sustained than on solid ground. If we are resolved to have such a population, and that they shall live by food, we must suppose that the waters contain at least so much solid matter as is requisite for the sustenance of the lowest classes; for the higher classes of animals will probably find their food in consuming the lower. I do not know whether the advocates of peopled worlds will think such a population as this worth contending for: but I think the only doubt can be, between such a population, and none. If Jupiter be a mere mass of water, with perhaps a few cinders at the center, and an envelope of clouds around it, it seems very possible that he may not be the seat of life at all. But if life be there, it does not seem in any way likely, that the living things can be anything higher in the scale of being, than such boneless, watery, pulpy creatures as I have imagined.

20 Perhaps it may occur to some one to ask, if this planet, which presents so glorious an aspect to our eyes, be thus the abode only of such imperfect and embryotic lumps of vitality as I have described; to what purpose was all that gorgeous array of satellites appended to him, which would

* It may be thought fanciful to suppose that because there is little or no solid matter (of any kind known to us) in Jupiter, his animals are not likely to have solid skeletons. The analogy is not very strong; but also, the weight assigned to it in the argument is small. *Valeat quantum valere debet.*

present, to intelligent spectators on his surface, a spectacle far more splendid than any that our skies offer to us: four moons, some as great, and others hardly less, than our moon, performing their regular revolutions in the vault of heaven. To which it will suffice, at present, to reply, that the use of those moons, under such a supposition, would be precisely the same, as the use of our moon, during the myriads of years which elapsed while the earth was tenanted by corals and madrepores, shell-fish and belemnites, the cartilaginous fishes of the Old Red Sandstone, or the Saurian monsters of the Lias; and in short, through all the countless ages which elapsed, before the last few thousand years: before man was placed upon the earth 'to eye the blue vault and bless the *useful* light:' to reckon by it his months and years: to discover by means of it, the structure of the universe, and perhaps, the special care of his Creator for him alone of all his creatures. The moons of Jupiter, may in this way be of use, as our own moon is. Indeed we know that they have been turned to most important purposes, in astronomy and navigation. And knowing this, we may be content not to know how, either the satellites of Jupiter, or the satellite of the Earth, tend to the advantage of the brute inhabitants of the waters.

21 There is another point, connected with this doctrine of the watery nature of Jupiter, which I may notice, though we have little means of knowledge on the subject. Jupiter being thus covered with water, is the water ever converted into ice? The planet is more than 5 times as far from the sun as the earth is: the heat which he receives is, on that account, 25 times less than ours. The veil of clouds which covers a large part of his surface, must diminish the heat still further. What effect the absence of land produces, on the freezing of the ocean, it is not easy to say. We cannot, therefore, pronounce with any confidence whether his waters

226|

are ever frozen or not. In the next considerable planet, Mars, astronomers conceive that they do trace the effects of frost; but in Mars we have also appearances of land. In Jupiter, we are left to mere conjecture; whether continents and floating islands of ice still further chill the fluids of the slimy tribes whom we have been led to regard as the only possible inhabitants; or whether the watery globe is converted into a globe of ice; retaining on its surface, of course, as much fluid as is requisite, under the evaporating power of the sun, to supply the currents of vapour which form the belts. In this case, perhaps, we may think it most likely that there are no inhabitants of these shallow pools in a planet of ice: at any rate, it is not worth while to provide any new speculations for such a hypothesis.

22 We may turn our consideration from Jupiter to Saturn; for in many respects the two planets are very similar. But in almost every point, which is of force against the hypothesis of inhabitants, the case is much stronger in Saturn than it is in Jupiter. Light and heat, at his distance, are only one ninetieth of those at the Earth. None but a very low degree of vitality can be sustained under such sluggish influences. The density of his mass is hardly greater than that of cork; much less than that of water: so that, it does not appear what supposition is left for us, except that a large portion of the globe, which we see as his, is vapour. That the outer part of the globe is vapour, is proved, in Saturn as in Jupiter, by the existence of several cloudy streaks or belts running round him parallel to his equator. Yet his mass, taken altogether, is considerable, on account of his great size; and gravity would be greater, at his outer surface, than it is at the earth's. For such reasons, then, as were urged in the case of Jupiter, we must either suppose that he has no inhabitants; or that they are aqueous, gelatinous creatures; too sluggish, almost, to be deemed |227

alive, floating in their ice-cold waters, shrouded for ever by their humid skies.

23 Whether they have eyes or no, we cannot tell; but probably if they had, they would never see the Sun; and therefore we need not commiserate their lot in not seeing the host of Saturnian satellites; and the Ring, which to an intelligent Saturnian spectator, would be so splendid a celestial object. The Ring is a glorious object for man's view, and his contemplation; and therefore is not altogether without its use. Still less need we (as some appear to do) regard as a serious misfortune to the inhabitants of certain regions of the planet, a solar eclipse of fifteen years' duration, to which they are liable by the interposition of the Ring between them and the Sun*.

24 The cases of Uranus and Neptune are similar to that of Saturn, but of course stronger, in proportion to their smaller light and heat. For Uranus, this is only 1-360th, for Neptune, as we have already said, 1-900th of the light and heat at the earth. Moreover, these two new planets agree with Jupiter and with Saturn, in being of very large size and of very small density: and also we may remark, one of them, probably both, in revolving with great rapidity, and in nearly the same period, namely, about 10 hours: at least, this has been the opinion of astronomers with regard to Uranus. The arguments against the hypothesis of these two planets being inhabited, are of course of the same kind as in the case of Jupiter and Saturn, but much increased in strength; and the supposition of the probably watery nature and low vitality of their inhabitants must be commended to the consideration of those who contend for inhabitants in those remote regions of the solar system.

25 We may now return towards the Sun, and direct

* Herschel, 522.

our attention to the planet Mars. Here we have some approximation to the condition of the Earth, in circumstances, as in position. It is true, his light and heat, so far as distance from the Sun affects them, are less than half those at the Earth. His density appears to be nearly equal to that of the Earth, but his mass is so much smaller, that gravity at his surface is only one-half of what it is here. Then, as to his physical condition, so far as we can determine it, astronomers discern in his face* the outlines of continents and seas. The ruddy colour by which he is distinguished, the red and fiery aspect which he presents, arise, they think, from the colour of the land, while the seas appear greenish. Clouds often seem to intercept the astronomer's view of the globe, which with its continents and oceans thus revolves under his eye; and that there is an atmosphere on which such clouds may float, appears to be further proved, by brilliant white spots at the poles of the planet, which are conjectured to be snow; for they disappear when they have been long exposed to the sun, and are greatest when just emerging from the long night of their polar winter; the snow-line then extending to about six degrees (reckoned upon the meridian of the planet) from the pole. Moreover, Mars agrees with the earth, in the period of his rotation; which is about 24 hours; and in having his axis inclined to his orbit, so as to produce a cycle of long and short days and nights, a return of summer and winter, in every revolution of the planet.

26 We have here a number of circumstances which speak far more persuasively for a similarity of condition, in this planet and the Earth, than in any of the cases previously discussed. It is true, Mars is much smaller than the earth, and has not been judged worthy of the attendance of a satellite, although further from the Sun; but still, he may

* Herschel, 510.

have been judged worthy of inhabitants by his Creator. Perhaps we are not quite certain about the existence of an atmosphere; and without such an appendage, we can hardly accord him tenants. But if he have inhabitants, let us consider of what kind they must be conceived to be, according to any judgment which we can form. The force of his gravity is so small, that we may allow his animals to be large, without fearing that they will break down by their own weight. In a planet so dense, they may very likely have solid skeletons. The ice about his poles will cumber the seas, cold even for the want of solar heat, as it does in our arctic and antarctic oceans; and we may easily imagine that these seas are tenanted, like those, by huge creatures of the nature of whales and seals, and by other creatures which the existence of these requires and implies. Or rather, since, as we have said, we must suppose the population of other planets to be more different from our existing population, than the population of other ages of our own planet, we may suppose the population of the seas and of the land of Mars, (if there be any, and if we are not carrying it too high in the scale of vital activity,) to differ from any terrestrial animals, in something of the same way in which the great land and sea saurians, or the iguanodon and dinotherium, differed from the animals which now live on the earth.

27 That we need not discuss the question, whether there are intelligent beings living on the surface of Mars, perhaps the reader will allow, till we have some better evidence that there are living things there at all; if he calls to mind the immense proportion which, on the earth, far better fitted for the habitation of the only intelligent creature which we know or can conceive, the duration of unintelligent life has borne to that of intelligent. Here, on this Earth, a few thousand years ago, began the life of a creature who

230|

can speculate about the past and the future, the near and the absent, the Universe and its Maker, duty and immortality. This began a few thousand years ago, after ages and myriads of ages, after immense varieties of lives and generations, of corals and mollusks, saurians, iguanodons, and dinotheriums. No doubt the Creator might place an intelligent creature upon a planet, without all this preparation, all this preliminary life. He has not chosen to do so on the earth, as we know; and that is by much the best evidence attainable by us, of what His purposes are. It is also possible that He should, on another planet, have established creatures of the nature of corals and mollusks, saurians and iguanodons, without having yet arrived at the period of intelligent creatures: especially if that other planet have longer years, a colder climate, a smaller mass, and perhaps no atmosphere. It is also possible that He should have put that smaller planet near the Earth, resembling it in some respects, as the Moon does, but without any inhabitants, as she has none; and that Mars may be such a planet. The probability against such a belief can hardly be considered as strong, if the arguments already offered be regarded as effective against the opinion of inhabitants in the other planets, and in the Moon.

28 The numerous tribe of small bodies, which revolve between Jupiter and Mars, do not admit of much of the kind of reasoning, which we have applied to the larger planets. They have, with perhaps one exception (Vesta) no disk of visible magnitude; they are mere dots, and we do not even know that their form is spherical. The near coincidence of their orbits has suggested, to astronomers, the conjecture that they have resulted from the explosion of a larger body, and from its fracture into fragments. Perhaps the general phenomena of the universe suggest rather the notion of a collapse of portions of sidereal matter, than of a |231

sudden disruption and dispersion of any portion of it; and these small bodies may be the results of some imperfectly effected concentration of the elements of our system; which, if it had gone on more completely and regularly, might have produced another planet, like Mars or Venus. Perhaps they are only the larger masses, among a great number of smaller ones, resulting from such a process: and it is very conceivable, that the meteoric stones which, from time to time, have fallen upon the earth's surface, are other results of the like process:—bits of planets which have failed in the making, and lost their way, till arrested by the resistance of the earth's atmosphere. A remarkable circumstance in these bodies is, that though thus coming apparently from some remote part of the system, they contain no elements but such as had already been found to exist in the mass of the earth; although some substances, as nickel and chrome, which are somewhat rare in the earth's materials, are common parts of the composition of meteoric stones. Also they are of crystalline structure, and exhibit some peculiarities in their crystallization. Such as these strange visitors are, they seem to show that the other parts of the solar system contain the same elementary substances, and are subject to the same laws of chemical synthesis and crystalline force, which obtain in the terrestrial region. The smallness of these specimens is a necessary condition of their reaching us; for if they had been more massive, they would have followed out the path of their orbits round the sun, however excentric these might be. The great excentricity of the smaller planets, their great deviation from the zodiacal path, which is the highway of the large planets, their great number, probably by no means yet exhausted by the discoveries of astronomers; all fall in with the supposition that there are, in the solar system, a vast multitude of such abnormal 232| planetoidal lumps. As I have said, we do not even know that they are approximately spherical; and if they are of the nature of meteoric stones, they are mere crude and irregularly crystallized masses of metal and earth. It will therefore, probably, be deemed unnecessary to give other reasons why these planetoids are not inhabited. But if it be granted that they are not, we have here, in addition to the moon, a large array of examples, to prove how baseless is the assumption, that all the bodies of the solar system are the seats of life.

29 We have thus performed our journey from the extremest verge of the Universe, so far as we have any knowledge of it, to the orbit of our own planet; and have found, till we came into our own most immediate vicinity, strong reasons for rejecting the assumption of inhabited worlds like our own; and indeed, of the habitation of worlds in any sense. And even if Mars, in his present condition, may be some image of the Earth, in some of its remote geological periods, it is at least equally possible that he may be an image of the Earth, in the still remoter geological period before life began. Of peculiar fitnesses which make the earth suited to the sustentation of life, as we know that it is, we shall speak hereafter: and at present, pass on to the other planets, Venus and Mercury. But of these, there is, in our point of view, very little to say. Venus, which, when nearest to us, fills a larger angle than any other celestial body, except the Sun and the Moon, might be expected to be the one of which we know most. Yet she is really one of the most difficult to scrutinize with our telescopes. Astronomers cannot discover in her, as in Mars, any traces of continents and seas, mountains and vallies: at least with any certainty*. Her illuminated part

* According to Bessel, Schrœter *once* saw one bright point on the dark ground, near the boundary of light in Venus. This was taken as proving a mountain, estimated at 60000 feet high. *Pop. Vorles.* p. 86.

shines with an intense lustre which dazzles the sight*: yet she is of herself perfectly dark; and it was the discovery, that she presented the phases of the Moon, made by the telescope of Galileo, which gave the first impulse to planetary research. She is almost as large as the earth; almost as heavy. The light and heat which she receives from the Sun must be about double those which come to the earth. We discern no traces of a gaseous or watery atmosphere surrounding her. Perhaps if we could see her better, we might find that she had a surface like the moon: or perhaps, in the nearer neighbourhood of the sun, she may have cooled more slowly and quietly, like glass which is annealed in the fire; and hence, may have a smooth surface, instead of the furrowed and pimpled visage which the Moon presents to us. With this ignorance of her conditions, it is hard to say what kind of animals we could place in her, if we were disposed to people her surface; except perhaps the microscopic creatures, with siliceous coverings, which, as modern explorers assert, are almost indestructible by heat. To believe that she has a surface like the earth, and tribes of animals, like terrestrial animals, and like man, is an exercise of imagination, which not only is quite gratuitous, but contrary to all the information which the telescope gives us; and with this remark, we may dismiss the hypothesis.

30 Of Mercury we know still less. He receives seven times as much light and heat as the Earth; is much smaller than the earth, but perhaps more dense; and has not, so far as we can tell, any of the conditions which make animal existence conceivable. If it is so difficult to find suitable inhabitants for Venus, the difficulty for Mercury is immensely greater.

31 So far then, we have traversed the Solar System,

* Herschel, 509.

and have found even here, the strongest grounds that there can be no animal existence, like that which alone we can conceive as animal existence, except in the planet next beyond the earth, Mars; and there, not without great modifications. But we may make some further remarks on the condition of the several planets, with regard to what appears to us to be the necessary elements of animal life.

CHAPTER X.

Theory of the Solar System.

---

1 WE have given our views respecting the various planets which constitute the Solar System;—views established, it would seem, by all that we know, of the laws of heat and moisture, density and attraction, organization and life. We have examined and reasoned upon the cases of the different planets separately. But it may serve to confirm this view, and to establish it in the reader's mind, if we give a description of the system which shall combine and connect the views which we have presented, of the constitution and peculiarities, as to physical circumstances, of each of the planets. It will help us in our speculations, if we can regard the planets not only as a collection, but as a scheme;—if we can give, not an enumeration only, but a theory. Now such a scheme, such a theory, appears to offer itself to us.

2 The planets exterior to Mars, Jupiter and Saturn especially, as the best known of them, appear, by the best judgment which we can form, to be spheres of water, and of aqueous vapour, combined, it may be, with atmospheric air, in which their cloudy belts float over their deep oceans. Mars seems to have some portion at least of aqueous atmosphere; the earth, we know, has a considerable atmosphere of air, and of vapour; but the Moon, so near to her mistress, has none. On Venus and Mercury, we see nothing of a gaseous or aqueous atmosphere; and they, and Mars, do not differ much in their density from the Earth. Now does not this look as if the water and the vapour, which belong to the solar system, were driven off into the outer regions of

its vast circuit; while the solid masses which are nearest to the focus of heat, are all approximately of the same nature? And if this be so, what is the peculiar physical condition which we are led to ascribe to the Earth? Plainly this: that she is situated just in that region of the system, where the existence of matter, both in a solid, a fluid, and a gaseous condition, is possible. Outside the Earth's orbit, or at least outside Mars and the small Planetoids, there is, in the planets, apparently, no solid matter; or rather, if there be, there is a vast preponderance of watery and vaporous matter. Inside the Earth's orbit, we see, in the planets, no traces of water or vapour, or gas; but solid matter, about the density of terrestrial matter. The Earth, alone, is placed at the border where the conditions of life are combined; ground to stand upon; air to breathe; water to nourish vegetables, and thus, animals; and solid matter to supply the materials for their more solid parts: and with this, a due supply of light and heat, a due energy of the force of weight. All these conditions are, in our conception, requisite for life: that all these conditions meet, elsewhere than in the neighbourhood of the Earth's orbit, we see strong reasons to disbelieve. The Earth, then, it would seem, is the abode of life, not because all the globes which revolve round the Sun may be assumed to be the abodes of life; but because the Earth is fitted to be so, by a curious and complex combination of properties and relations, which do not at all apply to the others. That the Earth is inhabited, is not a reason for believing that the other Planets are so, but for believing that they are not so.

3 Can we see any physical reason, for the fact which appears to us so probable, that all the water and vapour of the system is gathered in its outward parts? It would seem that we can. Water and aqueous vapour are driven from the Sun to the outer parts of the solar system, or are allowed to be permanent there only, as they are driven off and retained at |237

a distance by any other source of heat;—to use a homely illustration, as they are driven from wet objects placed near the kitchen-fire: as they are driven from the hot sands of Egypt into the upper air: as they are driven from the tropics to the poles. In this latter case, and generally, in all cases, in which vapour is thus driven from a hotter region, when it comes into a colder, it may again be condensed in water, and fall in rain. So the cold of the air in the temperate zone condenses the aqueous vapours which flow from the tropics; and so, we have our clouds and our showers. And as there is this rainy region, indistinctly defined, between the torrid and the frigid zones on the earth; so is there a region of clouds and rain, of air and water, much more precisely defined, in the solar system, between the central torrid zone and the external frigid zone which surrounds the Sun at a greater distance.

4 *The Earth's Orbit is the Temperate Zone of the Solar System.* In that Zone only is the play of Hot and Cold, of Moist and Dry, possible, The Torrid Zone of the Earth is not free from moisture; it has its rains, for it has its upper colder atmosphere. But how much hotter are Venus and Mercury than the Torrid Zone? There, no vapours can linger; they are expelled by the fierce solar energy; and there is no cool stratum to catch them and return them. If they were there, they must fly to the outer regions; to the cold abodes of Jupiter and Saturn, if on their way, the Earth did not with cold and airy finger outstretched afar, catch a few drops of their treasures, for the use of plant, and beast, and man. The solid stone only, and the metallic ore which can be fused and solidified with little loss of substance, can bear the continual force of the near solar fire, and be the material of permanent solid planets in that region. But the lava pavement of the Inner Planets bears 238| no superstructure of life; for all life would be scorched away

along with water, its first element. On the Earth first, can this superstructure be raised; and there, through we know not what gradation of forms, the waters were made to bring forth abundantly things that had life; plants, and animals nourished by plants, and conspiring with them, to feed on their respective appointed elements, in the air which surrounded them. And so, nourished by the influences of air and water, plants and animals lived and died, and were entombed in the scourings of the land, which the descending streams carried to the bottom of the waters. And then, these beds of dead generations were raised into mountain ranges; perhaps by the yet unextinguished forces of subterraneous fires. And then a new creation of plants and animals succeeded; still living under the fostering influence of the united pair, Air and Water, which never ceased to brood over the World of Life, their Nurseling; and then, perhaps, a new change of the limits of land and water, and a new creation again: till at last, Man was placed upon the Earth; with far higher powers, and far different purposes, from any of the preceding tribes of creatures: and with this, for one of his offices;—that there might be an intelligent being to learn how wonderfully the scheme of creation had been carried on, and to admire, and to worship the Creator.

5 But we have a few more remarks to make on the structure of the Solar System, in this point of view. When we say that the water and vapour of the System were driven to the outer parts, or retained there, by the central heat of the Sun, perhaps it might be supposed to be most simple and natural, that the aqueous vapour, and the water, should assume its place in a distinct circle, or rather a spherical shell, of which the Sun was the center; thus making an elemental sphere about the center, such as the ancients imagined in their schemes of the Universe. Nor will we venture to say that such an arrangement of elements might |239

not be; though perhaps it might be shown that no stable equilibrium of the system would be, in this way, mechanically possible. But this at least we may say; that a rotatory motion of all the parts of the universe appears to be a universal law prevalent in it, so far as our observation can reach: and that, by such rotation of the separate masses, the whole is put in a condition which is everywhere one of stable equilibrium. It was, then, agreeable to the general scheme, that the excess of water and vapour, which must necessarily be carried away, or stored up, in the outer regions of the System, should be put into shapes in which it should have a permanent place and form. And thus, it is suitable to the general economy of creation, that this water and vapour should be packed into rotating masses, such as are Jupiter and Saturn, Uranus and Neptune. When once collected in such rotating masses, the attraction of its parts would gather it into spheroidal forms; oblate by the effect of rotation, as Jupiter, or perhaps into annular forms, like the Ring of Saturn*; for such also is a mechanically possible form of equilibrium, for a fluid mass. And these spheroids once formed, the water would form a central nucleus, over which would hang a cover of vapour, raised by the evaporating power of the Sun, and forming clouds, where the rarity of the upper strata of vapour allowed the cold of the external space to act; and these clouds, spun into belts by the rotation of the sphere. And thus, the vapour, which would otherwise have wandered loose about the atmosphere, was neatly wound into balls; which, again, were kept in their due place, by being made to revolve in nearly circular orbits about the Sun.

6 And thus, according to our view, water and gases, clouds and vapours, form mainly the planets in the outer

* Other speculators also have regarded Saturn's Ring as a ring of cloud or water. See *Cosmos*, III. 527 and 553.

part of the solar system; while masses such as result from the fusion of the most solid materials, lie nearer the sun, and are found principally within the orbit of Jupiter*. To conceive planetary systems as formed by the gradual contraction of a nebular mass, and by the solidification of some of its parts, is a favourite notion of several speculators. If we adopt this notion, we shall, I think, find additional proofs in favour of our view of the system. For, in the first place, we have the zodiacal light, a nebulous appendage to the Sun, as Herschel conceives, extending beyond the orbits of Mercury and Venus. These planets, then, have not yet fully emerged from the atmosphere in which they had their origin:—the *mother-light* and *mother-fire*, in which they began to crystallize, as crystals do in their mother-water. Though they are already opaque, they are still immersed in luminous vapour: and bearing such traces of their chaotic state being not yet ended, we need not wonder, if we find no evidence of their having inhabitants, and some evidence to the contrary. They are within a nebular region, which may easily be conceived to be uninhabitable. And where this nebular region, marked by the zodiacal light, terminates, the world of life begins, namely at the Earth.

7 But further, outside this region of the Earth, what do we find in the solar system? Of solid matter, if our views are right, we find nothing but an immense number of small bodies; namely first, Mars, who, as we have said, is only about one eighth the earth in mass: the twenty-six small planetoids, (or whatever number may have been discovered when these pages meet the reader's eye†,) between Mars and Jupiter; the four satellites of Jupiter; the eight satellites of

* Humboldt has already remarked (*Cosmos*, I. 95, and III. 427), that the inner planets as far as Mars, and the outer ones beginning with Jupiter, form two groups having different properties. Also Encke. (See Humboldt's Note.)

† Printed Oct. 19, 1853.

Saturn; the six (if that be the true number,) satellites of Uranus; and the one satellite of Neptune, already detected. It is very remarkable, that all this array of small bodies begins to be found just outside the Earth's orbit. Supposing, as we have found so much reason to suppose, that Jupiter, and the other exterior planets, are not solid bodies, but masses of water and of vapour; the existence of great solid planetary masses, such as exist in the region of the Earth's orbit, is succeeded externally by the existence of a vast number of smaller bodies. The real quantity of matter in these smaller bodies we cannot in general determine. Perhaps the largest of them, (after Mars,) may be Jupiter's third satellite; which* is reckoned, by Laplace, to have a mass less than 1-10,000th of that of Jupiter himself; and thus, since Jupiter, as we have seen, has a mass 333 times that of the Earth, the satellite would be above 1-30th of the Earth's mass†. That none but masses of this size, and many far below this, are found outside of Mars, appears to indicate, that the *planet-making* powers which were efficacious to this distance from the sun, and which produced the great globe of the Earth, were, beyond this point, feebler; so that they could only give birth to smaller masses; to planetoids, to satellites, and to meteoric stones. Perhaps we may describe this want of energy in the planet-making power, by saying, that at so great a distance from the central fire, there was not heat enough to melt together these smaller fragments into a larger globe‡; or rather, when they existed

* Herschel, 540.

† It is probable, from the small density of Jupiter's satellites, that they also consist in a great measure of water and vapour. Only one of them is denser than Jupiter himself.—*Cosmos.*

‡ It has, in our own day, even in the present year, been regarded as a great achievement of man to direct the fiery influences which he can command, so as to cast a colossal statue in a single piece, instead of casting it in several portions.

in a nebular, perhaps in a gaseous state, that there was not heat enough to keep them in that state, till the attraction of the parts of all of them had drawn them into one mass, which might afterwards solidify into a single globe. The tendency of nebular matter to separate into distinct portions, which may afterwards be more and more detached from each other, so as to break the nebulous light into patches and specks, appears to be seen in the structure of the resolvable nebulæ, as we have already had occasion to notice. And according to the view we are now taking, we may conceive such patches, by further cooling and concentration, to remain luminous as comets, and perhaps shooting stars; or to become opaque as planets, planetoids, satellites, or meteoric stones. And here we may call to mind what we have already said, that the meteoric stones consist of the same elements as those of the earth, combined by the same laws; and thus appear to bring us a message from the other solid planets, that they also have the same elements and the same chemical forces as the earth has.

8 It has already been supposed, by many astronomers, that shooting stars, and meteoric stones, are bodies of connected nature and origin; and that they are cosmical, not terrestrial bodies;—parts of the solar system, not merely appendages to the earth. It has been conceived, that the luminous masses, which appear as shooting stars, when they are without the sphere of terrestrial influences, may, when they reach our atmosphere, collapse into such solid lumps as have from time to time fallen upon the earth's surface: many of them, with such sudden manifestations of light and heat, as implied some rapid change taking place in their chemical constitution and consistence. If shooting stars are of this nature, then, in those cases in which a great number of them appear in close succession, we have evidence that there is a region in which there is a large collection of matter of a

|243

nebulous kind, collected already into small clouds, and ready, by any additional touch of the powers that hover round the earth, to be further consolidated into planetary matter. That the earth's orbit carries her through such regions, in her annual course, we have evidence, in the curious fact, now so repeatedly observed, of showers of shooting stars, seen at particular seasons of every year; especially about the 13th of November, and the 10th of August. This phenomenon has been held, most reasonably, to imply that at those periods of the year, the earth passes through a crowd of such meteor-planets, which form a ring round the sun; and revolving round him, like the other planets, retain their place in the system from year to year*. It may be that the orbits of these meteor-planets are very elliptical. That they are to a certain extent elliptical, appears to be shown, by our falling in with them only once a year, not every half year, as we should do, if their orbit, being nearly circular, met the earth's orbit in two opposite points. That the shooting stars, thus seen in great numbers when the earth is at certain points of her orbit, are really planetoidal bodies, appears to be further proved by this;—that they all seem to move nearly in the same direction†. They are, each of them, visible for a short time only, (indeed commonly only for a few seconds,) while they are nearest the earth; much in the same way in which a comet is visible only for a small portion of its path: and this portion is described in a short time, because they move near the earth. They are so small that a little change of distance removes them beyond our vision.

9 Perhaps these revolving specks of nebulæ are the outriders of the zodiacal light; portions of it, which, being external to the permanently nebulous central mass, have broken into patches, and are seen as stars for the moment that we are near to them. And if this be true, we have to

244| * Herschel, 900—905. † Ibid. 901.

correct, in a certain way, what we have previously said of the zodiacal light;—that no one had thought of resolving it into stars: for it would thus appear, that in its outer region, it resolves itself into stars, visible, though but for a moment, to the naked eye.

10 And thus, all these phenomena concur in making it appear probable, that the Earth is placed in that region of the solar system in which the planet-forming powers are most vigorous and potent;—between the region of permanent nebulous vapour, and the region of mere shreds and specks of planetary matter, such as are the satellites and the planetoidal group. And from these views, finally it follows, that the Earth is really the largest planetary body in the Solar System. The vast globes of Jupiter and Saturn, Uranus and Neptune, which roll far above her, are still only huge masses of cloud and vapour, water and air; which, from their enormous size, are ponderous enough to retain round them a body of small satellites, perhaps, in some degree at least, solid; and which have perhaps a small lump, or a few similar lumps, of planetary matter at the center of their watery globe. The Earth is really the domestic hearth of this Solar System; adjusted between the hot and fiery haze on one side, the cold and watery vapour on the other. This region only is fit to be a domestic hearth, a seat of habitation; and in this region is placed the largest solid globe of our system; and on this globe, by a series of creative operations, entirely different from any of those which separated the solid from the vaporous, the cold from the hot, the moist from the dry, have been established, in succession, plants, and animals, and man. So that the habitation has been occupied; the domestic hearth has been surrounded by its family; the fitnesses so wonderfully combined have been employed; and the Earth alone, of all the parts of the frame which revolves round the Sun, has become a World.

|245

11 Perhaps it may tend still further to illustrate, and to fix in the reader's mind, the view of the constitution of the solar system here given, if we remark an analogy which exists, in this respect, between the Earth in particular, and the Solar System in general. The earth, like the central parts of the system, is warmed by the sun; and hence, drives off watery vapours into the circumambient space, where they are condensed by the cold. The upper regions of the atmosphere, like the outer regions of the solar system, form the vapours thus raised into clouds, which are really only water in minute drops; while in the solar system, the cold of the outer regions, and the rotation of the masses themselves, maintain the water, and the vapour, in immense spheres. But Jupiter and Saturn may be regarded as, in many respects, immense clouds; the continuous water being collected at their centers, while the more airy and looser parts circulate above. They are the permanent receptacles of the superfluous water and air of the system. What is not wanted on the Earth, is stored up there, and hangs above us, far removed from our atmosphere; but yet, like the clouds in our atmosphere, an example, what glorious objects accumulations of vapour and water, illuminated by the rays of the sun, may become in our eyes.

12 These views are so different from those hitherto generally entertained, and considered as having a sort of religious dignity belonging to them, that we may fear, at first at least, they will appear to many, rash and fanciful, and almost, as we have said, irreverent. On the question of reverence, we may hereafter say a few words; but as to the rashness of these views, we would beg the reader, calmly and dispassionately, to consider the very extraordinary number of points in the solar system, hitherto unexplained, which they account for, or, at least reduce into consistency and connexion, in 246| a manner which seems wonderful. The Theory, as we may

perhaps venture to call it, brings together all these known phenomena:—the great size and small density of the exterior planets;—their belts and streaks;—Saturn's ring;—Jupiter's oblateness;—the great number of satellites of the exterior planets;—the numerous group of planetoid bodies between Jupiter and Mars;—the appearance of definite shapes of land and water on Mars;—the showers of shooting stars which appear at certain periods of the year;—the Zodiacal Light;—the appearance of Venus as different from Mars;—and finally, the material composition of meteoric stones.

13 Perhaps there are other phenomena which more readily find an explanation in this theory, than in any other: for instance, the recent discovery of a dim half-transparent ring, as an appendage to the luminous ring of Saturn, which has hitherto alone been observed. Perhaps this is the ring of vapour which may naturally be expected to accompany the ring of water. It is the annular atmosphere of the aqueous annulus. But, the discovery of this faint ring being so new, and hitherto not fully unfolded, we shall not further press the argument, which, hereafter, perhaps, may be more confidently derived from its existence.

14 There are some other facts in the Solar System, which, we can hardly doubt, must have a bearing upon the views which we have urged; though we cannot yet undertake to explain that bearing fully. Not only do all the planetary bodies of the solar system, as well as the Sun himself, revolve upon their axes; but there is a very curious fact relative to these revolutions, which appears to point out a further connexion among them. So far as has yet been ascertained, all those which we, in our theory, regard as solid bodies, Mercury, Venus, the Earth, and Mars, revolve in very nearly the same time: namely, in about twenty-four hours. All those larger masses, on the other hand, which

we, in our theory, hold to be watery planets, Jupiter, Saturn, Uranus, revolve, not in a longer time, as would perhaps have been expected, from their greater size, but in a shorter time; in less than half the time; in about ten hours. The near agreement of the times of revolution in each of these two groups, is an extremely curious fact; and cannot fail to lead our thoughts to the probability of some common original cause of these motions. But no such common cause has been suggested, by any speculator on these subjects. If, in this blank, even of hypotheses, one might be admitted, as at least a mode of connecting the facts, we might say, that the compound collection of solid materials, water, and air, of which the solar system consists, and of which our earth alone, perhaps, retains the combination, being, by whatever means, set a spinning round an axis, at the rate of one revolution in 24 hours, the solid masses which were detached from it, not being liable to much contraction, retained their rate of revolution; while the vaporous masses which were detached from the fluid and airy part, contracting much, when they came into a colder region, increased their rate of revolution on account of their contraction. That such an acceleration of the rate of revolution would be the result of contraction, is known from mechanical principles; and indeed, is evident: for the contraction of a circular ring of such matter into a narrower compass, would not diminish the linear velocity of its elements, while it would give them a smaller path to describe in their revolutions. Such an hypothesis would account, therefore, both for the nearly equal times of revolution of all the solid planets, and for the smaller period of rotation, which the larger planets show.

15 In what manner, however, portions are to be detached from such a rotating mass, so as to form solid planets on the one side, and watery planets on the other, and how

248|

these planets, so detached, are to be made to revolve round the Sun, in orbits nearly circular, we have no hypothesis ready to explain. And perhaps we may say, that no satisfactory, or even plausible, hypothesis to explain these facts, has been proposed: for the Nebular Hypothesis, the only one which is likely to be considered as worthy any notice on this subject, is too imperfectly worked out, as yet, to enable us to know, what it will or will not account for. According to that hypothesis, the nebular matter of a system, having originally a rotatory motion, gradually contracts; and separating, at various distances from the center, forms rings; which again, breaking at some point of their circumference, are, by the mutual attraction of their parts, gathered up into one mass; which, when cooled down, so as to be opaque, becomes a planet; still revolving round the luminous mass which remains at the center. That such a process, if we suppose the consistency, and other properties, of the nebulous matter to be such as to render it possible, would produce planetary masses revolving round a sun in nearly circular orbits, and rotating about their own axes, seems most likely; though it does not appear that it has been very clearly shown*. But no

* Besides the curious relation of the times of rotation of the planets, just noticed, there is another curious relation, of their distance from the Sun, which any one, wishing to frame an hypothesis on the origin of our Solar System, ought by all means to try to account for.

The distances from the Sun, of the planets, Mercury, Venus, Earth, Mars, the Planetoids, Jupiter, Saturn, Uranus, are nearly as the numbers,

4, 7, 10, 16, 28, 52, 100, 196:

now the excesses of each of these numbers above the first are,

3, 3, 6, 12, 24, 48, 96:

a series in which each term (after the first,) is double of the preceding one. Hence, the distances of the planets conform to a series following this law, (*Bode's law*, as it is termed.) And though the law is by no means exact, yet it was so far considered a probable expression of a general fact, that the deviation from this law, in the interval between Mars and Jupiter, was the principal cause which led first to successful attempt has been made to deduce any laws of the distances from the center, times of rotation, or other properties of such planets; and therefore, we cannot say that the nebular hypothesis is yet in any degree confirmed.

16 The Theory which we have ventured to propose, of the Solar System, agrees with the Nebular Hypothesis, so far as that hypothesis goes; if we suppose that there is, at the center of the exterior planets, Jupiter, Saturn, Uranus, and Neptune, a solid nucleus, probably small, of the same nature as the other planets. Such an addition to our theory is, perhaps, on all accounts, probable: for that circumstance would seem to determine, to particular points, the accumulation of water and vapours, to which we hold that those planets owe the greater part of their bulk. Those planets then, Jupiter, Saturn, and the others, are really small solid planets, with enormous oceans and atmospheres. The Nebular Hypothesis, in that case, is that part of our Hypothesis, which relates to the condensation of luminous nebular matter; while *we* consider, further, the causes which, scorching the inner planets, and driving the vapours to the outer orbs, would make the region of the earth the only habitable part of the system.

17 The belief that other planets, as well as our own, are the seats of habitation of living things, has been entertained, in general, not in consequence of physical reasons, but in spite of physical reasons; and because there were

the suspicion of a planet interposed in the seemingly vacant space; and thus led to the discovery of the planetoids, which really occupy that region. It is true, that the law is found not to hold, in the case of the newly-discovered planet Neptune; for his distance from the Sun, which according to this law, should be 388, is really only 300, 30 times the Earth's distance, instead of 39 times. Still, Bode's law has a comprehensive approximate reality in the Solar System, sufficient to make it a strong recommendation of any hypothesis of the origin of the system, that it shall account for this law. This, however, the nebular hypothesis does not.

250|

conceived to be other reasons, of another kind, theological or philosophical, for such a belief. It was held that Venus, or that Saturn, was inhabited, not because any one could devise, with any degree of probability, any organized structure which would be suitable to animal existence on the surfaces of those planets; but because it was conceived that the greatness or goodness of the Creator, or His wisdom, or some other of His attributes, would be manifestly imperfect, if these planets were not tenanted by living creatures. The evidences of design, of which we can trace so many, and such striking examples, in our own sphere, the sphere of life, must, it was assumed, exist, in the like form, in every other part of the universe. The disposition to regard the Universe in this point of view, is very general; the disinclination to accept any change in our belief which seems, for a time, to interfere with this view, is very strong; and the attempt to establish the necessity of new views discrepant from these has, in many eyes, an appearance as if it were unfriendly to the best established doctrines of Natural Theology. All these apprehensions will, we trust, be shown, in the sequel, to be utterly unfounded: and in order that any such repugnance to the doctrines here urged, may not linger in the reader's mind, we shall next proceed to contemplate the phenomena of the universe in their bearing upon such speculations.

CHAPTER XI.

The Argument from Design.

---

1 THERE is no more worthy or suitable employment of the human mind, than to trace the evidences of Design and Purpose in the Creator, which are visible in many parts of the Creation. The conviction thus obtained, that man was formed by the wisdom, and is governed by the providence, of an intelligent and benevolent Being, is the basis of Natural Religion, and thus, of all Religion. We trust that some new lights will be thrown upon the traces of Design which the Universe offers, even in the work now before the reader: and as our views, regarding the plan of such Design, are different, in some respects, and especially as relates to the Planets and Stars, from those which have of late been generally entertained, it will be proper to make some general remarks, mainly tending to show, that the argument remains undisturbed, though the physical theory is changed.

2 It cannot surprize any one who has attended to the history of science, to find that the views, even of the most philosophical minds, with regard to the plan of the universe, alter, as man advances from falsehood to truth: or rather, from very imperfect truth to truth less imperfect. But yet such a one will not be disposed to look, with any other feeling than profound respect, upon the reasonings by which the wisest men of former times ascended from their erroneous views of nature to the truth of Natural Religion. It cannot seem strange to us that man, at any point, and perhaps at every point, of his intellectual progress, should have an imperfect insight into the plan of the Universe: but, in the

most imperfect condition of such knowledge, he has light enough from it, to see vestiges of the Wisdom and Benevolence of the Creating Deity; and at the highest point of his scientific progress, he can probably discover little more, by the light which physical science supplies. We can hardly hope, therefore, that any new truths with regard to the material universe, which may now be attainable, will add very much to the evidence of creative design; but we may be confident, also, that they will not, when rightly understood, shake or weaken such evidence. It has indeed happened, in the history of mankind, that new views of the constitution of the universe, brought to light by scientific researches, and established beyond doubt, in the conviction of impartial persons, have disturbed the thoughts of religious men; because they did not fall in with the view then entertained, of the mode in which God effects his purposes in the universe. But in these cases, it soon came to be seen, after a season of controversy, reproach, and alarm, that the old argument for design was capable of being translated into the language of the new theory, with no loss of force; and the minds of men were gradually tranquillized and pacified. It may be hoped that the world is now so much wiser than it was two or three centuries ago, that if any modification of the current arguments for the Divine Attributes, drawn from the aspect of the universe, become necessary, in consequence of the rectification of received errors, it will take place without producing pain, fear, or anger. To promote this purpose, we proceed to make a few remarks.

3 The proof of Design, as shown in the works of Creation, is seen most clearly, not in mere physical arrangements, but in the structure of organized things;—in the constitution of plants and animals. In those parts of nature, the evidences of intelligent purpose, of wise adaptation, of skilful selection of means to ends, of provident contrivance, are, in |253

many instances, of the most striking kind. Such, for example, are the structure of the human eye, so curiously adapted for its office of seeing; the muscles, cords, and pullies by which the limbs of animals are moved, exceeding far the mechanical ingenuity shown in human inventions; the provisions which exist, before the birth of offspring, for its sustenance and well-being when it shall have been born;—these are lucid and convincing proofs of an intelligent Creator, to which no ordinary mind can refuse its conviction. Nor is the evidence, which we here recognize, deprived of its force, when we see that many parts of the structure of animals, though adapted for particular purposes, are yet framed as a portion of a system which does not seem, in its general form, to have any bearing on such purposes*. The beautiful contrivances which exist in the skeleton of man, and the contrivances, possessing the same kind of beauty, in the skeleton of a sparrow, do not appear to any reasonable person less beautiful, because the skeleton of a man, and of a sparrow, have an agreement, bone for bone, for which we see no reason, and which appears to us to answer no purpose. The way in which the human hand and arm are made capable of their infinite variety of use, by the play of the radius and ulna, the bones of the wrist and the fingers, is not the less admirable, because we can trace the representatives and rudiments of each of these bones, in cases where they answer no such ends;—in the foreleg of the pig, the ox, the horse, or the seal. The provision for feeding the young creature, which is made, with such bounteous liberality, and such opportune punctuality, by the breasts of the mother, has not any doubt thrown upon its reality, by the teats of male animals and the

* The greatest anatomists, and especially Mr Owen, have recently expressed their conviction, that researches on the structure of animals must be guided by the principle of *unity of composition* as well as the principle of *final causes*. See Owen *On the Nature of Limbs*.

paps of man, which answer no such purpose. That in these cases there is manifested a wider plan, which does not show any reference to the needs of particular cases; as well as peculiar contrivances for the particular cases, does not disturb our impression of design in each case. Why should so large a portion of the animal kingdom, intended, as it seems, for such different fields of life and modes of living;—beasts, birds, fishes;—still have a skeleton of the same plan, and even of the same parts, bone for bone; though many of the parts, in special cases, appear to be altogether useless (namely, the vertebrate plan)? We cannot tell. Our naturalists and comparative anatomists, it would seem, cannot point out any definite end, which is answered by making so many classes of animals on this one vertebrate plan. And since they cannot do this, and since we cannot tell why animals are so made, we must be content to say that we do not know; and therefore, to leave this feature in the structure of animals out of our argument for design. Hence we do not say that the making of beasts, birds, and fishes, on the same vertebrate plan, proves design in the Creator, in any way in which we can understand design. That plan is not of itself a proof of design; it is something in addition to the proofs of design; a general law of the animal creation, established, it may be, for some other reason. But this common plan being given, we can discern and admire, in every kind of animal, the manner in which the common plan is adapted to the particular purposes which the animal's kind of life involves*. The general law is not all; there is also, in every instance, a special care for the species. The general law may seem, in many cases, to remove further from us the proof of providential care; by showing that the elements of the benevolent

* This has been termed by physiologists *The Law of the Development from the General to the Special.*

contrivance are not provided in the cases alone where they are needed, but in others also. But yet this seeming, this obscuration of the evidence of design, by interposing the form of general law, cannot last long. If the general law supplies the elements, still a special adaptation is needed to make the elements answer such a purpose; and what is this adaptation, but design? The radius and ulna, the carpal and metacarpal bones, are all in the general type of the vertebrate skeleton. But does this fact make it the less wonderful, that man's arm and hand and fingers should be constructed so that he can make and use the spade, the plough, the loom, the pen, the pencil, the chisel, the lute, the telescope, the microscope, and all other instruments? Is it not, rather, very wonderful that the bones which are to be found rudimentally, in the leg-bone of a horse, or the hoof of an ox, should be capable of such a curious and fertile development and modification? And is not such development and modification a work, and a proof, of design and intention in the Creator? And so in other cases. The teats of male animals, the nipples of man, may arise from this, that the general plan of the animal frame includes paps, as portions of it; and that the frame is so far moulded in the embryo, before the sex of the offspring is determined. Be it so. Yet still this provision of paps in the animal form in general, has reference to offspring; and the development of that part of the frame, when the sex *is* determined, is evidence of design, as clear as it is possible to conceive in the works of nature. The general law is moulded to the special purpose, at the proper stage; and this play of general laws, and special contrivances, into each other's provinces, though it may make the phenomena a little more complex, and modify our notion as to the mode of the Creator's working, will not, in philosophical minds, disturb the conviction that there is design in the special adaptations:

besides which, some other feature of the operation of the Creative Mind may be suggested by the prevalence of general laws in the Creation.

4 There is, however, one caution suggested by this view. Since, besides, and mixed with the examples of Design which the creation offers, there are also results of General Laws, in which we cannot trace the purpose and object of the law; we may fall into error, if we fasten upon something which is a result of such mere general laws, and imagine that we can discern its object and purpose. Thus, for instance, we might possibly persuade ourselves that we had discovered the use and purpose of the teats of male animals; or of the trace of separation into parts which the leg-bone of a horse offers; or of the false toes of a pig: all which are, as we have seen, the rudiments of a plan more general than is developed in the particular case. And if, when we had made such a fancied discovery, it were found that the uses and purposes which we had imagined to belong to these parts or features, were not really served by them; at first, perhaps, we might be somewhat disturbed, as having lost one of the evidences of the design of the Creator, all which are precious to a reverent mind. But it is not likely that any disturbance of a reverent mind on such grounds as this, would continue long, or go far. We should soon come to recollect, how light and precarious, perhaps, how arbitrary and ill-supported by our real knowledge, were the grounds on which we had assigned such uses to such parts. We should turn back from them to the more solid and certain evidences, not shaken, nor likely to be shaken, by any change in prevalent zoological or anatomical doctrines, which those who love to contemplate such subjects habitually dwell upon; and, holding ourselves ready to entertain any speculations by which the bearing of those general Laws upon Natural Religion could be shown, in such a way as to convince our reason, we should rest in |257

the confident and tranquil persuasion that no success or failure in such speculations could vitally affect our belief in a wise and benevolent Deity:—that though additional illustrations of his attributes might be interesting and welcome, no change of our scientific point of view could make his being or action doubtful.

5 This is, it would seem, the manner in which a reasonable and reverent man would regard the proof of a Supreme Creator and Governor, which is derived from Design, as seen in the organic creation; and the mode in which such proof would be affected by changes in the knowledge which we may acquire of the general laws by which the organic creation is constituted and governed. And hence, if it should be found to be established by the researches of the most comprehensive and exact philosophy, that there are, in any province of the universe, resemblances, gradations, general laws, indications of the mode in which one form approaches to another, and seems to pass into and generate another, which tend to obliterate distinctions which at first appeared broad and conspicuous; still the argument, from the design which appears in the parts of which we most clearly see the purpose, would not lose its force. If, for instance, it should be made apparent, by geological investigations of the extinct fossil creation, that the animal forms which have inhabited the earth, have gradually approached to that type in which the human form is included, passing from the rudest and most imperfect animal organizations, mollusks, or even organic monads, to vertebrate animals, to warm-blooded animals, to monkeys, and to men; still, the evidences of design in the anatomy of man are not less striking than they were, when no such gradation was thought of. And what is more to the purpose of our argument, the evidences of the peculiar nature and destination of man, as shown in other characters than his 258| anatomy,—his moral and intellectual nature, his history and

capacities,—stand where they stood before; nor is the vast chasm which separates man, as a being with such characters as these latter, from all other animals, at all filled up or bridged over.

6 The evidence of design in the inorganic world,—in the relation of earth, air, water, heat and light,—is, to most persons, less striking and impressive, than it is in the organic creation. But even among these mere physical elements of the world, when we consider them with reference to living things, we find many arrangements which, on a reflective view, excite our admiration, by the beneficial effect, and seemingly beneficent purpose. Our condition is furnished with the solid earth, on which we stand, and in which we find the materials of man's handiworks; stone and metal, clay and sand;—with the atmosphere which we breathe, and which is the vehicle of oral intercourse between man and man;—with revolutions of the sun, by which are brought round the successions of day and night, through all their varying lengths, and of summer and winter;—with the clouds above us, which pour upon the earth their fertilizing showers. All this furniture of the earth, so marvellously adapting it for the abode of living creatures, and especially of man, may well be regarded as a collection of provisions for his benefit:—as *intended* to do him the good, which they do. Nor would this impression be removed, or even weakened, if we were to discover that some of these arrangements, instead of being produced by a machinery confined to that single purpose, were only partial results of a more general plan. For instance; we learn that the varying lengths of days and nights through the year, and the varying declination of the sun, are produced, not, as was at first supposed, by the sun moving round the earth, in a complex diurnal and annual path, but by the earth revolving in an annual orbit round the sun; while at the same time she has a diurnal rotation about her |259

own axis, which axis, by the laws of mechanics, remains always parallel to itself. When we learn that this is so, we see that the effect is produced by a mechanical arrangement far more simple than any which the imagination of man had devised; but in this case, the effect is plainly rather an increased admiration at the simplicity of the mechanism, than a wavering belief in the reality of the purpose. In like manner when, instead of supposing water to exist in a continuous reservoir in a firmament above the earth, and to fall in the earlier and in the latter rain, by some special agency for that purpose; men learnt to see that the water in the upper regions of the air must exist in clouds and in vapours only, and must fall in showers by the condensing influence of cold currents of air; they needed not to cease to admire the kindness of the Creator, in providing the rain to water the earth, and the wind to dry it; although the mechanism by which the effect was produced was of a larger kind than they had before imagined. And even if this mechanism extend through the solar system: if the arrangement by which the Earth's atmosphere is the special region in which there are winds hot and cold, clouds compact or dissolving,—be an arrangement which extends its influence to other planets, as well as to ours;—if this mixed atmosphere be placed, not only at the meeting point of clear aqueous vapour above, and warmer airs below, but also at the meeting point of a hot central region surrounding the Sun, and a cold exterior zone in which water and vapour can exist in immense collected masses, such as are Jupiter and Saturn;—still it would not appear, to a reasonable view, that this larger expansion of the machinery by which the effect is produced, makes the machinery less remarkable; or can at all tend to diminish the belief that it was *intended* to produce the effect which it does produce. Hot and cold, moist and dry, are constantly
260| mixed together for the support of vegetable and animal life;

and not the less so, if we believe that, though elements of this kind pervade the whole solar system, it is only at the Earth that they are combined so as to foster and nourish living things.

7 But it will perhaps be said, that to suppose the whole Solar System to be a machine merely operating for the benefit of the Earth and its population, is to give to the Earth and its population an importance in the scheme of creation which is quite extravagant and improbable:—it is to make the greater orbs, Jupiter and Saturn, minister to the less; instead of having their own purpose, and their own population, which their size naturally leads us to expect. To this we reply, that, in the first place, we have shown good reason for believing that the Earth is really the largest dense solid globe which exists in the solar system, and that the size of Jupiter and Saturn arises from their being composed mainly of water and vapour. And with regard to the difficulty of the greater ministering to the less;—if by *greater*, mere size and extent be understood, it appears to be the universal law of creation, that the greater, in that sense, *should* minister to the less, when the less includes living things. Even if the planets be all inhabited, the sun, which is greater far than all of them together, ministers light and heat to all of them. Even on this supposition, the vast spaces by which the planets are separated have no use, that we can discern, except to place them at suitable distances from the sun. Even on this supposition, their solid globes within, their atmospheres without, are all merely subservient to the benefit of a thin and scattered stratum of population on the surface. The space occupied by man and animals on the earth's surface, even taking into account the highest buildings and the deepest seas, is only a few hundreds, or a thousand feet. The benefit of this minute shell, interrupted in many places for vast distances, everywhere loosely and |261

sparely filled, is ministered to by the solidity and attraction of a mass below it 20 millions of feet deep; by the influence of an atmosphere above it 200 thousand feet high at least, and it may be, much more. And this being so, if we increase the depth of the center 20 thousand times; if we carry the extreme verge of air and vapour to thirty times the radius of the earth's orbit from us, how does the construction of the machine become more improbable, or the disproportion of its size to its purpose more incongruous? Is mere size,—extent of brute matter or blank space,—so majestic a thing? Is not infinite space large enough to admit of machines of any size without grudging? But if we thus move the center of the Earth's peopled surface 20 thousand times further off, we reach the Sun. If we carry the limit of air and vapour to the distance of 30 times the radius of the Earth's orbit, we arrive at Neptune. Are these new numbers monstrous, while the old ones were accepted without scruple? Is number such an alarming feature in the description of the Universe? Does not the description of every part and every aspect of it, present us with numbers so large, that wonder and repugnance, on that ground, are long ago exhausted? Surely this is so: and if the evidence really tend to prove to us that all the solar system ministers to the earth's population; the mere size of the system, compared with the space occupied by the population, will not long stand in the way of the reception of such a doctrine.

8 But the objection will perhaps be urged in another form. It will be said that the other Planets have so many points of resemblance with the Earth, that we must suppose their nature and purpose the same. They, like the Earth, revolve in circles round the sun, rotate on their own axes, have, several of them, satellites, are opaque bodies, deriving light and probably heat from the sun. To an external 262| spectator of the Solar System, they would not be distin-

guishable from the Earth. Such a spectator would never be tempted to guess that the Earth alone, of all these, neither the greatest nor the least, neither the one with the most satellites, nor the fewest, neither the innermost nor the outermost of the planets, is the only one inhabited; or at any rate the only one inhabited by an intelligent population. And to this we reply; that the largest of the other planets, if we judge rightly, are *not* like the Earth in one most essential respect, their density; and none of them, in having a surface consisting of land and water; except perhaps Mars: that if the supposed external spectator could see that this was so, he might see that the earth was different from the rest; and he might be able to see the vaporous nature of the outer planets, so that he would no more dream of peopling them, than we do, of peopling the grand Alpine ridges and vallies which we see in the clouds of a summer-sky.

9 But even if the supposed spectator attended only to the obvious and superficial resemblances between one of the planets and another, he might still, if he were acquainted with the general economy of the Universe, have great hesitation in inferring that, if one of them were inhabited, the others also must be inhabited. For, as we have said, in the plan of creation, we have a profusion of examples, where similar visible structures do not answer a similar purpose; where, so far as we can see, the structure answers no purpose in many cases; but exists, as we may say, for the sake of similarity: the similarity being a general Law, the result, it would seem, of a creative energy, which is wider in its operation than the particular purpose. Such examples are, as we have said, the finger-bones which are packed into the hoofs of a horse, or the paps and nipples of a male animal. Now the spectator, recollecting such cases, might say: I know that the earth is inhabited; no doubt Mars and Jupiter are a good deal like the Earth; but are they inhabited? They look

like the terrestrial breast of Nature: but are they really nursing breasts? Do they, like that, give food to living offspring? Or are they mere images of such breasts? male teats, dry of all nutritive power? sports, or rather over-works of nature; marks of a wider law than the needs of Mother Earth require? many sketches of a design, of which only one was to be executed? many specimens of the preparatory process of making a Planet, of which only one was to be carried out into the making of a World? Such questions, might naturally occur to a person acquainted with the course of creation in general; even before he remarked the features which tend to show that Jupiter and Saturn, that Venus and Mercury, have not been developed into peopled worlds, like our Earth.

10 Perhaps it may be said, that to hold this, is to make Nature work in vain; to waste her powers; to suppose her to produce the framework, and not to build; to make the skeleton, and not to clothe it with living flesh; to delude us with appearances of analogy and promises of fertility, which are fallacious. What can we reply to this?

11 We reply, that to work in vain, in the sense of producing means of life which are not used, embryos which are never vivified, germs which are not developed; is so far from being contrary to the usual proceedings of nature, that it is an operation which is constantly going on, in every part of nature. Of the vegetable seeds which are produced, what an infinitely small proportion ever grow into plants! Of animal ova, how exceedingly few become animals, in proportion to those that do not; and that are wasted, if this be waste! It is an old calculation, which used to be repeated as a wonderful thing, that a single female fish contains in its body 200 millions of ova, and thus, might, of itself alone, replenish the seas, if all these were fostered into life. But in truth, this, though it may excite wonder, cannot excite

264|

wonder as anything uncommon. It is only one example of what occurs everywhere. Every tree, every plant, produces innumerable flowers, the flowers innumerable seeds, which drop to the earth, or are carried abroad by the winds, and perish, without having their powers unfolded. When we see a field of thistles shed its downy seeds upon the wind, so that they roll away like a cloud, what a vast host of possible thistles are there! Yet very probably none of them become actual thistles. Few are able to take hold of the ground at all; and those that do, die for lack of congenial nutriment, or are crushed by external causes before they are grown. The like is the case with every tribe of plants*. The like with every tribe of animals. The possible fertility of some kinds of insects is as portentous as anything of this kind can be. If allowed to proceed unchecked, if the possible life were not perpetually extinguished, the multiplying energies perpetually frustrated, they would gain dominion over the largest animals, and occupy the earth. And the same is the case, in different degrees, in the larger animals. The female is stocked with innumerable ovules, capable of becoming living things: of which incomparably the greatest number end as they began, mere ovules;—marks of mere possibility, of vitality frustrated. The universe is so full of such rudiments of things, that they far outnumber the things which outgrow their rudiments. The marks of possibility are much more numerous than the tale of actuality.

* Every reader of physiological works knows how easy it would be to multiply examples of this kind to any extent. Thus it is held by physiologists, that the sporules of fungi are universally diffused through the atmosphere, ready to vegetate whenever an opportunity presents itself: and that a single individual produces not less than ten millions of germs. It is held also that innumerable seeds of plants still capable of vegetation, lie in strata far below the earth's surface, finding the occasion to vegetate only by the rarest and most exceptional occurrences.—Carpenter, *Manual of Physiology.* 1851, Art. 44.

The vitality which is frustrated is far more copious than the vitality which is consummated. So far, then, as this analogy goes, if the earth alone, of all the planetary harvest, has been a fertile seed of creation;—if the terrestrial embryo have alone been evolved into life, while all the other masses have remained barren and dead:—we have, in this, nothing which we need regard as an unprecedented waste, an improbable prodigality, an unusual failure in the operations of nature: but on the contrary, such a single case of success among many of failure, is exactly the order of nature in the production of life. It is quite agreeable to analogy, that the Solar System, of which the *flowers* are not many, should have borne but one *fertile* flower. One in eight, or in twice eight, reared into such wondrous fertility as belongs to the Earth, is an abundant produce, compared with the result in the most fertile provinces of Nature. And even if any number of the Fixed Stars were also found to be barren flowers of the sky; objects, however beautiful, yet not sources of life or development, we need not think the powers of creation wasted or frustrated, thrown away or perverted. One such fertile result as the Earth, with all its hosts of plants and animals, and especially with Man, an intelligent being, to stand at the head of those hosts, is a worthy and sufficient produce, so far as we can judge of the Creator's ways by analogy, of all the Universal Scheme.

12 But when we follow this analogy, so far as to speak of the mere material mass of a planet as an *embryo world;*—a barren flower;—a seed which has never been developed into a plant;—we are in danger of allowing the analogy to mislead us. For a planet, as to its brute mass, has really nothing in common with a seed or an embryo. It has no organization, or tendency to organization; no principle of life, however obscure. So far as we can judge, no progress of time, or operation of mere natural influences, would clothe a

266|

brute mass with vegetables, or stock it with animals. No species of living thing would have its place upon the surface, by the mere order of unintelligent nature. So much is this so, according to all that our best knowledge teaches, that those geologists who must most have desired, for the sake of giving completeness and consistency to their systems, to make the production of vegetable and animal species from brute matter, a part of the order of nature, (inasmuch as they have explained everything else by the order of nature,) have not ventured to do so. They allow, generally at least, each separate species to require a special act of creative power, to bring it into being. They make the peopling of the earth, with its successive races of inhabitants, a series of events altogether different from the operation of physical laws in the sustentation of existing species. The creation of life is, they allow, something out of the range of the ordinary laws of nature. And therefore, when we speak of uninhabited planets, as cases in which vital tendencies have been defeated; in which their apparent destiny, as worlds of life, has been frustrated; we really do injustice to our argument. The planets had no vital tendencies: they could have had such given, only by an additional act, or a series of additional acts, of Creative Power. As mere inert globes, they had no settled destiny to be seats of life: they could have such a destiny, only by the appointment of Him who creates living things, and puts them in the places which he chooses for them. If, when a planetary mass had come into being, (in virtue of the same general physical law, suppose, which produced the earth,) the Creator placed a host of living things upon the earth, and none upon the other planet; there was still no violation of analogy, no seeming change of purpose, no unfinished plan. In the solar system, we can see what seem to be good reasons why he did this; but if we could not see such reasons, still we should be yet further from

|267

being able to see reasons why he necessarily must place inhabitants upon the other planet.

13 It is sometimes said, that it is agreeable to the goodness of God, that all parts of the creation should swarm with life: that life is enjoyment; and that the benevolence of the Supreme Being is shown in the diffusion of such enjoyment into every quarter of the universe. To leave a planet without inhabitants, would, it is thought, be to throw away an opportunity of producing happiness. Now we shall not here dwell upon the consideration, that the enjoyment thus spoken of, is, in a great degree, the enjoyment which the mere life of the lower tribes of animals implies;—the enjoyment of madrepores and oysters, cuttle-fish and sharks, tortoises and serpents; but we reply more broadly, that it is not the rule followed by the Creator, to fill all places with living things. To say nothing of the vast intervals between planet and planet, which, it is presumed, no one supposes to be occupied by living things; how large a portion of the surface of the earth is uninhabited, or inhabited only in the scantiest manner. Vast desert tracts exist in Africa and in Asia, where the barren sand nourishes neither animal nor vegetable life. The highest regions of mountain-ranges, clothed with perpetual snow, and with far-reaching sheets of glacier ice, are untenanted, except by the chamois at their skirts. There are many uninhabited islands; and were formerly many more. The ocean, covering nearly three-fourths of the globe, is no seat of habitation for land animals or for man; and though it has a large population of the fishy tribes, is probably peopled in smaller numbers than if it were land, as well as by inferior orders. We see, in the Earth then, which is the only seat of life of which we really know anything, nothing to support the belief that every field in the material universe is tenanted by living inhabitants.

268| 14 That vegetables and animals, being once placed upon

the earth, have multiplied or are multiplying, so as to occupy every part of the land and water which is suited for their habitation, we can see much reason to believe. Philosophical natural-historians have been generally led to the conviction that each species has had an original center of dispersion, where it was first native, and that from this center it has been diffused in all directions, as far as the circumstances of climate and soil were favourable to its production. But we can see also much reason to believe that this general diffusion of vegetable and animal life from centers, is a part of the order of nature which may often be made to give way to other and higher purposes;—to the diffusion, over the whole surface of the earth, of a race of intelligent, moral agents. This process may often interfere with the general law of diffusion: as for instance, when man exterminates noxious animals. And whatever may be the laws which tend to replenish the earth, on which such centers of the diffusion of life exist for animals and plants; according to all analogy, these laws can have no force on any other planet, till such origins and centers of life are established on their surfaces. And even if any of the species which have ever tenanted the earth were so established on any other planet, we have the strongest reason to believe that they could not survive to a second generation.

15 Perhaps it may be said that we unjustifiably limit the power and skill of the Supreme Creator, if we deny that he could frame creatures fitted to live on any of the other planets, as well as in the Earth:—that the wonderful variety, and unexpected resource, of the ways in which animals are adapted for all kinds of climates, habitations, and conditions, upon the earth, may give us confidence that, under conditions still more extended, in habitations still further removed, in climates going beyond the terrestrial extremes, still the same

wisdom and skill may well be supposed to have devised possible modes of animal life.

16 To this we reply, that we are so far from saying that the Creator could not place inhabitants in the other planets, that we have attempted to show what kind of inhabitants would be most likely to be placed there, by considering the way in which animals are accommodated to special conditions in their habitation. In judging of such modes of accommodating animals to an abode on other planets, as well as the earth, we have reasoned from what we know, of the mode in which animals are accommodated to their different habitations on the earth. We believe this to be the only safe and philosophical way of treating the question. If we are to reason at all about the possibility of animal life, we must suppose that heat and light, gravity and buoyancy, materials and affinities, air and moisture, produce the same effect, require the same adaptations, in Jupiter or in Venus, as they do on the Earth. If we do not suppose this, we run into the error which so long prevented many from accepting the Newtonian system:—the error of thinking that matter in the heavens is governed by quite different laws from matter on the earth. We must adopt that belief, if we hold that animals may live under relations of heat and moisture, materials and affinities, in Jupiter or Venus, under which they could not live on our planet. And that belief, as we have said, appears to us contrary to all the teaching which the history of science offers us.

17 And not only is it contrary to the teaching of the history of science, to suppose the laws, which connect elemental and organic nature, to be different in the other planets from what they are on ours; but moreover the supposition would not at all answer the purpose, of making it probable that the planets are inhabited. For if we begin to imagine new and unknown laws of nature for those abodes, what is there to

limit or determine our assumptions in any degree? What extravagant mixtures of the attributes and properties of mind and matter may we not then accept as probable truths? We know how difficult the poets have found it to describe, with any degree of consistency, the actions and events of a world of angels, or of evil spirits, souls or shades, embodied in forms so as to admit of description, and yet not subject to the laws of human bodies. Virgil, Tasso, Milton, Klopstock, and many others, have struggled with this difficulty :—no one of them, it will be probably agreed, with any great success; at least, regarding his representation as a hypothesis of a possible form of life, different from all the forms which we know. Yet if we are to reject the laws which govern the known forms of life, in order that we may be able to maintain the possibility of some unknown form in a different planet, we must accept some of these hypotheses, or find a better. We must suppose that weight and cohesion, wounds and mutilations, wings and plumage, would have, either the effect which the poets represent them as having, or some different effect; and in either case, it will be impossible to give any sufficient reason why we should confine the population to the surface of a planet. If gravity have not, upon any set of beings, the effect which it has upon us, such beings may live upon the surface of Saturn, though it be mere vapour: but then, on that supposition, they may equally well live in the vast space between Saturn and Jupiter, without needing any planet for their mansion. If we are ready to suppose that there are, in the solar system, conscious beings, not subject to the ordinary laws of life, we may go on to imagine creatures constituted of vaporous elements, floating in the fiery haze of a nebula, or close to the body of a sun; and cloudy forms which soar as vapours in the regions of vapour. But such imaginations, besides being rather fitted for the employment of poets than of philosophers, will not, as we

|271

have said, find a population for the planets; since such forms may just as easily be conceived swimming round the sun in empty space, or darting from star to star, as confining themselves to the neighbourhood of any of the solid globes which revolve about the central sun.

18 We should not, then, add anything to the probability of inhabitants on the other planets of our system, even if we were arbitrarily to assume unlimited changes in the laws of nature, when we pass from our region to theirs. But probably, all readers will be of opinion that such assumptions are contrary to the whole scheme and spirit of such speculations as we are here presuming:—that if we speculate on such subjects at all, it must be done by supposing that the same laws of nature operate in the same manner, in planetary, as in terrestrial spaces;—and that as we suppose, and prove, gravity and attraction, inertia and momentum, to follow the same rules, and produce the same effects, on brute matter there, which they do here; so, both these forces, and others, as light and heat, moisture and air, if, in the planets, they go beyond the extremes which limit them here, yet must imply, in any organized beings which exist in the planets, changes, though greater in amount, of the same kind as those which occur in approaching the terrestrial extremes of those elementary agents. And what kind of a population that would lead us to suppose in Jupiter or Saturn, Mars or Venus, the reader has already seen our attempt to determine; and may thence judge whether, when we go so far beyond the terrestrial extremes of heat and cold, light and dimness, vapour and water, air and airlessness, any population at all is probable.

19 Perhaps some persons, even if they cannot resist the force of these reasons, may still yield to them with regret; and may feel as if, having hitherto believed that the planets were inhabited, and having now to give up that

272

belief, their view of the solar system, as one of the provinces of God's creation, were made narrower and poorer than it was before. And this feeling may be still further increased, if they are led to believe also that many of the fixed stars are not the centers of inhabited systems; or that very few, or none are. It may seem to them, as if, by such a change of belief, the field of God's greatness, benevolence, and government, were narrowed and impoverished, to an extent painful and shocking;—as if, instead of being the Maker and Governor of innumerable worlds, of the most varied constitution, we were called upon to regard him as merely the Master of the single world in which we live:—as if, instead of being the object of reverence and adoration to the intelligent population of these thousand spheres, he was recognized and worshipped on one only, and on that, how scantily and imperfectly!

20 It is not to be denied that there may be such a regret and disturbance naturally felt, at having to give up our belief that the planets and the stars probably contain servants and worshippers of God. It must always be a matter of pain and trouble, to be urged with tenderness, and to be performed in time, to untwine our reverential religious sentiments from erroneous views of the constitution of the universe with which they have been involved. But the change once made, it is found that religion is uninjured, and reverence undiminished. And therefore we trust that the reader will receive with candour and patience the argument which we have to offer with reference to this view, or rather, this sentiment.

21 We remark, in the first place, that however repugnant it may be to us to believe a state of any part of the universe in which there are not creatures who can know, obey and worship God; we are compelled, by geological evidence, to admit that such a state of things has existed upon the

earth, during a far longer period than the whole duration of man's race. If we suppose that the human race, if not by their actual knowledge, obedience, and worship of God, yet at least by their faculties for knowing, obeying, and worshipping, are a sufficient reason why there should be such a province in God's empire; still in fact, this race has existed only for a few thousand years, out of the, perhaps, millions of years of the earth's existence; and during all the previous period, the earth, if tenanted, was tenanted by brute creatures, fishes and lizards, beasts and birds, of which none had any faculty, intellectual, moral, or religious. By the same analogy, therefore, on which we have already insisted, we may argue that there is reason to believe, that if other planets, and other stars, are the seats of habitation, it is rather of such habitation as has prevailed upon the earth during the millions, than during the six thousand years; and that if we have, in consequence of physical reasons, to give up the belief of a population in the other planets, or in the stars; we are giving up, not anything on which we might dwell with religious pleasure—hosts of fellow-servants and fellow-worshippers of the Divine Author of all:—but the mere brute tribes, of the land and of the water, things that creep and crawl, prowl and spring;—none that can lift its visage to the sky, with a feeling that it is looking for its Maker and Master. There have not existed upon the Earth, during the immense ages of its præhuman existence, beings who could recognize and think of the Creator of the world: and if astronomy introduces us, as geology has done, to a new order of material structures, thus barren of an intelligent and religious population, we must learn to accept the prospect, in the one case, as in the other. Nor need we fear that on a further contemplation of the universe, we shall find every part of it ministering, though perhaps not in the way our first thoughts had guessed, to senti-

274|

ments of reverence and adoration towards the Maker of the universe.

22 The truth is, as the slightest recollection of the course of opinion about the stars may satisfy us, that men have had repeatedly to give up the notions which they had adopted, of the manner in which the material heavens, the stars and the skies, are to minister to man's feeling of reverence for the Creator. It was long ago said, that the heavens declare the glory of God, and the firmament sheweth his handiwork: that day and night, sun and moon, clouds and stars, unite in impressing upon us this sentiment. And this language still finds a sympathetic echo, in the breasts of all religious persons. Nor will it ever cease to do so, however our opinions of the structure and nature of the heavenly bodies may alter. When the new aspects of things become familiar, they will show us the handiwork of God, and declare his glory, as plainly as the old ones. But in the progress of opinions, man has often had to resign what seemed to him, at the time, visions so beautiful, sublime, and glorious, that they could not be dismissed without regret. The Universal Lord was at one time conceived as directing the motions of all the spheres by means of Ruling Angels, appointed to preside over each. The prevalence of proportion and number, in the dimensions of these spheres, was assumed to point to the existence of harmonious sounds, accompanying their movements, though unheard by man; as proportion and number had been found to be the accompaniments and conditions of harmony upon earth. The time came, when these opinions were no longer consistent with man's knowledge of the heavenly motions, and of the wide-spreading causes by which they are produced. Then 'Ruling Angels from their spheres were hurled,' as a matter of belief; though still the poets loved to refer to imagery in which so many lofty and reverent thoughts had so long been clothed. |275

The aspect of the stars was most naturally turned to a lesson of cheerful and thoughtful piety, by the adoption of such a view of their nature and office; and thus, the midnight contemplator of an Italian sky teaches his companion concerning the starry host;

> Sit, Jessica; look how the floor of heav'n
> Is thick inlaid with patterns of bright gold.
> There's not the meanest orb, which thou behold'st,
> But in his motion like an angel sings,
> Still quiring to the young-eyed cherubims;
> Such harmony is in immortal souls.

Meaning, apparently, the harmony between the immortal spirits that govern each star, and the cherubims that sing before the throne of God. But however beautiful and sublime may be this representation, the philosopher has had to abandon it in its literal sense. He may have adopted, instead, the opinion that each of the stars is the seat, or the center of a group of seats, of choirs of worshippers: but this again, is still to suppose the nature of those orbs to be entirely different from that of this earth; though in many respects, we know that they are governed by the same laws. And if he will be content to know no more than he has the means of knowing, or even to know only according to his best means of knowing, he must be prepared, if the force of proof so requires, to give up this belief also; at least for the present.

23 Indeed those who have not been content with this, and have sought to combine with the visible splendour of the skies, some scheme, founded upon astronomical views, which shall people them with intelligent beings and worshippers, have drawn upon their fancy quite as much as Lorenzo in his lesson to Jessica: or rather, they have done what he and those from whom his lore was derived, had done before. They have taken the truths which astronomers have discovered and taught, and made the objects and regions so revealed,

276|

the scenes and occasions of such sentiments of piety as they themselves have, or feel that they ought to have. Even in Shakespeare, the stars are already *orbs*, each orb has his *motion*, and in his motion produces the music of the spheres. More recent preachers, following sounder views of the nature of these orbs and motions, have been equally poetical when they come to their religious reflections. When the poet of the *Night Thoughts* says,

> "Each of these stars is a religious house;
> I saw their altars smoke, their incense rise,
> And heard hosannas ring through every sphere."

he is no less imaginative than the poet of that *Midsummer Night's Dream*, which we have in the *Merchant of Venice*. And we are compelled, by all the evidence which we can discern, to say the same of the preacher who speaks, from the pulpit, of these orbs or worlds, and tells us of the stars which 'give animation to other systems*;' when he says† 'worlds roll in these distant regions; and these worlds must be the centers of life and intelligence;' when he speaks of the earth‡ as 'the humblest of the provinces of God's empire.' But then we must recollect that these thoughts still prove the religious nature of man; they show how he is impelled to endeavour to elevate his mind to God by every part of the universe; and it is not too much to say, that through the faculties of man, thus regarding the starry heavens, every star does really testify to the greatness of God, and minister to His worship.

24 We may trust that this mere material magnificence does not require inhabitants, to make it lift man's heart towards the Universal Creator, and to make him accept it as a sublime evidence of His greatness. The grandest objects in nature are blank and void of life;—the mountain-peaks that stand, ridge beyond ridge, serene in the region of perpetual

* Chalmers, p. 35. † Ibid. p. 21. ‡ Ibid. p. 119.

snow;—the summer-clouds, images of such mountain-tracts, even upon a grander scale, and tinted with more gorgeous colours;—the thunder-cloud with its dazzling bolt;—the stormy ocean with its mountainous waves;—the Aurora Borealis, with its mysterious pillars of fire;—all these are sublime; all these elevate the soul, and make it acknowledge a mighty Worker in the elements, in spite of any teaching of a material philosophy. And if we have to regard the planets as merely parts of the same great spectacle of nature, we shall not the less regard them with an admiration which ministers to pious awe. Even merely as a spectacle, Saturn, made visible in his real shape, only by a vast exertion of human skill, yet shining like a star, in form so curiously complex, symmetrical and seemingly artificial, will never cease to be an object of the ardent and contemplative gaze of all who catch a sight of him. And however much the philosopher may teach that he is merely a mass of water and vapour, ice and snow, he must be far more interesting to the eye than the Alps, or the clouds that crown them, or the ocean with its icebergs; where the same elements occur in forms comparatively shapeless and lawless, irregular and chaotic.

25 But perhaps there is in the minds of many persons, a sentiment connected with this regular and symmetrical form of the heavenly bodies; that being thus beautifully formed and finished, they must have been the objects of especial care to the Creator. These regular globes, these nearly circular orbits, these families of satellites, they too so regular in their movements; this ring of Saturn; all the adjustments by which the planetary motions are secured from going wrong, as the profoundest researches into the mechanics of the universe show;—all these things seem to indicate a peculiar attention bestowed by the Maker on each part of the machine. So much of law and order, of symmetry and

278|

beauty in every part, implies, it may be thought, that every part has been framed with a view to some use;—that its symmetry and its beauty are the marks of some noble purpose.

26 To reply to this argument, so far as it is requisite for us to do so, we must recur to what we have already said; that though we see, in many parts of the universe, inorganic as well as organic, marks which we cannot mistake, of design and purpose; yet that this design and purpose are often effected by laws which are of a much wider sweep than the design, so far as we can trace its bearing. These laws, besides answering the purpose, produce many other effects, in which we can see no purpose. We have now to observe further, that these laws, thus ranging widely through the universe, and working everywhere, as if the Creator delighted in the generality of the law, independently of its special application, do often produce innumerable results of beauty and symmetry, as if the Creator delighted in beauty and symmetry, independently of the purpose answered.

27 Thus, to exemplify this reflexion: the powers of aggregation and cohesion, which hold together the parts of solid bodies, as metals and stones, salts and ice,—which solidify matter, in short,—we can easily see, to be necessary, in order to the formation and preservation of solid terrestrial bodies. They are requisite, in order that man may have the firm earth to stand upon, and firm materials to use. But let us observe, what a wonderful and beautiful variety of phenomena grows out of this law, with no apparent bearing upon that which seems to us its main purpose. The power of aggregation of solid bodies is, in fact, the force of crystallization. It binds together the particles of bodies by molecular forces, which not only hold the particles together, but are exerted in special directions, which form triangles, squares, hexagons, and the like. And hence we have all the variety of crystalline forms which sparkle in gems, ores, |279

earths, pyrites, blendes; and which, when examined by the crystallographers, are found to be an inexhaustible field of the play of symmetrical complexity. The diamond, the emerald, the topaz, have got each its peculiar kind of symmetry. Gold and other metals have, for the basis of their forms, the cube, but run from this into a vastly greater variety of regular solids than ever geometer dreamt of. Some single species of minerals, as calc-spar, present hundreds of forms, all rigorously regular, and have been alone the subject of volumes. Ice crystallizes by the same laws as other solid bodies; and our Arctic voyagers have sometimes relieved the weariness of their sojourn in those regions, by collecting some of the innumerable forms, resembling an endless collection of hexagonal flowers, sporting into different shapes, which are assumed by flakes of snow*. In these and many other ways, the power of crystallization produces an inexhaustible supply of examples of symmetrical beauty. And what are we to conceive to be the object and purpose of this? As we have said, that part of the purpose which is intelligible to us is, that we have here a force holding together the particles of bodies, so as to make them solid. But all these pretty shapes add nothing to this intelligible use. Why then are they there? They are there, it would seem, for their own sake;—because they are pretty;—symmetry and beauty are there on their own account; or because they are universal adjuncts of the general laws by which the Creator works. Or rather, we may say, combining different branches of our knowledge, that crystallization is the mark and accompaniment of chemical composition: and that as chemical composition takes place according to definite numbers, so crystalline aggregation takes place according to definite forms. The

* Dr Scoresby, in his *Account of the Arctic Regions* (1820) Vol. II., has given figures of 96 such forms, selected for their eminent regularity from many more.

symmetrical relations of space in crystals correspond to the simple relations of number in synthesis; and thus, because there is rule, there is regularity, and regularity assumes the form of beauty.

28 This, which thus shows itself throughout the mineral kingdom, or, speaking more widely and truly, throughout the whole range of chemical composition, is still more manifest in the vegetable domain. All the vast array of flowers, so infinitely various, and so beautiful in their variety, are the results of a few general laws; and show, in the degree of their symmetry, the alternate operation of one law and another. The rose, the lily, the cowslip, the violet, differ in something of the same way, in which the crystalline forms of the several gems differ. Their parts are arranged in fives or in threes, in pentagons or in hexagons, and in these regular forms, one part or another is expanded or contracted, rendered conspicuous by colour or by shape, so as to produce all the multiplicity of beauty which the florist admires. Or rather, in the eye of the philosophical botanist, the whole of the structure of plants, with all their array of stems and leaves, blossoms and fruits, is but the manifestation of one Law; and all these members of the vegetable form, are, in their natures, the same, developed more or less in this way or in that. The daisy consists of a close cluster of flowers of which each has, in its form, the rudiments of the valerian. The peablossom is a rose, with some of its petals expanded into butterfly-like wings. Even without changing the species, this general law leads to endless changes. The garden-rose is the common hedge-rose with innumerable filaments changed into glowing petals. By the addition of whorl to whorl, of vegetable coronet over coronet, green and coloured, broad and narrow, filmy and rigid, every plant is generated, and the glory of the field and of the garden, of the jungle and of the forest, is brought forth in all its magnificence. Here, then, we have an immeasurable |281

wealth of beauty and regularity, brought to view by the operation of a single Law. And to what use? What purpose do these beauties answer? What is the object for which the lilies of the field are clothed so gaily and gorgeously? Some plants, indeed, are subservient to the use of animals and of man: but how small is the number in which we can trace this, as an intelligible purpose of their existence! And does it not, in fact, better express the impression which the survey of this province of nature suggests to us, to say, that they grow because the Creator willed that they should grow? Their vegetable life was an object of His care and contrivance, as well as animal and human life. And they are beautiful, also because He willed that they should be so:—because He delights in producing beauty:—and, as we have further tried to make it appear, because He acts by general law, and law produces beauty. Is not such a tendency here apparent, as a part of the general scheme of Creation?

29 We have already attempted to show, that in the structure of animals, especially that large class best known to us, vertebrate animals, there is also a general plan which, so far as we can see, goes beyond the circuit of the special adaptation of each animal to its mode of living: and is a rule of creative action, in addition to the rule that the parts shall be subservient to an intelligible purpose of animal life. We have noticed several phenomena in the animal kingdom, where parts and features appear, rudimentary and inert, discharging no office in their economy, and speaking to us, not of purpose, but of law:—consistent with an end which is visible, but seemingly the results of a rule whose end is in itself.

30 And do we not, in innumerable cases, see beauties of colour and form, texture and lustre, which suggest to us irresistibly the belief that beauty and regular form are rules of the Creative agency, even when they seem to us, looking

282|

at the creation for uses only, idle and wanton expenditure of beauty and regularity. To what purpose are the host of splendid circles which decorate the tail of the peacock, more beautiful, each of them, than Saturn with his rings? To what purpose the exquisite textures of microscopic objects, more curiously regular than anything which the telescope discloses? To what purpose the gorgeous colours of tropical birds and insects, that live and die where human eye never approaches to admire them? To what purpose the thousands of species of butterflies with the gay and varied embroidery of their microscopic plumage, of which one in millions, if seen at all, only draws the admiration of the wandering schoolboy? To what purpose the delicate and brilliant markings of shells, which live, generation after generation, in the sunless and sightless depths of the ocean? Do not all these examples, to which we might add countless others, (for the world, so far as human eye has scanned it, is full of them,) prove that beauty and regularity are universal features of the work of Creation, in all its parts, small and great: and that we judge in a way contrary to a vast range of analogy, which runs through the whole of the Universe, when we infer that, because the objects which are presented to our contemplation are beautiful in aspect and regular in form, they must, in each case, be means for some special end, of those which we commonly fix upon, as the main ends of the Creation, the support and advantage of animals or of man?

31 If this be so, then the beautiful and regular objects which the telescope reveals to us; Jupiter and his Moons, Saturn and his Rings, the most regular of the Double Stars, Clusters and Nebulæ; cannot reasonably be inferred, because they are beautiful and regular, to be also fields of life, or scenes of thought. They may be, as to the poet's eye they often appear, the gems of the robe of Night, the flowers of the celestial fields. Like gems and like flowers, they are

|283

beautiful and regular, because they are brought into being by vast and general laws. These laws, although, in the mind of the Creator, they have their sufficient reason, as far as they extend, may have, in no other region than that which we inhabit, the reason which we seek to discover everywhere, the sustentation of a life like ours. That we should connect with the existence of such laws, the existence of Mind like our own mind, is most natural; and, as we might easily show, is justifiable, reasonable, even necessary. But that we should suppose the results of such laws are so connected with Mind, that wherever the laws gather matter into globes, and whirl it round the central body, *there* is also a local seat of minds like ours; is an assumption altogether unwarranted; and is, without strong evidence, of which we have as yet no particle, quite visionary.

32 But finally, it may be said that by this our view of the universe, we diminish the greatness of the work of creation, and the majesty of the Creator. Such a view appears to represent the other planets as mere fragments, which have flown off in the fabrication of this our earth, and of the mechanism by which it answers its purpose. Instead of a vast array of completed worlds, we have one world, surrounded by abortive worlds and inert masses. Instead of perfection everywhere, we have imperfection everywhere, except at one spot; if even there the workmanship be perfect.

33 To this, the reply is contained in what we have already said: but we may add, that it cannot be wise or right, to prop up our notions of God's greatness, by physical doctrines which will not bear discussion. God's greatness has no need of man's inventions for its support. The very conviction that the Creation must be such as to confirm our belief in the greatness of God, shows that such a belief is more deeply seated than any special views of the structure of the universe, and will triumphantly survive the removal of

284|

error in such views. We may add, that till within a few thousand years, this earth, compared with what it now is, having upon it no intelligent beings, might be regarded as an abortive world; that all the parts of the solar system which we can best scrutinize, the moon, and meteoric stones, are inert masses; and further, that there is everywhere the perfection which results from the operation of law, and that *that* seems to be the perfection with which the Creator is contented.

34 And perhaps, when the view of the universe which we here present has become familiar, we may be led to think that the aspect which it gives to the mode of working of the Creator, is sufficiently grand and majestic. Instead of manufacturing a multitude of worlds on patterns more or less similar, He has been employed in one great work, which we cannot call imperfect, since it includes and suggests all that we can conceive of perfection. It may be that all the other bodies, which we can discover in the universe, show the greatness of this work, and are rolled into forms of symmetry and order, into masses of light and splendour, by the vast whirl which the original creative energy imparted to the luminous element. The planets and the stars are the lumps which have flown from the potter's wheel of the Great Worker;—the shred-coils which, in the working, sprang from His mighty lathe:—the sparks which darted from His awful anvil when the solar system lay incandescent thereon;—the curls of vapour which rose from the great cauldron of creation when its elements were separated. If even these superfluous portions of the material are marked with universal traces of regularity and order, this shows that universal rules are his implements, and that Order is the first and universal Law of the heavenly work.

35 And, that we may see the full dignity of this work, we must always recollect that Man is a part of it, |285

and the crowning part. The workmanship which is employed on mere matter is, after all, of small account, in the eyes of intellectual and moral creatures, when compared with the creation and government of intellectual and moral creatures. The majesty of God does not reside in planets and stars, in orbs and systems; which are, after all, only stone and vapour, materials and means. If, as we believe, God has not only made the material world, but has made and governs man, we need not regret to have to depress any portion of the material world below the place which we had previously assigned to it: for, when all is done, the material world *must* be put in an inferior place, compared with the world of mind. If there be a World of Mind, *that*, according to all that we can conceive, must have been better worth creating, must be more worthy to exist, as an object of care in the eyes of the Creator, than thousands and millions of stars and planets, even if they were occupied by a myriad times as many species of brute animals as have lived upon the earth since its vivification. In saying this, we are only echoing the common voice of mankind, uttered, as so often it is, by the tongues of poets. One such speaks thus of stellar systems:

> Behold this midnight splendour, worlds on worlds;
> Ten thousand add and twice ten thousand more,
> Then weigh the whole: one soul outweighs them all,
> And calls the seeming vast magnificence
> Of unintelligent creation, poor.

And as this is true of intelligence, with the suggestion, which that faculty so naturally offers, of the inextinguishable nature of mind, so is it true of the moral nature of man. No accumulation of material grandeur, even if it fill the universe, has any dignity in our eyes, compared with moral grandeur: as poetry has also expressed:

> Look then abroad through nature, to the range
> Of planets, suns, and adamantine spheres,
> Wheeling unshaken through the void immense,
> And speak, O man! Can this capacious scene

With half that kindling majesty exalt
Thy strong conception, as when Brutus rose
Refulgent from the stroke of Cesar's fate
Amid the band of patriots; and his arm
Aloft extending, like eternal Jove
When guilt calls down the thunder, call'd aloud
On Tully's name, and shook his crimson steel,
And bade the Father of his Country, Hail!
For lo! the tyrant prostrate in the dust,
And Rome again is free.

This action being taken, as it is here meant to be conceived, for one of the highest examples of moral greatness. And however we may judge of this action, we must allow that the characters which are implied in this praise of it,—the loftiest kinds of moral excellence,—are more suitable to the highest idea of the object and purpose of a Deity creating worlds, than would be any mere material structure of planets and suns, whether kept in their places by adamantine spheres, wheeling unshaken through the void immense, or themselves wheeling unshaken by the power of a universal law. The thoughts of Rights and Obligations, Duty and Virtue, of Law and Liberty, of Country and Constitution, of the Glory of our Ancestors, the Elevation of our Fellow-Citizens, the Freedom and Happiness and Dignity of Posterity,—are thoughts which belong to a world, a race, a body of beings, of which any one individual, with the capacities which such thoughts imply, is more worthy of account, than millions of millions of mollusks and belemnites, lizards and fishes, sloths and pachyderms, diffused through myriads of worlds.

36 We might illustrate this argument further, by taking actions of the moral character of which there will be less doubt. If we look at the great acts which render Greece illustrious and interesting in our eyes,—such as the death of Socrates, for instance, the triumph of a reverence for Law

and a love of country;—can we think it any real diminution of the glory of the universe, if we are reduced to the necessity of rejecting the belief in a multitude of worlds, which though, it may be, peopled with lower animals, contain none endowed with any higher principle than hunger and thirst?

37 That the human race possesses a worth in the eyes of Reason beyond that which any material structure, or any brute population can possess, might be maintained on still higher and stronger grounds; namely, on religious grounds: but we do not intend here to dwell on that part of the subject. If man be, not merely (and he alone of all animals) capable of Virtue and Duty, of Universal Love and Self-Devotion, but be also immortal; if his being be of infinite duration, his soul created never to die; then, indeed, we may well say that one soul outweighs the whole unintelligent creation. And if the Earth have been the scene of an action of Love and Self-Devotion for the incalculable benefit of the whole human race, in comparison with which the death of Socrates fades into a mere act of cheerful resignation to the common lot of humanity; and if this action, and its consequences to the whole race of man, in his temporal and eternal destiny, and in his history on earth before and after it, were the main object for which man was created, the cardinal point round which the capacities and the fortunes of the race were to turn; then indeed we see that the Earth has a pre-eminence in the scheme of creation, which may well reconcile us to regard all the material splendour which surrounds it, all the array of mere visible luminaries and masses which accompany it, as no unfitting appendages to such a drama. The elevation of millions of intellectual, moral, religious, spiritual creatures, to a destiny so prepared, consummated, and developed, is no unworthy occupation of all the capacities of

space, time, and matter. And, so far as any one has yet shown, to regard this great scheme as other than the central point of the divine plan; to consider it as one part among other parts, similar, co-ordinate, or superior; involves those who so speculate, in difficulties, even with regard to the plan itself, which they strive in vain to reconcile; while the assumption of the subjects of such a plan, in other regions of the universe, is at variance with all which we, looking at the analogies of space and time, of earth and stars, of life in brutes and in man, have found reason to deem in any degree probable.

38 And thus that conjecture of the Plurality of Worlds, to which a wide and careful examination of the physical constitution of the Universe supplied no confirmation, derives also little support from a contemplation of the Design which the Creator may be supposed to have had in the work of the Creation; when such Design is regarded in a comprehensive manner, and in all its bearings. Such a survey seems to speak rather in favour of the Unity of the World, than of a Plurality of Worlds. A further consideration of the intellectual, moral, and religious nature of man may still further illustrate this view; and with that object, we shall make a few additional remarks.

CHAPTER XII.

The Unity of the World.

1 THE two doctrines which we have here to weigh against each other are the Plurality of Worlds, and the Unity of the World. In so saying, we include, as in our present view, a necessary part of the conception of a *World*, a collection of intelligent creatures: for even if the suppositions to which we have been led, respecting the kind of unintelligent living things which may inhabit other parts of the Universe, be conceived to be probable; such a belief will have little interest for most persons, compared with the belief of other worlds, where reside intelligence, perception of truth, recognition of moral Law, and reverence for a Divine Creator and Governor. In looking outwards at the Universe, there are certain aspects which suggest to man, at first sight, a conjecture that there may be other bodies like the Earth, tenanted by other creatures like man. This conjecture, however, receives no confirmation from a closer enquiry, with increased means of observation. Let us now look inwards, at the constitution of man; and consider some characters of his nature, which seem to remove or lessen the difficulties which we may at first feel, in regarding the Earth as, in a unique and special manner, the field of God's Providence and Government.

2 In the first place, the Earth, as the abode of man, the intellectual creature, contains a being, whose mind is, in some measure, of the same nature as the Divine Mind of the Creator. The Laws which man discovers in the Creation must be Laws known to God. The truths,—for

instance, the truths of geometry,—which man sees to be true, God also must see to be true. That there were, from the beginning, in the Creative Mind, Creative Thoughts, is a doctrine involved in every intelligent view of Creation.

3 This doctrine was presented by the ancients in various forms: and the most recent scientific discoveries have supplied new illustrations of it. The mode in which Plato expressed the doctrine which we are here urging was, that there were in the Divine Mind, before or during the work of creation, certain archetypal Ideas, certain exemplars or patterns of the world and its parts, according to which the work was performed: so that these Ideas or Exemplars existed in the objects around us, being in many cases discernible by man, and being the proper objects of human reason. If a mere metaphysician were to attempt to revive this mode of expressing the doctrine, probably his speculations would be disregarded, or treated as a pedantic resuscitation of obsolete Platonic dreams. But the adoption of such language must needs be received in a very different manner, when it proceeds from a great discoverer in the field of natural knowledge: when it is, as it were, forced upon *him*, as the obvious and appropriate expression of the result of the most profound and comprehensive researches into the frame of the whole animal creation. The recent works of Mr Owen, and especially one work, *On the Nature of Limbs*, are full of the most energetic and striking passages, inculcating the doctrine which we have been endeavouring to maintain. We may take the liberty of enriching our pages with one passage bearing upon the present part of the subject.

'If the world were made by any antecedent Mind or Understanding, that is by a Deity, then there must needs be an Idea and Exemplar of the whole world before it was made, and consequently actual knowledge, both in the order of Time and Nature, before Things. But conceiving of knowledge as it was got by their own finite minds, and ignorant of any evidence of an ideal Archetype for the world or any part of it, they [the Democritic Philosophers who denied a Divine Creative Mind] affirmed that there was none, and concluded that there could be no knowledge or mind before the world was, as its cause.' Plato's assertion of Archetypal Ideas was a protest against this doctrine, but was rather a guess, suggested by the nature of mathematical demonstration, than a doctrine derived from a contemplation of the external world.

'Now however,' Mr Owen continues, 'the recognition of an ideal exemplar for the vertebrated animals proves that the knowledge of such a being as Man must have existed before Man appeared. For the Divine Mind which planned the Archetype also foreknew all its modifications. The Archetypal Idea was manifested in the flesh under divers modifications upon this planet, long prior to the existence of those animal species which actually exemplify it. To what natural or secondary causes the orderly succession and progression of such organic phenomena may have been committed, we are as yet ignorant. But if without derogation to the Divine Power, we may conceive such ministers and personify them by the term *Nature*, we learn from the past history of our globe that she has advanced with slow and stately steps, guided by the archetypal light amidst the wreck of worlds, from the first embodiment of the vertebrate idea, under its old ichthyic vestment, until it became arrayed in the glorious garb of the human form.'

4 Law implies a Lawgiver, even when we do not see the object of the Law; even as Design implies a Designer, when we do not see the object of the Design. The Laws of Nature are the indications of the operation of the Divine Mind; and are revealed to us, as such, by the operations of our minds, by which we come to discover them.

292|

They are the utterances of the Creator, delivered in language which we can understand; and being thus Language, they are the utterances of an Intelligent Spirit.

5 It may seem to some persons too bold a view, to identify, so far as we thus do, certain truths as seen by man, and as seen by God*:—to make the Divine Mind thus cognizant of the truths of geometry, for instance. If any one has such a scruple, we may remark that truth, when of so luminous and stable a kind as are the truths of geometry, must be alike *Truth* for all minds, even for the highest. The mode of arriving at the knowledge of such truths, may be very different, even for different human minds;—deduction for some;—intuition for others. But the intuitive apprehension of necessary truth is an act so purely intellectual, that even in the Supreme Intellect, we may suppose that it has its place. Can we conceive otherwise, than that God does contemplate the universe as existing in space, since it really does so;—and subject to the relations of space, since these are as real as space itself? We are well aware that the Supreme Being must contemplate the world under many other aspects than this;—even man does so. But that does not prevent the truths, which belong to the aspect of the world, contemplated as existing in space, from being truths, regarded as such, even by the Divine Mind.

6 If these reflections are well founded, as we trust they will, on consideration, be seen to be, we may adopt many of the expressions by which philosophers heretofore have attempted to convey similar views; for in fact, this view, in its general bearing at least, is by no means new. The Mind of Man is a partaker of the thoughts of the Divine

* Among the most recent expositors of this doctrine we may place M. Henri Martin, whose *Philosophie Spiritualiste de la Nature* is full of striking views of the universe in its relation to God. (Paris. 1849.)

Mind. The Intellect of Man is a spark of the Light by which the world was created. The Ideas according to which man builds up his knowledge, are emanations of the archetypal Ideas according to which the work of creation was planned and executed. These, and many the like expressions, have been often used: and we now see, we may trust, that there is a great philosophical truth, which they all tend to convey; and this truth shows at the same time, how man may have some knowledge respecting the Laws of Nature, and how this knowledge may, in some cases, seem to be a knowledge of necessary relations, as in the case of space *.

7 Now the views to which we have been led, bear very strongly upon that argument. For if man, when he attains to a knowledge of such Laws, is really admitted, in some degree, to the view with which the Creator himself beholds his creation;—if we can gather, from the conditions of such knowledge, that his intellect partakes of the Nature

* Most readers who have given any attention to speculations of this kind will recollect Newton's remarkable expressions concerning the Deity: 'Æternus est et infinitus, omnipotens et omnisciens; id est, durat ab æterno in æternum, et adest ab infinito in infinitum . . . Non est æternitas et infinitas, sed æternus et infinitus; non est duratio et spatium, sed durat et adest. Durat semper et adest ubique, et existendo semper et ubique durationem et spatium constituit.'

To say that God by existing always and everywhere *constitutes duration and space,* appears to be a form of expression better avoided. Besides that it approaches too near to the opinion, which the writer rejects, that He *is* duration and space, it assumes a knowledge of the nature of the Divine existence, beyond our means of knowing, and therefore rashly. It appears to be safer, and more in conformity with what we really know, to say, not that the existence of God constitutes time and space; but that God has constituted *man,* so that *he* can apprehend the works of creation, only as existing in time and space. That God has constituted time and space as conditions of man's knowledge of the creation, is certain: that God has constituted time and space as results of His own existence in any other way, *we* cannot know.

294|

of the Supreme Intellect;—if his Mind, in its clearest and largest contemplations, harmonizes with the Divine Mind;—we have, in this, a reason which may well seem to us very powerful, why, even if the Earth alone be the habitation of intelligent beings, still, the great work of Creation is not wasted. If God have placed upon the earth a creature who can so far sympathize with Him, if we may venture upon the expression;—who can raise his intellect into some accordance with the Creative Intellect; and that, not once only, nor by few steps, but through an indefinite gradation of discoveries, more and more comprehensive, more and more profound; each, an advance, however slight, towards a Divine Insight;—then, so far as intellect alone (and we are here speaking of intellect alone) can make Man a worthy object of all the vast magnificence of Creative Power, we can hardly shrink from believing that he is so.

8 We may remark further, that this view of God, as the Author of the Laws of the Universe, leads to a view of all the phenomena and objects of the world, as the work of God; not a work made, and laid out of hand, but a field of his present activity and energy. And such a view cannot fail to give an aspect of dignity to all that is great in creation, and of beauty to all that is symmetrical, which otherwise they could not have. Accordingly, it is by calling to their thoughts the presence of God as suggested by scenes of grandeur or splendour, that poets often reach the sympathies of their readers. And this dignity and sublimity appear especially to belong to the larger objects, which are destitute of conscious life; as the mountain, the glacier, the pine-forest, the ocean; since in these, we are, as it were, alone with God, and the only present witnesses of His mysterious working.

9 Now if this reflection be true, the vast bodies which hang in the sky, at such immense distances from us, and roll on their courses, and spin round their axles with such exceed-

ing rapidity; Jupiter and his array of Moons, Saturn with his still larger host of Satellites, and with his wonderful Ring, and the other large and distant Planets, will lose nothing of their majesty, in our eyes, by being uninhabited; any more than the summer-clouds, which perhaps are formed of the same materials, lose their dignity from the same cause;—any more than our Moon, one of the tribe of satellites, loses her soft and tender beauty, when we have ascertained that she is more barren of inhabitants than the top of Mount Blanc. However destitute the planets and moons and rings may be of inhabitants, they are at least vast scenes of God's presence, and of the activity with which he carries into effect, everywhere, the laws of nature. The light which comes to us from them is transmitted according to laws which He has established, by an energy which He maintains. The remotest planet is not devoid of life, for God lives there. At each stage which we make, from planet to planet, from star to star, into the regions of infinity, we may say, with the patriarch, 'Surely God is here, and I knew it not.' And when those who question the habitability of the remote planets and stars are reproached as presenting a view of the universe, which takes something from the magnificence hitherto ascribed to it, as the scene of God's glory, shown in the things which He has created; they may reply, that they do not at all disturb that glory of the creation which arises from its being, not only the product, but the constant field of God's activity and thought, wisdom and power; and they may perhaps ask, in return, whether the dignity of the Moon would be greatly augmented if her surface were ascertained to be abundantly peopled with lizards; or whether Mount Blanc would be more sublime, if millions of frogs were known to live in the crevasses of its glaciers.

10 Again: the Earth is a scene of Moral Trial. Man 296| is subject to a Moral Law: and this Moral Law is a

Law of which God is the Legislator. It is a Law which man has the power of discovering, by the use of the faculties which God has given him. By considering the nature and consequences of actions, man is able to discern, in a great measure, what is right and what is wrong;—what he ought and what he ought not to do;—what is duty and virtue, what is crime and vice. Man has a Law on such subjects, written on his heart, as the Apostle Paul says. He has a conscience which accuses or excuses him; and thus, recognizes his acts as worthy of condemnation or approval. And thus, man is, and knows himself to be, the subject of Divine Law, commanding and prohibiting; and is here, in a state of probation, as to how far he will obey or disobey this Law. He has impulses, springs of action, which urge him to the violation of this Law. Appetite, Desire, Anger, Lust, Greediness, Envy, Malice, impel him to courses which are vicious. But these impulses he is capable of resisting and controlling;—of avoiding the vices and practising the opposite virtues;—and of rising from one stage of Virtue to another, by a gradual and successive purification and elevation of the desires, affections and habits, in a degree, so far as we know, without limit.

11 Now in considering the bearing of this view upon our original subject, we have, in the first place, to make this remark: that the existence of a body of creatures, capable of such a Law, of such a Trial, and of such an Elevation as this, is, according to all that we can conceive, an object infinitely more worthy of the exertion of the Divine Power and Wisdom, in the Creation of the universe, than any number of planets occupied by creatures having no such lot, no such law, no such capacities, and no such responsibilities. However imperfectly the moral law may be obeyed; however ill the greater part of mankind may respond to the appointment which places them here in a state of moral probation; how-

|297

ever few those may be who use the capacities and means of their moral purification and elevation;—still, that there is such a plan in the creation, and that any respond to its appointments,—is really a view of the Universe which we can conceive to be suitable to the nature of God, because we can approve it, in virtue of the moral nature which He has given us. One school of moral discipline, one theatre of moral action, one arena of moral contests for the highest prizes, is a sufficient center for innumerable hosts of stars and planets, globes of fire and earth, water and air, whether or not tenanted by corals and madrepores, fishes and creeping things. So great and majestic are those names of *Right* and *Good, Duty* and *Virtue*, that all mere material or animal existence is worthless in the comparison.

12 But further: let us consider what is this moral progress of which we have spoken;—this purification and elevation of man's inner being. Man's intellectual progress, his advance in the knowledge of the general laws of the Universe, we found reason to believe that we were not describing unfitly, when we spoke of it as bringing us nearer to God;—as making our thoughts, in some degree, resemble His thoughts;—as enabling us to see things as He sees them. And on that account, we held that the placing man, with his intellectual powers, in a condition in which he was impelled, and enabled, to seek such knowledge, was of itself a great thing, and tended much to give to the Creation a worthy end. Now the moral elevation of man's being is the elevation of his sentiments and affections towards a standard or idea, which God, by his Law, has indicated as that point towards which man ought to tend. We do not ascribe *Virtue* to God, adapting to Him our notions taken from man's attributes, as we do when we ascribe Knowledge to God: for Virtue implies the control and direction of 298| human springs of action;—implies human efforts and human

habits. But we ascribe to God infinite Goodness, Justice, and Truth, as well as infinite Wisdom and Power; and Goodness, Justice, Truth, form elements of the character at which man also is, by the Moral Law, directed to aim. So far, therefore, man's moral progress is a progress towards a likeness with God; and such a progress, even more than a progress towards an intellectual likeness with God, may be conceived as making the soul of man fit to endure for ever with God; and therefore, as making this earth a prefatory stage of human souls, to fit them for eternity;—a nursery of plants which are to be fully unfolded in a celestial garden.

13 And to this, we must add that, on other accounts also, as well as on account of the capacity of the human soul for moral and intellectual progress, thoughtful men have always been disposed, on grounds supplied by the light of nature, to believe in the existence of human souls after this present earthly life is past. Such a belief has been cherished in all ages and nations, as the mode in which we naturally conceive that which is apparently imperfect and deficient in the moral government of the world, to be completed and perfected. And if this mortal life be thus really only the commencement of an infinite Divine Plan, beginning upon earth and destined to endure for endless ages after our earthly life; we need no array of other worlds in the universe to give sufficient dignity and majesty to the scheme of the Creation.

14 We may make another remark which may have an important bearing upon our estimate of the value of the moral scheme of the world which occupies the earth. If, by any act of the Divine Government, the number of those men should be much increased, who raise themselves towards the moral standard which God has appointed, and thus, towards a likeness to God, and a prospect of a future eternal union with him;—such an act of Divine Government would do far more towards making the Universe a scene in which God's |299

goodness and greatness were largely displayed, than could be done by any amount of peopling of planets with creatures who were incapable of moral agency; or with creatures whose capacity for the developement of their moral faculties was small, and would continue to be small till such an act of Divine Government were performed. The Interposition of God, in the history of man, to remedy man's feebleness in moral and spiritual tasks, and to enable those who profit by the Interposition, to ascend towards a union with God, is an event entirely out of the range of those natural courses of events which belong to our subject: and to such an Interposition, therefore, we must refer with great reserve; using great caution, that we do not mix up speculations and conjectures of our own, with what has been revealed to man concerning such an Interposition. But this, it would seem, we may say:—that such a Divine Interposition for the moral and spiritual elevation of the human race, and for the encouragement and aid of those who seek the purification and elevation of their nature, and an eternal union with God, is far more suitable to the Idea of a God of Infinite Goodness, Purity, and Greatness, than any supposed multiplication of a population, (on our planet or on any other,) not provided with such means of moral and spiritual progress.

15 And if we were, instead of such a supposition, to imagine to ourselves, in other regions of the Universe, a moral population purified and elevated without the aid or need of any such Divine Interposition; the supposed possibility of such a moral race would make the sin and misery, which deform and sadden the aspect of our earth, appear more dark and dismal still. We should therefore, it would seem, find no theological congruity, and no religious consolation, in the assumption of a Plurality of Worlds of Moral Beings: while, to place the seats of such worlds in the Stars and the Planets, would be, as we have already shown, a step

300|

discountenanced by physical reasons; and discountenanced the more, the more the light of science is thrown upon it.

16 Perhaps it may be said, that all which we have urged to show that other animals, in comparison with man, are less worthy objects of creative design, may be used as an argument to prove that other planets are tenanted by men, or by moral and intellectual creatures like men; since, if the creation of *one* world of such creatures exalts so highly our views of the dignity and importance of the plan of creation, the belief in *many* such worlds must elevate still more our sentiments of admiration and reverence of the greatness and goodness of the Creator; and must be a belief, on that account, to be accepted and cherished by pious minds.

17 To this we reply, that we cannot think ourselves authorized to assert cosmological doctrines, selected arbitrarily by ourselves, on the ground of their exalting our sentiments of admiration and reverence for the Deity, *when the weight of all the evidence which we can obtain respecting the constitution of the universe is against them.* It appears to us, that to discern one great scheme of moral and religious government, which is the spiritual center of the universe, may well suffice for the religious sentiments of men in the present age; as in former ages such a view of creation was sufficient to overwhelm men with feelings of awe, and gratitude, and love; and to make them confess, in the most emphatic language, that all such feelings were an inadequate response to the view of the scheme of Providence which was revealed to them. The thousands of millions of inhabitants of the Earth to whom the effects of the Divine Plan extend, will not seem, to the greater part of religious persons, to need the addition of more, to fill our minds with sufficiently vast and affecting contemplations, so far as we are capable of pursuing such contemplations. The possible extension of God's spiritual kingdom upon the earth will probably appear to them a far

|301

more interesting field of devout meditation, than the possible addition to it of the inhabitants of distant stars, connected in some inscrutable manner with the Divine Plan.

18 To justify our saying that the weight of the evidence is against such cosmological doctrines, we must recall to the reader's recollection the whole course of the argument which we have been pursuing.

It is a possible conjecture, at first, that there may be other Worlds, having, as this has, their moral and intellectual attributes, and their relations to the Creator. It is also a possible conjecture, that this World, having such attributes, and such relations, may, on that account, be necessarily unique and incapable of repetition, peculiar, and spiritually central. These two opposite possibilities may be placed, at first, front to front, as balancing each other. We must then weigh such evidence and such analogies as we can find on the one side or on the other. We see much in the intellectual and moral nature of man, and in his history, to confirm the opinion that the human race is thus unique, peculiar and central. In the views which Religion presents, we find much more, tending the same way, and involving the opposite supposition in great difficulties. We find, in our knowledge of what we ourselves are, reasons to believe that if there be, in any other planet, intellectual and moral beings, they must not only be *like* men, but must *be* men, in all the attributes which we can conceive as belonging to such beings. And yet to suppose other groups of the human species, in other parts of the universe, must be allowed to be a very bold hypothesis, to be justified only by some positive evidence in its favour. When from these views, drawn from the attributes and relations of man, we turn to the evidence drawn from physical conditions, we find very strong reason to believe that, so far as the Solar System is concerned, the Earth *is*, with regard to the conditions of life, in a peculiar

302|

and central position; so that the conditions of any life approaching at all to human life, exist on the Earth alone. As to other systems which may circle other suns, the possibility of their being inhabited by men, remains, as at first, a mere conjecture, without any trace of confirmatory evidence. It was suggested at first by the supposed analogy of other stars to our sun; but this analogy has not been verified in any instance; and has been, we conceive, shown in many cases, to vanish altogether. And that there may be such a plan of creation,—one in which the moral and intelligent race of man is the climax and central point to which innumerable races of mere unintelligent species tend,—we have the most striking evidence, in the history of our own earth, as disclosed by geology. We are left therefore with nothing to cling to, on one side, but the bare possibility that some of the stars are the centers of systems like the Solar System;—an opinion founded upon the single fact, shown to be highly ambiguous, of those stars being self-luminous: and to this possibility, we oppose all the considerations, flowing from moral, historical, and religious views, which represent the human race as unique and peculiar. The force of these considerations will, of course, be different in different minds, according to the importance which each person attaches to such moral, historical, and religious views: but whatever the weight of them may be deemed, it is to be recollected that we have on the other side a bare possibility, a mere conjecture; which, though suggested at first by astronomical discoveries, all more recent astronomical researches have failed to confirm in the smallest degree. In this state of our knowledge, and with such grounds of belief, to dwell upon the Plurality of Worlds of intellectual and moral creatures, as a highly probable doctrine, must, we think, be held to be eminently rash and unphilosophical.

19 On such a subject, where the evidences are so imperfect, and our power of estimating analogies so small, far be it from us to speak positively and dogmatically. And if any one holds the opinion, on whatever evidence, that there are other spheres of the Divine Government than this earth, —other regions in which God has subjects and servants,—other beings who do his will, and who, it may be, are connected with the moral and religious interests of man;—we do not breathe a syllable against such a belief; but, on the contrary, regard it with a ready and respectful sympathy. It is a belief which finds an echo in pious and reverent hearts*; and it is, of itself, an evidence of that religious and spiritual character in man, which is one of the points of our argument. But the discussion of such a belief does not belong to the present occasion, any further than to observe, that it would be very rash and unadvised,—a proceeding unwarranted, we think, by Religion, and certainly at variance with all that Science teaches,—to place those other, extra-human spheres of Divine Government, in the Planets and in the Stars. With regard to the planets and the stars, if we reason at all, we must reason on physical grounds; we must suppose, as to a great extent we can prove, that the laws and properties of terrestrial matter and motion apply to them also. On such grounds, it is as improbable that visitants from Jupiter or from Sirius can come to the Earth, as that men can pass to those stars: as unlikely that inhabitants of those stars know and take an interest in human affairs, as that we can learn what they are doing. A belief in the Divine Government of other races of

* "For doubt not that in other worlds above
There must be other offices of love,
That other tasks and ministries there are,
Since it is promised that His servants, there,
Shall serve Him still."—TRENCH.

spiritual creatures besides the human race, and in Divine Ministrations committed to such beings, cannot be connected with our physical and astronomical views of the nature of the stars and the planets, without making a mixture altogether incongruous and incoherent; a mixture of what is material and what is spiritual, adverse alike to sound religion and to sound philosophy.

20 Perhaps again, it may be said, that in speaking of the shortness of the time during which man has occupied the earth, in comparison with the previous ages of irrational life, and of blank matter, we are taking man at his present period of existence on the earth:—that we do not know that the race may not be destined to continue upon the earth for as many ages as preceded the creation of man. And to this we reply, that in reasoning, as we must do, at the present period, we can only proceed upon that which has happened up to the present period. If we do not know how long man will continue to inhabit the earth, we cannot reason as if we did know that he will inhabit it longer than any other species has done. We may not dwell upon a mere possibility, which, it is assumed, may at some indefinitely future period, alter the aspect of the facts now before us. For it would be as easy to assume possibilities which may come hereafter to alter the aspect of the facts, in favour of the one side, as of the other*. What the future destinies of our race, and of the earth, may be, is a subject which is, for us, shrouded in deep darkness. It would be very rash to assume that they will be such as to alter the impression derived from what we now know, and to alter it in a cer-

* For instance, we may assume that in two or three hundred years, by the improvement of telescopes, or by other means, it may be ascertained that the other planets of the Solar System are not inhabited, and that the other Stars are not the centers of regular systems.

tain preconceived manner. But yet it is natural to form conjectures on this subject; and perhaps we may be allowed to consider for a moment what kind of conjectures the existing state of our knowledge suggests, when we allow ourselves the licence of conjecturing. The next Chapter contains some remarks bearing upon such conjectures.

CHAPTER XIII.

The Future.

---

1 WE proceed then to a few reflections to which we cannot but feel ourselves invited by the views which we have already presented in these pages. What will be the future history of the human race, and what the future destination of each individual, most persons will, and most wisely, judge on far other grounds than the analogies which physical science can supply. Analogies derived from such a quarter can throw little light on those grave and lofty questions. Yet perhaps a few thoughts on this subject, even if they serve only to show how little the light thus attainable really is, may not be an unfit conclusion to what has been said; and the more so, if these analogies of science, so far as they have any specific tendency, tend to confirm some of the convictions, with regard to those weighty and solemn points,—the destiny of Man, and of Mankind,—which we derive from other and higher sources of knowledge.

2 Man is capable of looking back upon the past history of himself, his Race, the Earth, and the Universe. So far as he has the means of doing so, and so far as his reflective powers are unfolded, he cannot refrain from such a retrospect. As we have seen, man has occupied his thoughts with such contemplations, and has been led to convictions thereupon, of the most remarkable and striking kind. Man is also capable of looking forwards to the future probable or possible history of himself, his race, the earth, and the universe. He is irresistibly tempted to do this, and to endeavour to shape his conjectures on the Future, by what he

knows of the Past. He attempts to discern what future change and progress may be imagined or expected, by the analogy of past change and progress, which have been ascertained. Such analogies may be necessarily very vague and loose; but they are the peculiar ground of speculation, with which we have here to deal. Perhaps man cannot discover with certainty any fixed and permanent laws which have regulated those past changes which have modified the surface and population of the earth; still less, any laws which have produced a visible progression in the constitution of the rest of the universe. He cannot, therefore, avail himself of any close analogies, to help him to conjecture the future course of events, on the earth or in the universe; still less can he apply any known laws, which may enable him to predict the future configurations of the elements of the world; as he can predict the future configurations of the planets for indefinite periods. He can foresee the astronomical revolutions of the heavens, so long as the known laws subsist. He cannot foresee the future geological revolutions of the earth, even if they are to be produced by the same causes which have produced the past revolutions, of which he has learnt the series and order. Still less can he foresee the future revolutions which may take place in the condition of man, of society, of philosophy, of religion; still less, again, the course which the Divine Government of the world will take, or the state of things to which, even as now conducted, it will lead.

3 All these subjects are covered with a veil of mystery, which science and philosophy can do little in raising. Yet these are subjects to which the mind turns, with a far more eager curiosity, than that which it feels with regard to mere geological or astronomical revolutions. Man is naturally, and reasonably, the greatest object of interest to man. What shall happen to the human race, after thousands of 308| years, is a far dearer concern to him, than what shall happen

to Jupiter or Sirius: and even, than what shall happen to the continents and oceans of the globe on which he lives, except so far as the changes of his domicile affect himself. If our knowledge of the earth and of the heavens, of animals and of man, of the past condition and present laws of the world, is quite barren of all suggestion of what may or may not hereafter be the lot of man, such knowledge will lose the charm which would have made it most precious and attractive in the eyes of mankind in general. And if, on such subjects, any conjectures, however dubious,—any analogies, however loose,—can be collected from what we know, they will probably be received as acceptable, in spite of their insecurity; and will be deemed a fit offering from the scientific faculty, to those hopes and expectations,—to that curiosity and desire of all knowledge,—which gladly receive their nutriment and gratification from every province of man's being.

4 Now if we ask, what is likely to be the future condition of the population of the earth as compared with the present; we are naturally led to recollect, what has been the past condition of that population as compared with the present. And here, our thoughts are at once struck by that great fact, to which we have so often referred; which we conceive to be established by irrefragable geological evidence, and of which the importance cannot be overrated:—namely, the fact that the existence of man upon the earth has been for only a few thousand years:—that for thousands, and myriads, and it may be for millions of years, previous to that period, the earth was tenanted, entirely and solely, by brute creatures, destitute of reason, incapable of progress, and guided merely by animal instincts, in the preservation and continuation of their races. After this period of mere brute existence, in innumerable forms, had endured for a vast series of cycles, there appeared upon the earth a creature, |309

even in his organization, superior far to all; but still more superior, in his possession of peculiar endowments;—reason, language, the power of indefinite progress, and of raising his thoughts towards his Creator and Governor: in short, to use terms already employed, an intellectual, moral, religious, and spiritual creature. After the ages of intellectual darkness, there took place this creation of intellectual light. After the long continued play of mere appetite and sensual life, there came the operation of thought, reflection, invention, art, science, moral sentiments, religious belief and hope; and thus, life and being, in a far higher sense than had ever existed, even in the slightest degree, in the long ages of the earth's previous existence.

5 Now this great and capital fact cannot fail to excite in us many reflections, which, however vaguely and dimly, carry us to the prospect of the future. The present being *so* related to the past, how may we suppose that the future will be related to the present?

In the first place, *this* is a natural reflection. The terrestrial world having made this advance from brute to human life, can we think it at all likely, that the present condition of the earth's inhabitants is a final condition? Has the vast step from animal to human life, exhausted the progressive powers of nature? or to speak more reverently and justly, has it completed the progressive plan of the Creator? After the great revolution by which man became what he is, can and will nothing be done, to bring into being something better than man now is; however that future creature may be related to man? We leave out of consideration any supposed progression, which may have taken place in the animal creation previous to man's existence; any progression by which the animal organization was made to approximate, gradually or by sudden steps, to the human organization; partly, because such successive approximation

310|

is questioned by some geologists; and is, at any rate, obscure and perplexed: but much more, because it is not really to our purpose. Similarity of organization is not the point in question. The endowments and capacities of man, by which he is Man, are the great distinction, which places all other animals at an immeasurable distance below him. The closest approximation of form or organs, does nothing to obliterate this distinction. It does not bring the monkey nearer to man, that his tongue has the same muscular apparatus as man's, so long as he cannot talk; and so long as he has not the thought and ideas which language implies, and which are unfolded indefinitely in the use of language. The step, then, by which the earth became a *human* habitation, was an immeasurable advance on all that existed before; and therefore there is a question which we are, it seems, irresistibly prompted to ask, Is this the last such step? Is there nothing beyond it? Man is the head of creation, in his present condition; but is that condition the final result and ultimate goal of the progress of creation in the plan of the Creator? As there was found and produced something so far beyond animals, as man is, may there not also, in some course of the revolutions of the world, be produced something far beyond what man is? The question is put, as implying a difficulty in believing that it should be so; and this difficulty must be very generally felt. Considering how vast the resources of Creative Power have been shown to be, it is difficult to suppose that they are exhausted. Considering how great things have been done, in the progress of the work of creation, we naturally think that even greater things than these, still remain to be done.

6 But then, on the other hand, there is an immense difficulty in supposing, even in imagining, any further change, at all commensurate in kind and degree, with the step which carried the world from a mere brute population, to a human

|311

population. In a proportion in which the two first terms are *brute* and *man*, what can be the third term? In the progress from mere Instinct to Reason, we have a progress from blindness to sight: and what can we do more than see? When pure Intellect is evolved in man, he approaches to the nature of the Supreme Mind: how can a creature rise higher? When mere impulse, appetite, and passion are placed under the control and direction of duty and virtue, man is put under a Divine Government: what greater lot can any created being have?

7 And the difficulty of conceiving any ulterior step at all analogous to the last and most wonderful of the revolutions which have taken place in the condition of the earth's inhabitants, will be found to grow upon us, as it is more closely examined. For it may truly be said, the change which occurred when man was placed on the earth, was not one which could have been imagined and constructed beforehand, by a speculator merely looking at the endowments and capacities of the creatures which were previously living. Even in the way of organization, could any intelligent spectator, contemplating anything which then existed in the animal world, have guessed the wonderful new and powerful purposes to which it was to be made subservient in man? Could such a spectator, from seeing the *rudiments of a Hand*, in the horse or the cow, or even from seeing the hand of a quadrumanous animal, have conjectured, that the Hand was, in man, to be made an instrument by which infinite numbers of new instruments were to be constructed, subduing the elements to man's uses, giving him a command over nature which might seem supernatural, taming or conquering all other animals, enabling him to scrutinize the farthest regions of the universe, and the subtlest combinations of material things?

312| 8 Or again; could such a spectator, by dissecting the

tongues of animals, have divined that the Tongue, in man, was to be the means of communicating the finest movements of thought and feeling; of giving one man, weak and feeble, an unbounded ascendancy over robust and angry multitudes; and, assisted by the (writing) hand, of influencing the intimate thoughts, laws, and habits of the most remote posterity?

9 And again, could such a spectator, seeing animals entirely occupied by their appetites and desires, and the objects subservient to their individual gratification, have ever dreamt that there should appear on earth a creature who should desire to know, and should know, the distances and motions of the stars, future as well as present; the causes of their motions, the history of the earth, and his own history; and even should know truths by which all possible objects and events not only are, but must be regulated?

10 And yet again, could such a spectator, seeing that animals obeyed their appetites with no restraint but external fear, and knew of no difference of good and bad except the sensual difference, ever have imagined that there should be a creature acknowledging a difference of right and wrong, as a distinction supreme over what was good or bad to the sense; and a rule of duty which might forbid and prevent gratification by an internal prohibition?

11 And finally, could such a spectator, seeing nothing but animals with all their faculties thus entirely immersed in the elements of their bodily being, have supposed that a creature should come, who should raise his thoughts to his Creator, acknowledge Him as his Master and Governor, look to His Judgment, and aspire to live eternally in His presence?

12 If it would have been impossible for a spectator of the præhuman creation, however intelligent, imaginative, bold and inventive, to have conjectured beforehand the endowments of such a creature as Man, taking only those which we have thus indicated; it may well be thought, that if there

is to be a creature which is to succeed man, as man has succeeded the animals, it must be equally impossible for us to conjecture beforehand, what kind of creature *that* must be, and what will be *his* endowments and privileges.

13 Thus a spectator who should thus have studied the præhuman creation, and who should have had nothing else to help him in his conjectures and conceptions, (of course, by the supposition of a præhuman period, not any knowledge of the operation of intelligence, though a most active intelligence would be necessary for such speculations,) would not have been able to divine the future appearance of a creature, so excellent as Man; or to guess at his endowments and privileges, or his relation to the previous animal creation: and just as little able may we be, even if there is to exist at some time, a creature more excellent and glorious than man, to divine what kind of creature he will be, and how related to man. And here, therefore, it would perhaps be best, that we should quit the subject; and not offer conjectures which we thus acknowledge to have no value. Perhaps, however, the few brief remarks which we have still to make, put forwards, as they are, merely as suggestions to be weighed by others, cannot reasonably give offense, or trouble even the most reverent thinker.

14 To suppose a higher developement of endowments which already exist in man, is a natural mode of rising to the imagination of a being nobler than man is: but we shall find that such hypotheses do not lead us to any satisfactory result. Looking at the first of those features of the superiority of man over brutes, which we have just pointed out, the Human Hand, we can imagine this superiority carried further. Indeed, in the course of human progress, and especially in recent times, and in our own country, man employs instead of, or in addition to the hand, innumerable instruments to make nature serve his needs and do his will. He

314|

works by Tools and Machinery, derivative hands, which increase a hundred-fold the power of the natural hand. Shall we try to ascend to a New Period, to imagine a New Creature, by supposing this power increased hundreds and thousands of times more, so that nature should obey man, and minister to his needs, in an incomparably greater degree than she now does? We may imagine this carried so far, that all need for manual labour shall be superseded; and thus, abundant time shall be left to the creature thus gifted, for developing the intellectual and moral powers which must be the higher part of its nature. But still, that higher nature of the creature itself, and not its command over external material nature, must be the quarter in which we are to find anything which shall elevate the creature above man, as man is elevated above brutes.

15 Or, looking at the second of the features of human superiority, shall we suppose that the means of Communication of their thoughts to each other, which exist for the human race, are to be immensely increased, and that this is to be the leading feature of a New Period? Already, in addition to the use of the tongue, other means of communication have vastly multiplied man's original means of carrying on the intercourse of thought:—writing, employed in epistles, books, newspapers; roads, horses and posting establishments; ships; railways; and, as the last and most notable step, made in our time, electric telegraphs, extending across continents and even oceans. • We can imagine this facility and activity of communication, in which man so immeasurably exceeds all animals, still further increased, and more widely extended. But yet so long as what is thus communicated is nothing greater or better than what is now communicated among men;—such news, such thoughts, such questions and answers, as now dart along our roads;—we could hardly think that the creature, whatever wonderful means of intercourse with

its fellow-creatures it might possess, was elevated above man, so as to be of a higher nature than man is.

16 Thus, such improved endowments as we have now spoken of, increased power over materials, and increased means of motion and communication, arising from improved mechanism, do little, and, we may say, nothing, to satisfy our idea of a more excellent condition than that of man. For such extensions of man's present powers are consistent with the absence of all intellectual and moral improvement. Men might be able to dart from place to place, and even from planet to planet, and from star to star, on wings, such as we ascribe to angels in our imagination: they might be able to make the elements obey them at a beck: and yet they might not be better, nor even wiser, than they are. It is not found generally, that the improvement of machinery, and of means of locomotion, among men, produces an improvement in morality, nor even an improvement in intelligence, except as to particular points. We must therefore look somewhat further, in order to find possible characters, which may enable us to imagine a creature more excellent than man.

17 Among the distinctions which elevate man above brutes, there is one which we have not mentioned, but which is really one of the most eminent. We mean, his faculty and habit of forming himself into Societies, united by laws and language for some common object, the furtherance of which requires such union. The most general and primary kind of such societies, is that Civil Society which is bound together by Law and Government, and which secures to men the Rights of property, person, family, external peace, and the like. That this kind of society may be conceived, as taking a more excellent character than it now possesses, we can easily see: for not only does it often very imperfectly

316| attain its direct object, the preservation of Rights, but it becomes the means and source of wrong. Not only does it often fail to secure peace with strangers, but it acts as if its main object were to enable men to make wars with strangers. If we were to conceive a Universal and Perpetual Peace to be established among the nations of the earth; (for instance by some general agreement for that purpose;) and if we were to suppose, further, that those nations should employ all their powers and means in fully unfolding the intellectual and moral capacities of their members, by early education, constant teaching, and ready help in all ways; we might then, perhaps, look forwards to a state of the earth in which it should be inhabited, not indeed by a being exalted above Man, but by Man exalted above himself as he now is.

18 That by such combinations of communities of men, even with their present powers, results may be obtained, which at present appear impossible, or inçonceivable, we may find good reason to believe; looking at what has already been done, or planned as attainable by such means, in the promotion of knowledge, and the extension of man's intellectual empire. The greatest discovery ever made, the discovery, by Newton, of the laws which regulate the motions of the cosmical system, has been carried to its present state of completeness, only by the united efforts of all the most intellectual nations upon earth; in addition to vast labours of individuals, and of smaller societies, voluntarily associated for the purpose. Astronomical observatories have been established in every land; scientific voyages, and expeditions for the purpose of observation, wherever they could throw light upon the theory, have been sent forth; costly instruments have been constructed, achievements of discovery have been rewarded; and all nations have shown a ready sympathy with every attempt to forward this part of knowledge. Yet the largest and wisest plans for the extension of human knowledge in other provinces of science by the like means, have remained |317

hitherto almost entirely unexecuted, and have been treated as mere dreams. The exhortations of Francis Bacon to men, to seek, by such means, an elevation of their intellectual condition, have been assented to in words; but his plans of a methodical and organized combination of society for this purpose, it has never been even attempted to realize. If the nations of the earth were to employ, for the promotion of human knowledge, a small fraction only of the means, the wealth, the ingenuity, the energy, the combination, which they have employed in every age, for the destruction of human life and of human means of enjoyment; we might soon find that what we hitherto knew, is little compared with what man has the power of knowing.

19 But there is another kind of Society, or another object of Society among men, which in a still more important manner aims at the elevation of their nature. Man sympathizes with man, not only in his intellectual aspirations, but in his moral sentiments, in his religious beliefs and hopes, in his efforts after spiritual life. Society, even Civil Society, has generally recognized this sympathy, in a greater or less degree; and has included Morality and Religion, among the objects which it endeavoured to uphold and promote. But any one who has any deep and comprehensive perception of man's capacities and aspirations, on such subjects, must feel that what has commonly, or indeed ever, been done by nations for such a purpose, has been far below that which the full developement of man's moral, religious, and spiritual nature requires. Can we not conceive a Society among men, which should have for its purpose, to promote this developement, far more than any human society has yet done?—a Body selected from all nations, or rather, including all nations, the purpose of which should be to bind men together by a universal feeling of kindness and mutual regard, to associate them in the acknowledgement of a common Divine Lawgiver, Governor, and

318|

Father;—to unite them in their efforts to divest themselves of the evil of their human nature, and to bring themselves nearer and nearer to a conformity with the Divine Idea; and finally, a Society which should unite them in the hope of such a union with God that the parts of their nature which seem to claim immortality, the Mind, the Soul, and the Spirit, should endure for ever in a state of happiness arising from their exalted and perfected condition? And if we can suppose such a Society, fully established and fully operative, would not this be a condition, as far elevated above the ordinary earthly condition of man, as that of man is elevated above the beasts that perish?

20 Yet one more question; though we hesitate to mix such suggestions from analogy, with trains of thought and belief, which have their proper nutriment from other quarters. We know, even from the evidence of natural science, that God *has* interposed in the history of this Earth, in order to place Man upon it. In that case, there was a clear, and, in the strongest sense of the term, a *supernatural interposition* of the Divine Creative Power. God interposed to place upon the earth, Man, the social and rational being. God thus directly instituted Human Society; gave man his privileges and his prospects in such society; placed him far above the previously existing creation; and, endowed him with the means of an elevation of nature entirely unlike any thing which had previously appeared. Would it then be a violation of analogy, if God were to interpose again, to institute a Divine Society, such as we have attempted to describe; to give to *its* members their privileges; to assure to them their prospects; to supply to them his aid in pursuing the objects of such a union with each other; and thus, to draw them, as they aspire to be drawn, to a spiritual union with Him?

It would seem that those who believe, as the records of |319 the earth's history seem to show, that the establishment of Man, and of Human Society, or of the germ of human society, upon the earth, was an interposition of Creative Power beyond the ordinary course of nature; may also readily believe that another supernatural Interposition of Divine Power might take place, in order to plant upon the earth the Germ of a more Divine Society; and to introduce a period in which the earth should be tenanted by a more excellent creature than at present.

21 But though we may thus prepare ourselves to assent to the possibility, or even probability, of such a Divine Interposition, exercised for the purpose of establishing upon earth a Divine Society: it would be a rash and unauthorized step,—especially taking into account the vast differences between material and spiritual things,—to assume that such an Interposition would have any resemblance to the commencement of a New Period in the earth's history, analogous to the Periods by which that history has already been marked. What the manner and the operation of such a Divine Interposition would be, Philosophy would attempt in vain to conjecture. It is conceivable that such an event should produce its effect, not at once, by a general and simultaneous change in the aspect of terrestrial things, but gradually, by an almost imperceptible progression. It is possible also that there may be such an Interposition, which is only one step in the Divine Plan;—a preparation for some other subsequent Interposition, by which the change in the Earth's inhabitants is to be consummated. Or it is possible that such a Divine Interposition in the history of man, as we have hinted at, may be a preparation, not for a new form of terrestrial life, but for a new form of human life;—not for a new peopling of the Earth, but for a new existence of Man. These possibilities are so vague and doubtful, so 320| far as any scientific analogies lead, that it would be most

unwise to attempt to claim for them any value, as points in which Science supplies support to Religion. Those persons who most deeply feel the value of Religion, and are most strongly convinced of its truths, will be the most willing to declare, that religious belief is, and ought to be, independent of any such support, and must be, and may be, firmly established on its own proper basis.

22 We find no encouragement, then, for any attempt to obtain, from Science, by the light of the analogy of the past, any definite view of a future condition of the Creation. And that this is so, we cannot, for reasons which have been given, feel any surprise. Yet the reasonings which we have, in various parts of this Essay, pursued, will not have been without profit, even in their influence upon our religious thoughts, if they have left upon our minds these convictions :—That if the analogy of science proves any thing, it proves that the Creator of man can make a Creature as far superior to Man, as Man, when most intellectual, moral, religious, and spiritual, is superior to the brutes :—and again, That Man's Intellect is of a divine, and therefore of an immortal nature. Those persons who can, on any basis of belief, combine these two convictions, so as to feel that they have a personal interest in both of them ;—those who have such grounds as Religion, happily appealed to, can furnish, for hoping that their imperishable element may, hereafter, be clothed with a new and more glorious apparel by the hand of its Almighty Maker ;—may be well content to acknowledge that Science and Philosophy could not give them this combined conviction, in any manner in which it could minister that consolation, and that trust in the Divine Power and Goodness, which human nature, in its present condition, requires.

THE END.

PART TWO

Pages excised by Whewell from the first edition just before the book was printed

THE EXCISED CHAPTERS

As mentioned in the introduction, at the urging of Sir James Stephen who thought them too specialized for the *Essay,* there were five chapters removed by Whewell from the original edition. A condensed, and not-altogether-coherent alternative chapter was substituted. However, in the Trinity College Library there is a version of the *Essay* with the original chapters, and it is these which are printed here publicly for the first time.

As implied in my introduction, it is the first of these excised chapters, "The Argument from Law," which is particularly interesting, for it is here that Whewell develops his claim that the very fact of law proves a Designer, and suggests that we here on Earth have the intellectual task of understanding the nature of the Divine Mind. The following briefer chapters develop some further ideas, building on this major argument. In "The Omnipresence of the Deity," Whewell takes up the issue of God's continual involvement in His creation. Whewell wants to steer clear of suspicions of deism (no doubt with the consequent dangers of evolutionism), and to avoid suggesting that God has created the universe and its laws and now sits back and lets things run their course. For Whewell, God is always involved in His creation, sustaining it. Whewell's argument here is much dependent on his assumption (shared by other philosophers of his generation, particularly Sir John Herschel) that force and will are closely linked, and hence inasmuch as the universe is governed by Newtonian forces, a sustaining will is not far behind. (A young scientist, mentored by Whewell and much impressed by Herschel's writings, Charles Darwin, also mixed up force and will, and spent much time arguing that although artificial selection is a force guided by intelligence, this is not necessarily the case for natural selection—although as it happened, at the time when Darwin made his discovery of natural selection, he saw it as a force guided by God.)

"Man's Intellectual Task" further develops Whewell's claim that we are put on this earth to follow the work of the Creator. In spelling out his position, Whewell uses some of the apparatus and terminology of the *Philosophy of the Inductive Sciences*, especially his Kantian distinction between the Ideas of the Mind and the Phenomena of Nature. Perhaps Stephen gave Whewell good advice when he urged that a discussion of this philosophical sophistication was not needed in an *Essay* such as Whewell was now writing. The complementary "Man's Moral Trial" seeks to show that the traditional task set for us humans—obeying God's law—is in no wise diminished by the overall thrust of Whewell's argument. Indeed, the claim is made that a moral trial only really makes sense in the light of the Atonement, which is threatened by a much-populated universe.

Finally we have "The Design of Animal Springs of Action," and if any pre-Darwinian writing shows that the mid-Victorian science/religion mixture was a mess, it is this chapter. In passages which might fit without comment in the *Origin of Species*, Whewell stresses just how miserable is much of animal life and how there is an ongoing bloody struggle for being: "Pain and death, and the fear of pain and death, have, throughout, been the sources of life, and of the means of life and enjoyment" (322). A huge amount of unconvincing wriggling is required to show that the animal strife for life does not exist just in its own right, but for humans as well. In a bastardized version of Owen's thinking about homology, Whewell suggests that there is a general plan to animal life, and in order for us to have our own emotions and actions, which leave the way open for peace as well as war, animals apparently have to share in this pattern. No doubt this kind of thinking much comforts the deer when faced with the wolf! Stressing again that we should not read Whewell's *Essay* simply in the light of what is to come later in the decade, here above all we have truth to the claim that people change their views due both to the failure of the old and the success of the new.

38 We shall not dwell further on these reasonings here. But there is one part of man's pre-eminence in creation which is so connected with the speculations which we have been pursuing, that it may be of use to follow it somewhat further than we have done. We will not, in the present essay, attempt to trace the importance which man's moral, religious, and spiritual character impress on all that is subservient to those parts of his nature, or that flows from them. But man's *intellectual* nature is so closely bound up with the discoveries which he has been able to make with regard to the scheme of the material universe, and the laws which govern it;—the amount of truth on those subjects, which, by the exercise of his reason, he has been able to acquire, so directly leads us up to the consideration of the source of such truth, and the reasons why he can acquire it; —the manifestation of his mind in this way so plainly has a relation to the manifestations of the Divine Mind;—that we are only following out speculations which have already engaged us here, when we enquire what that relation is. If man can, in any degree, understand what God does, we may conceive that God created him for that purpose, at least among other purposes; and we may believe that the creation

of a being who could do this, was an object of the act of creation, in addition to all the purposes of the mere material creation; and an object to which the scheme of the material creation, and its laws, may reasonably be held to be subservient.

On these subjects we shall therefore try to make a few plain reflections.

CHAPTER XII.

The Argument from Law.

---

1 IN the course of the views here presented, we have seen that there exist, and operate in the universe, many Laws of nature, producing regularity, order, symmetry, in the forms and motions of its parts. Or rather, we have seen that the whole universe is merely a collection of such Laws. For if we trace the subject into its elements, we find that we cannot distinguish between the Laws, and the Things which are subject to the Laws. The planets, for instance, and all the bodies of the Solar System, obey the Laws of Motion, and the Law of Universal Gravitation. But when we say that *bodies* obey these Laws, what do we mean by *bodies?* Bodies are substances held together by certain forces of cohesion; which, in the case of solid, or rather we may say, rigid bodies, is the force of crystallization, acting according to its Laws, more or less completely. Such bodies exclude other bodies from the space which they themselves occupy; and such exclusion may be ascribed to a Law of Repulsion. Here, then, is another Law. As another distinctive character of bodies, bodies have *Inertia;* and Inertia is expressed by one of the Laws of Motion. But if we take from a body Inertia, Resistance to pressure, Rigidity, or whatever other Consistence it has, what do we leave! The body itself is gone; or remains a mere imaginary solid space, such as geometers suppose in their demonstrations; such solids as can interpenetrate each other, and occupy the same space without any restriction. Such a geometrical solid is not a physical body;—is not matter: and thus body or matter is really only a collection of Laws.

2 But for the sake of our present argument, it is not necessary to go so far as this. At any rate, no one who gives attention to the constitution of the universe, can doubt that there are, disclosed by its phenomena, many Laws producing regularity, symmetry, order. Now by contemplating such Laws, man's thoughts are naturally directed to a Lawgiver. Regularity, symmetry, order, imply a Creative Mind, which has impressed these characters on its creation. As it is one of the highest operations of our intellect, to discover such Laws in Nature, and to trace their consequences; so we are irresistibly led to suppose that these Laws must have been present to the Divine Intellect, before they were apprehended by the human Intellect. That we can understand such Laws, is for us, much: they must then have been devised by an Understanding greater than ours. To our Reason these Laws are often, especially at our first apprehension of them, obscure, complicated, partial; and by further study, we find they grow clearer, simpler, more general. There must then, it would seem, have been a Supreme Reason, in which these Laws were always perfectly clear, simple, and universal. What we read so slowly, with so much difficulty, and yet in the end, so confidently, must be writing, must be a language, must imply Intelligence and Purpose—cannot be the work of Chance and Necessity. What we can thus clearly interpret, must have a meaning; and in that case, *Whose* meaning? Where we find Geometry and Mechanics, there must have been a creative Geometer and Mechanist. Where we are compelled to employ all the resources of mathematical calculation and reasoning, in order to explain the phenomena, a Mathematical Spirit must have produced the phenomena. Where, by a laborious analysis, we find the elements of the phenomena to be the relations of space, number, and the like, there must have been some one to make the synthesis of these elements;—some Divine Thought to combine these

330|

relations. And thus, the existence of Laws of Nature, governing and producing the phenomena of the universe, makes manifest to us the existence and operation of God.

3 We have stated this inference from the Laws of Nature to a Lord of Nature in various forms, because the conclusion appears to some minds clearest in one shape and to some in another. The Argument from Law, which we have to treat of at present, is different from the Argument from Design, of which we were speaking in the last Chapter; and is, to the minds of most persons, less clear and obvious, though we hope, with due attention, it may be made convincing and satisfactory. The Argument from Design, as generally apprehended, assumes the properties which bodies have, as something fixed and given; and contemplates these properties as being used, applied, and contrived, to answer an end; in something of the same way in which man uses, applies, and combines the properties which he finds bodies to possess, in order to promote his ends. When we consider the structure of the human eye, as contrived in order to produce vision, we regard this contrivance, as consisting in the selection of transparent bodies, in adjusting them according to their powers of refracting light; in giving them due shapes and positions: much as we should consider the similar selection of transparent bodies, their adjustment according to their refractive powers, their formation into proper shapes, and their collocation in due positions, in the construction of a telescope or a microscope. The transparency, the refractive power, are the given elements, out of which the contrivance is produced, by acts of thought and invention. The thought and invention are regarded as employed in combining, rather than in creating such elements. So in contemplating the actions of the muscles, by which the limbs are moved, we regard the bones and tendons as bars and cords, which are combined, as bars and

cords are combined in a machine made by man. We do not direct our attention to the forces of cohesion, by which the parts of a bone, or of a muscle, are held together and have their special textures given them. These properties are supposed to be given, as the means to be used by the Divine Mechanist, as they are by the human mechanist. What strikes us as proving design, and therefore as manifesting a Divine Mechanist, is the way in which these means are used. So again, in the machinery by which the rotation of the earth produces the succession of seasons, this rotation once begun, is conceived to continue by the Laws of Motion; and these Laws of Motion we regard as applied, not instituted, for the purpose. So, in the provisions for the permanence of the motions of the solar system,—abstruse and subtle as are the calculations by which mathematicians show that the existing arrangements of the system will produce such permanence by the action of the Law of Universal Gravitation;—this Law is, in this case, not conceived as established for this purpose. The arrangements only,—the nearly circular orbits, the circular motions all in nearly the same direction (from west to east,) the smallness of the planets;—which arrangements are the conditions of the permanent motion never deviating far from regularity; are conceived as something devised subsequently to the Law of Force, by which all bodies attract each other; and are conceived, by most philosophers, to be, therefore, a special contrivance, having such permanence and regularity for its end. We are accustomed to regard bodies as having certain original properties of themselves, naturally, it may be, necessarily; and we look for (and we find) in the combination and appreciation of these properties, wonderful proofs of Design, exercised in every part of creation. This is the way in which man is accustomed to work:—the only way in
332| which he can work. And therefore, this is the way in which,

at first at least, he conceives the workmanship of God. God is thus a Mighty Workman, employing a power, a skill, an amount of resources, which go far beyond what man can employ;—which he can only with effort, and to a small extent, apprehend: but which are sufficiently seen, to produce conviction and admiration, in those who give due attention to such subjects.

4 And this power, this skill, these resources, are applied to promote an end, which man can easily conceive as an end;—with which he can, to a great extent, sympathize;—that end being the life, comfort, and enjoyment of sentient creatures;—the enabling them to perform those functions which are requisite to their life, sustenance, and perpetuation. A design to attain and further such purposes, can easily be conceived by man, as design. As man finds and prepares his food, clothes himself, builds himself dwellings, makes himself weapons, so God has prepared food, clothing, dwellings, weapons, for animals in general; or has given them the means of acquiring those necessaries and comforts. As man provides for his children, so the Universal Father has provided for all creatures, all being his children. As man changes his contrivances to obtain food, warmth, shelter, security, according to the circumstances in which he is placed, so the Governor of the World changes the mode in which animals are provided with food, warmth, shelter, security, according to the circumstances in which they are placed. Man feels a benevolence towards man, and even towards animals, which prompts his exertions for their good: and hence, he can recognize the indications of a Universal Benevolence in the construction and conduct of the whole world of living things. From this aspect of the Creation, he is led to raise his thoughts to one all-powerful, all-wise, all-benevolent Creator.

5 This effect of the aspect of the Creation, this argu- |333

ment from the Design manifested in the Universe, is, as we have said, a habit of man's mind, in the highest degree suitable to his nature and position. To be able to see things in this aspect, to draw this inference from the phenomena of the world, is a high privilege of man; and is, in most minds, the beginning and foundation of all other religious thoughts, and therefore, of all religious privileges. And while we attempt to bring into view the mode in which another aspect of the Universe also, in its way, suggests to us, and impresses upon us, the conviction of a Supreme Mind, the Creator and Director of all the arrangements of the Universe, even of those in which we do not discern such a design as we have now spoken of; we still would leave upon the mind of the reader, unimpaired, and if it might be, augmented, the impression of the argument from such Design. That argument is more adapted to the common habits of human thought, than the argument from Law; and even in the most philosophical minds, those most disciplined by science, it loses none of its effect, by the introduction of the view of the General Laws, by which the universe is governed, if this view be rightly apprehended.

6 The argument from Design, and the argument from Law, in proof of the existence and activity of a Divine Creator and Governor, may, indeed, be stated briefly in similar modes. Those who urge the former argument, are accustomed to say, Design proves a Designer, and that Designer is God. In like manner, we may say, Law proves a Lawgiver, and that Lawgiver is God. But perhaps we may be able to unfold this argument a little further, and to bring into view some trains of thought, by which the notion of Law, as implying a Lawgiver, is more clearly developed in the mind. We may be able to show that the assumption which we noticed before, as pervading the common mode of apprehending the argument from design, is really an assumption,

334|

and is an assumption in which inquiring minds cannot rest; namely, the assumption that bodies have, of themselves, naturally, or necessarily, certain properties, and modes of being and of operating, which the Supreme Mind only uses, does not create;—the assumption that intelligence is shown in the application of such properties, but that there is no intelligence manifested in the existence of such properties.

7 Let us take some of the cases which we have just given, in order to exemplify this assumption; and let us endeavour to go a step further. In the structure of the eye, certain transparent media are adjusted with the most exact adaptation of their forms and the laws of their refractive powers, so as to produce on the retina a distinct image of external objects. The adjustment is an evident mark of design. But what are the Laws of the refractive powers of which we thus speak? The Laws of the Refractive Powers of all transparent bodies are mathematical Laws; of a certain degree of complexity, as may appear from this; that it required long-continued labour and thought, many experiments, great mathematical skill, to discover them. In the attempt, the Greeks failed, the Arabians failed; and it was only at last, when the spirit of scientific discovery was revived in modern Europe, that the laws were detected. The law, when reduced to its most simple shape, and expressed in technical terms, invented for the purpose, is simple enough. It is that in each transparent medium, the *sine of incidence* is to the *sine of refraction* in a constant ratio for all incidences: the ratio being different for different media. Upon the basis of this law, the eye is constructed; as upon the basis of this law, optical instruments are constructed. But whence came this Law? Shall we say that it exists naturally, necessarily, of itself? Has it no cause? no origin? Is there some necessity in nature, anterior to and independent of all cause, by which media are transparent, transmit

light, refract it, and refract it according to this Law? according to this mathematical Law? Must not such a mathematical law have a mathematical source? Must not that which it cost mind so much trouble to discover, have its origin in mind? Must not a Law which governs the phenomena according to such precise relations of angles and sines, be appointed by some Intelligence which comprehended geometrical relations? Is not the establishment of such a law an act of thought, no less than its application? If man must be a geometer to understand the Law, must there not be somewhere a Geometer who made the Law?

8 We may pursue the other cases in the same way. By the Laws of motion, the diurnal and annual movements of the earth, once impressed upon it, continue to go on, without pause or retardation. And thus, the earth, once launched into her orbit, and set spinning on her axis, may go on, it would seem, for ever, without further help. It required only the first impulse of the Creative Hand. But when we say that it will thus go on indefinitely, *by the Laws of Motion*, can we forbear to ask, who established the Laws of motion? Some speculative persons have attempted to prove that the Laws of motion exist necessarily; but what is this necessity? what is the bond by which it holds its subjects, and prescribes their actions? Other persons have attempted to show that the primary Laws of motion are not necessary truths;—that they might have been other than they are. Must not those who hold this, hold also, that some one has selected these from the other possible Laws: rejecting others and establishing these? According to the Law with which we are here especially concerned, a body once set in motion, and not stopped or retarded by any obstacle or force, will go on moving for ever, with undiminished velocity. Who has thus determined what shall take

336| place *for ever?* Can matter, of itself, thus have a native power of moving for an infinite time through an infinite space? Again, the laws of motion act according to mathematical rules: must not such rules be established by a mind cognizant of mathematical relations? The universe consists of matter and motion: is there matter,—matter with its properties,—is there motion,—motion with its Laws,—without our being able to say, that the properties of matter, that the Laws of motion, imply any creative agency, by which they are what they are? Must we stop here, without being able to see any working of a Divine Author in these fundamental conditions of the Universe?

9 We put the argument here in the form of these questions, because the questions imply that difficulty which the mind of man must feel, in stopping at this point. It may require abstruse thought and comprehensive views, to go further; but still, if we do not go further, we have a view which is arbitrarily limited; our train of thought is manifestly incomplete. The properties of matter must be given by the Creator of the world; the mathematical Laws of these properties imply a contemplation of mathematical relations by the Lawgiver. If we would go as far as the powers of the human intellect permit, in a knowledge of the operations of the Author of the Universe, we must take account of all the laws which bind together the parts of the universe; even the most abstruse and mathematical, even the most comprehensive and fundamental. We must not stop at a superficial and imperfect view, looking only at those operations which most resemble the work of man: we must go on to contemplate that operation which is peculiar to God. Man can employ the properties of matter, but he cannot create such properties. He can apply the laws of nature, but he cannot establish laws of nature. But this need not stop his speculations at once. Though he can-

not create properties of matter, he can investigate what they are. Though he cannot establish laws of nature, he can determine such laws. And having found that such properties and such laws are governed by the most subtle and general mathematical relations, which his reason is able to apprehend; he must needs ask himself, whether such properties and such Laws could have come into being, without being apprehended from the beginning by a Reason superior to his own.

10 Perhaps the argument may appear more evident, if we consider an example, in which the operation of Law appears mainly in geometrical relations;—in the relations of space and number. We have already noticed such an example; namely, the forms produced by the Laws of Crystallization. These forms, as we have said, are of the most multiplied and varied kind, yet pervaded by the most striking aspect of order and symmetry in their faces, edges, sides, and angles. The beauty of a collection of such forms must strike the most incurious observer: from the mutual play of resemblances and differences; the mode in which the most complex regular forms, are, by various steps, deduced from the most simple; the simplicity of the fundamental forms and laws, and the apparently inexhaustible variety of the results, all controlled by some Law of Symmetry. The diagrams which, in a work on crystallography, represent the various forms of crystals, strike the reader's eye as if they were a series of figures invented by geometers, in order to show what kinds of forms might be produced by combination and derivation of elementary regular solids, and in order to trace the geometrical properties of the figures so formed. And accordingly, a great number of very curious geometrical properties have been found to belong to the figures which crystallography requires us to study; and by the labours of geometers, the

discovery of such properties is still going on: one beautiful theorem after another is added to this new branch of applied mathematics.

11 Now shall we say that the Creative Power which produced all these crystals, and crystalline laws, was unaccompanied by any Creative Thought, which apprehended the laws and their consequences? Shall we say that while man can discover, and go on discovering, these beautiful theorems, —these truths which have existed and been realized ever since crystals began to exist,—that these truths had no existence for any mind, till the mind of man in these late times discovered them? Did these truths create themselves? or were they not rather, present to the Divine Creating Mind, as involved in the Laws and Principles from which they flow; —learnt by man, as a lesson concerning what God had long before done? Man's intellect has a geometrical faculty; and by the use of that faculty, he comes, though tardily and with difficulty, to a knowledge of the geometry which there is in the construction of the world. But is his geometrical thought, thus imperfect, slow, limited, the only Geometrical Thought which exists in the world? Must there not be, somewhere in the Universe, that Thought in a higher form, which he possesses in a lower? that Thought in a form free from imperfection, obstacle, and limit, which he possesses so imperfectly? If man, the creature, the student of creation, the slow Learner, the mistaker, yet the Lover of Truth, must pursue his study, and acquire his truth, by the use of his geometry; must not God, the Creator, the Lawgiver of Creation, the Teacher, the Infallible, the Author of Truth, also have done his work, by introducing into it geometry?

12 Of course, when we say this, we are not to imagine that the Divine Mind employs itself in geometrical demonstration, and deduction of truths from principles, as we are wont to do, in our geometrical investigations. To Him, all

the consequences of principles must be equally manifest, all theorems intuitively seen to be true. We have, even in the operations of human intellect, that which may help us to conceive this. Even among men, those who have been most highly gifted with mathematical power, have had an insight which has enabled them to gather up, at a single glance, the convictions which duller minds can reach only through long and laborious trains of deduction. Newton and Pascal are reported to have apprehended at once, as truths, the long series of truths and demonstrations which form the mass of the elementary books of geometry. Such minds see vast bodies of truth, relative to space and number, by a native insight. They have an intuitive power of grasping all the comprehensive consequences of the first principles of such truth. And this insight, this *intuition*, using for the sake of explanation, terms borrowed from the operations of the human mind, and venturing, with a due sense of their vast insufficiency, to apply them to the Divine Mind, must be supposed to exist, in an infinite degree, in the Mind of the Creator. To Him all the relations of space and of number, in their most complex and remote results, must lie bare and clear, as the plainest axioms do to us. With Him, there is, in Truth, no first and second step, no primary and secondary truth, no deduction and reasoning from Principles. With Him, All Truth is the Principle of being. The highest Reason supersedes reasoning. Everything which man, or which any finite and limited form of intelligence, can discover, lies manifest in its own light, to the eye of the Supreme Reason. And thus, we may in some measure understand, how all the complex relations of figures and combinations of figures, such for instance, as those which crystallography is employed about, were present to the Supreme Mind, which impressed on matter the Laws of Crystallization; and that 340| man, with the most laborious exertion of his mathematical faculties, detects, slowly and gradually, the truths of which a Divine Intellect inserted the germs into the intimate frame of matter; thus concealing for a time, from us, the truths which were to Him, from the first, manifest in their extremest developement.

13 This view is by no means new. Even in the very first stages of man's cultivation of geometry, in Greece, at the time of Plato, it was declared by thoughtful men, and especially by Plato himself, that God acts by Geometry:—Ὁ Θεὸς γεωμετρεῖ. This was said then, because men found that every discovery which they made respecting the motions of the stars and planets, (and they made very remarkable discoveries, which even yet hold an important place in astronomical science,) required them to reason concerning the geometrical properties of the machinery by which those celestial bodies seemed to them to be carried along their courses. The cycles and epicycles and excentrics, which, to so great an extent, explained the motions of the planets, could be made to minister to such explanation, only by those who possessed all the geometrical knowledge which had then been opened to man. The investigation of the motions of the heavens, at once showed the necessity of adding new provinces to that science which had its origin in the measurement of the earth:—new provinces larger than the old ones, so that, as Plato said, it is an inappropriate name to call this science *Geometry*, the measurement of earth, since its greatest office is in the heavens. We must add, he says, to Plane Geometry, a Solid Geometry, which shall enable us to reason about spheres; to the Geometry which treats of figures at rest, we must add a new kind of calculation, which shall investigate the phenomena produced by the motions of spheres. And this, accordingly, was done. Spherical Geometry was created by the Greek Mathematicians; and they were thus enabled to calculate the lengths of days and nights |341

through the varying seasons, the relations of Zones and Climates, the distances of places from each other on the globe of the earth, and of planets from each other in the visible vault of the sky. More than this; they were enabled to explain how it was that eclipses occurred, and to predict their occurrence, and their amount, beforehand. And thus, finding the heavens full of geometry, to which they had with effort attained, they judged at once, that this was the Geometry of God, the Maker of the Heavens. They conceived that what so tasked their intellects to discover in part, was the work of a greater, of a Supreme Intellect, which had so arranged the whole. Obtaining these imperfect, and after all, obscure glimpses of the Universe, they judged that it had been planned in the clearness and fulness of an Infinite Faculty of intellectual Vision. They were led to the conviction that the Principle of the Universe was Mind:—that all earlier guesses, which assumed matter, in any of its forms, to be an ultimate thing, with properties of its own, independent of an Intelligence which made them what they are, was short of the mark of philosophy:—was a solution of the problem of being, below even what man could ascend to:—was an answer to their questions, which left new questions, no less inevitable. The ultimate answer to the inquiry, how the universe came to be what it was, full of laws which intellect alone could apprehend, was, that it was created by an Intellect whose mode of operation is Law.

14 This view of the universe, which prevailed while the *geometry* of the heavens only was concerned, did not lose, but gained force, when the *mechanism* of the universe also came to be studied, and, in some degree, to be discovered. When the motions of the celestial spheres, and cycles, and epicycles, was assumed, it required the science of geometry alone, to trace their consequences. When the causes of these motions, or of the machinery, different from this, whatever

342|

it might be, by which the motions of the planets was produced, was further studied, the science of mechanics became necessary, in order to pursue the investigation. And as the science of Spherical Geometry was created in order to carry onwards the old investigation, so the science of Mechanics, which treats of the force which produces and changes motion, was created in order to carry on the new inquiry. It was only when this science had been created, that the truth so sought could be discovered. In the attempt at such discovery, Kepler failed, because he was ignorant of Mechanics: Descartes failed, because he had not courage and skill to apply what was then known of Mechanics: Newton succeeded because he first completed the science of Mechanics, as it governs motions near to us, and therefore known to us; and then applied it with undaunted courage, but with consummate caution, to all the bodies which exist in the universe. And in this, as in the former part of the history of astronomy, it was found, that one or two simple principles, when their consequences, near and remote, were traced by the light of a piercing and comprehensive intellect, led to all the phenomena, (including many phenomena of the most complex, minute, and seemingly irregular kind,) which astronomers had till then observed. Everything was found to depend upon the action of one single kind of force; urging, according to the same law, the earth and all the planets to the sun, the satellites to the planets; urging every mass of matter to every other, every particle of matter to every other; while the planets and the satellites were kept in their orbits by the operation of the universal Laws of Motion. And if, as we have already said, the Laws of Motion, which produce their effects according to mathematical relations, imply a Lawgiver, who has impressed such Laws, with a comprehension of their mathematical relations; no less did the new Law, the Law of force, the Law of gravitation, producing its multitudinous and complex effects according to mathematical relations, more abstruse and difficult to disentangle, also imply a Lawgiver, who has impressed that Law upon matter and the particles of matter; still with a comprehension of the relations, by which its effect is regulated, and determined to be what it is. What it required the intellect of Newton to discover, could not have been produced by blind chance or dull necessity, without any intervention of intellect. The curious and complex trains of demonstration, by which he proved that the Law of force, asserted by him, would explain the facts in the minutest particulars, could only be the human representatives of the Divine intuition by which such facts were foreseen in the act of creation. Even now, the clearest understandings do not follow these demonstrations without effort. That which the human mind can hardly do, a Supreme Mind may do with no effort: but how should matter without mind arrive so exactly at the result of the most difficult reasonings?

15 We may without impropriety, and with some advantage, represent to ourselves the phenomena of the universe, and especially the mathematical laws of the phenomena, as a kind of Language, in which the Supreme Mind speaks to our human minds. A few men only discover the meaning of this language; they communicate the meaning to other men, who thus learn also to understand it. What is thus discovered, what is thus learnt, is probably but little, compared with all that is really inscribed on creation: a few sentences, a few phrases. But still it *is* understood. We feel, and are immovably convinced, that we do understand some part of what is written:—that the Laws which discoverers disclose to us are really Laws of Nature:—really sentences inscribed on the face of the Creation. And then, understanding these inscriptions, we cannot doubt that they

344| are the expressions of the Thoughts of the Creator. They are a Language, which implies an Intelligent Utterer of language. They have a meaning, and this must be the meaning of the Creator. They tell us what he has willed and wills. Not only has he willed, but he has made the expression of his will, the law of his will, intelligible to us. We can discern that he has willed that matter should be governed by certain universal Laws of Motion:—that the particles of matter should be drawn together by a certain universal Force of Gravitation. These Laws of Motion, this Law of Gravitation, clear-sighted men have decyphered. No competent inquirer can doubt that the inscription is there—that it is rightly decyphered. For, having decyphered it in its simpler parts, having first read these, we find that we can go on reading to the end. Having first, for instance, explained by the Law of Gravitation, the larger motions of the solar system, the motions of the planets and satellites in their orbits, we find that we can go on to explain the most complex irregularities, the most curious facts, the most minute inequalities of motion:—for instance, the moon's many deviations from equable motion, discovered by astronomers long before Newton's time; the curious habits of Jupiter's satellites; the perturbations of the motions of the planets by occasion of their relative position. All these, and many more features, of the planetary motions, are at once read into significance, by the aid of the alphabet with which Newton has supplied us—the Law of Gravitation. How then can we question that we are really reading? How can we doubt that all this significance is real? But if we are reading, whose is the writing? If the significance be real, whose thought is signified? What can we see in this, but a message, a lesson, intended for us: or at least, a record which we are able to peruse and to interpret? And if so, does it not record the operation of the Intellect by which the creation was framed, and the Principles on which it was constructed?

|345

16 But there is another feature which sometimes occurs, in the progress of man's discoveries respecting laws of nature, and which appears still further to confirm our views. It has happened, not unfrequently, that a law which was discovered by studying one set of phenomena, and which satisfactorily and fully explains those, has been found, afterwards, to explain some other set of phenomena entirely different, and not contemplated in the original investigation, by which the law was discovered. The law so discovered performs, not only what was expected and desired of it, but more. It not only explains the facts from which it was collected, but another set of facts, seemingly unconnected with the former, which might have been expected to depend on some other principle. Thus, the law of gravitation was discovered by studying the motions of the planets and satellites in their orbits;—the motion of *translation* as it is called; that is, of the transfer of the whole masses from place to place. But there was another phenomenon, known to astronomers from the time of Hipparchus, the Precession of the Equinoxes; manifested by a slow motion of the visible pole of the heavens from star to star; but indicating really a motion of the earth's axis, by which it does not exactly remain parallel to itself, but slowly shifts with a conical motion (travelling round its circle in 26,000 years). This motion of the earth's axis appeared, at first sight, to have nothing to do with the motion of the planets and satellites in their orbits;—with the forces by which they were kept in their paths. But yet, when the law of universal gravitation had been discovered by the study of those orbital motions, Newton saw, and showed, that the same law would produce such a motion of the earth's axis as was then known to exist. For the rotation of the earth makes it protuberant at the equator; and the slight attraction of the sun and of the moon on this protuberant part—an ex-

ample of the universal law of attraction,—gives rise to the slow conical motion of the axis. And thus, the law of universal gravitation was confirmed, we may say, from an unexpected quarter. The discovery was clenched by an unforeseen coincidence. The rule, extracted from the motions of translation, accounted also for the conical motion of the axis of rotation. The explanation of the orbital motions, and of the precession, facts lying in different regions of the system, jumped together. The Induction of Law from the one agreed with the Induction of Law from the other. There was, what has been called, a *Consilience of Inductions.*

17 Other examples of such Consilience of Inductions might be given. What strong, overwhelming confirmation such events in the history of science, supply, to the discoveries so supported, we need not here urge. But it cannot, it would appear, be denied, that such steps in the discovery of Laws of Nature, also still further strengthen and impress the belief that those Laws are impressed upon Nature by an Intellect of infinite clearness and comprehensiveness. Here we find, that not only the consequences of such principles as we can discover, are already unfolded in all their mathematical consequences, where *we* trace those consequences with difficulty: but that the consequences of the same principles are unfolded elsewhere, where we had never looked for them. The operation of the Divine Mind is seen, not only in presenting mathematical results, which we too, to a certain extent, were searching for: but in presenting other mathematical results, which we were not seeking for, but which we must needs acknowledge to be true when they are presented to us. Or to recur to the illustration, in which we likened the discovery of laws of nature to the decyphering of a writing; we find here, that the alphabet which we discovered by the study of the inscription

before us, enables us not only to read that inscription to its uttermost sentence, and in every phrase; but that another inscription, on a remote part of the same building, and which at first seemed written in another alphabet, becomes, on closer inspection, legible by the same key which had served us before. If this were to happen, in decyphering inscriptions written by man, could we doubt that we really had the key to the writing? or that the writer, both of the one inscription and of the other, had really that power of intelligence which language implies? And just as little, it must seem, can we doubt that when we find this coincidence, this consilience, in the Interpretation of Nature, to which we are led, we have before us, really, the manifestation of the Will and Thought of the Divine Author of Nature.

18 We have been considering the impressions made upon our minds by contemplating Mathematical Laws of Nature; or those of which the consequences require to be traced (by us) through the aid of mathematical reasoning. But if we consider Laws of Nature of other kinds, we shall still find that a Law suggests to us the conviction of a Lawgiver;—that symmetry, regularity, order, even when not presented to us under mathematical forms, still lead to the conviction that they are the work of a Spirit which can consciously delight in symmetry, regularity, and order. We have already spoken of an example of such a Law of Nature, which is seen in the construction of plants. According to a principle which has been termed the Principle of Morphology, all the parts of plants, though apparently different, are really the same in their nature. Leaves, and the parts of the flowers, and of the fruit, the calyx, the corolla, the stamens, the styles, the seed-cells, are all only so many whorls of vegetable elements; which, according to their nutriment, developement, and other influences, assume various shapes, textures, and colours: and by this universal law

348|

of similarity, combined with certain principles of symmetry which pervade large classes of plants, are produced all the multiplicity of beauties, resemblances, and differences among flowers, which make them so graceful and splendid a portion of the creation. We have before adduced this example, in order to show that there may be large portions of the Creation in which we cannot trace any design for the good of sentient beings; and in which we must suppose that the symmetry and beauty and variety which exist, exist on their own account, and as a manifestation of Law directed to no purpose such as we can understand. But what we have now to remark is, that in this field of nature, though we cannot see the marks of Design, which commonly supply the argument for a Designer, yet that the marks of Law which are obvious, imply a Lawgiver:—the display of beauty and order reveals a Creative Spirit which delights in beauty and order. Of design indeed, so far as the preservation, sustenance, developement, and propagation of vegetable species is concerned, we have, in the structure and circumstances of vegetables, abundant evidences. Not to speak of the structure and economy of flowers and fruits,—the tendrils of the vine, the filaments with which the ivy clings to its support, and thousands of similar appendages, show the same kind of contrivance which appears in the limbs of animals. But the consideration of such contrivance is not to our present purpose. We have now rather to ask, whether we can conceive all this vast display of beauty and splendour, not unfolded in a lawless and irregular manner, like a floor on which bright colours have been casually shed, but governed by wide laws of symmetry and gradation, which make every form and every hue own an affinity with the others, while yet they are separated by boundary lines which are never transgressed;—whether this sameness so fertile of variety, this rule so flexible and yet so firm;—this symmetry so boldly extended, and yet never going beyond the limits of symmetry;—can be supposed, by any one who opens his mind to the contemplation of the law and its consequences, to proceed from any other source than a mighty Creative Mind, which also has contemplated the Law and its consequences; which had present to it, in the energy of Creative Power, the thought of this symmetry;—which delighted in exemplifying such a thought in endless forms;—which was pleased, in the working, with the beauty which man admires in the work. Undoubtedly, the general, we may say the natural impulse of the human soul, is to recognize, in such an aspect of the vegetable world, an evidence of the majesty and brightness of its Creator; of the beautiful and glorious ideas which predominated in its Creation; of an interest, if we may so speak, which the Supreme Mind takes in those aspects, even of the inanimate world, in which contemplative man takes so lively an interest. It is, we say, the natural tendency of the mind of man to assume such a kind of sympathy with him, in the Divine Mind; and it does not appear that, after any amount of philosophical discipline, any view of the Author of the Creation from which such an assumption is excluded, can at all satisfy him, as an intelligible or admissible account of the origin of this frame of things.

19 If we pass to the animal creation, and confine ourselves, in order that we may take the plainest case, to vertebrate animals, those which have an internal skeleton with backbone and limbs; we have, as we have already noticed, in addition to all the special contrivances by which each animal is fitted to its destination, a general plan, a certain collection and arrangement of bones; all which are present, or are represented, whether they are wanted or not for the purposes of the animal's life; but which are developed more or less, in this way or that, according to those purposes.

350|

Here also, as in the plan of the ordinary vegetables, we have symmetry and uniformity made the basis of infinite variety and endless modes of adaptation. The symmetry and uniformity, as we have said, are not seen by us to be necessary conditions of the adaptations. But do they not, of themselves, suggest the notion of a Law, and of a Lawgiver? Of a Lawgiver, whose plans are wider than those which the argument from design contemplates; but a Lawgiver whom, considering how narrow our knowledge of his plans is, we can easily imagine to have ample reasons for these wider laws; and who, being a Lawgiver, is a conscious and intelligent Being, manifesting in his Laws, the Ideas on which the creation is constructed, even where the object in working on such Ideas is concealed from us.

20 The mode in which Plato expressed the doctrine which we are here urging was, that there were in the Divine Mind, before or during the work of creation, certain archetypal Ideas, certain exemplars or patterns of the world and its parts, according to which the work was performed: so that these Ideas or Exemplars existed in the objects around us, being in many cases discernible by man, and being the proper objects of human reason. If a mere metaphysician were to attempt to revive this mode of expressing the doctrine, probably his speculations would be disregarded, or treated as a pedantic resuscitation of obsolete Platonic dreams. But the adoption of such language must needs be received in a very different manner, when it proceeds from a great discoverer in the field of natural knowledge; when it is, as it were, forced upon *him*, as the obvious and appropriate expression of the result of the most profound and comprehensive researches into the frame of the whole animal creation. The recent works of Mr Owen, and especially one work, *On the Nature of Limbs*, are full of the most energetic and striking passages, inculcating the doctrine

which we have been endeavouring to maintain. We may take the liberty of enriching our pages with one passage bearing upon the present part of the subject.

21 'If the world were made by any Antecedent Mind or Understanding, that is by a Deity, then there must needs be an Idea and Exemplar of the whole world before it was made, and consequently actual knowledge, both in the order of Time and Nature, before Things. But conceiving of knowledge as it was got by their own finite minds, and ignorant of any evidence of an ideal Archetype for the world or any part of it, they [the Democritic Philosophers who denied a Divine Creative Mind] affirmed that there was none, and concluded that there could be no knowledge or mind before the world was, as its cause.' Plato's assertion of Archetypal Ideas was a protest against this doctrine; but was rather a guess, suggested by the nature of mathematical demonstration, than a doctrine derived from a contemplation of the external world.

22 'Now however,' Mr Owen continues, 'the recognition of an ideal exemplar for the vertebrated animals proves that the knowledge of such a being as Man must have existed before Man appeared. For the Divine Mind which planned the Archetype also foreknew all its modifications. The Archetypal Idea was manifested in the flesh under divers modifications upon this planet, long prior to the existence of those animal species which actually exemplify it. To what natural or secondary causes the orderly succession and progression of such organic phenomena may have been committed, we are as yet ignorant. But if without derogation to the Divine Power, we may conceive such ministers and personify them by the term *Nature*, we learn from the past history of our globe that she has advanced with slow and stately steps, guided by the archetypal light amidst the

352| wreck of worlds, from the first embodiment of the vertebrate idea, under its old ichthyic vestment, until it became arrayed in the glorious garb of the human form.'

23 Law, then, implies a Lawgiver, even when we do not see the object of the Law; even as Design implies a Designer, when we do not see the object of the Design. The Laws of Nature are the indications of the operation of the Divine Mind; and are revealed to us, as such, by the operations of our minds, by which we come to discover them. They are the utterances of the Creator, delivered in language which we can understand; and being thus Language, they are the utterances of an Intelligent Spirit. We may imagine two objections made, from different quarters, to the line of speculation which we have been pursuing;—two objections, which we must consider together; since the answer to both must be given, by combining the views which they respectively urge. Each sees a difficulty, by regarding the subject from one side; and the difficulties are solved, by regarding it from both sides.

24 On the one hand, it may be said, that in urging the argument from Law, we have supposed too much the mind of man to be of the same nature as the Mind of God. We have reasoned on the supposition, that our apprehension of the relations of space, our knowledge of mathematical truth, is of the same nature as His:—that a conception of symmetry and order, such as we have, is a conception present also to the Supreme Intelligence:—that our knowledge of the law of universal gravitation is of the same kind as His knowledge of that law. To make such suppositions is, it may be said, to overlook too much the difference between the creature and the Creator:—between our finite, and his Infinite Mind;—between our dull and obscure and scanty knowledge, and his Omniscience. Are we, the objector may ask, to conceive God as knowing geometry as we know it? knowing the geometrical propositions which we know? tracing

the consequences of the law of the inverse squares which we trace? Is it not too low, too anthropomorphical a notion of the nature of the Infinite Spirit, to represent Him as delighting in what appears to us Symmetry and Beauty, because we delight in it? Is it not too bold, too presumptuous, a view of the nature of man, to regard the few scanty and imperfect glimpses which he can obtain of the Laws of Nature, as adequate specimens of the Ideas on which the plan of the Universe is based? Can Ideas belong to man and to God in any common sense?—in any identical meaning of the term?

25 This is the objection on one side, questioning the value of human Ideas in these speculations. But an objection may also be urged on the other side, derived from the necessary validity of certain of our Ideas. It may be said that the universe, in some of those aspects in which man has been most successful in discovering the relations which exist, cannot properly be said to be governed by Laws of Nature. For example, the relations of space, in which man has discovered such a large body of geometrical truth, cannot be said to be established by Laws of Nature. They are necessarily what they are: they could not be otherwise: they need no Law of Nature to make them be. It is not a Law of Nature, established by the Creator of the Universe, that two triangles which have all their sides equal to each other, shall have all their angles also equal. It is a property of space; and is true in the Universe, because the Universe is in space. We do not arrive at such knowledge by the study of natural facts and phenomena, as we arrive at the knowledge of the Laws of Nature;—we see, by intuition, such truths, or the principles of such truths. Such truths are described, by some speculators, as derived from our Idea of space; by others, as the result of hypotheses and definitions; but all agree, that they are *necessarily* true; and

354|

thus, distinct altogether from Laws of Nature, which are only known by experiment to be true; which can easily be conceived to be other than they are. We have no difficulty in conceiving that the Creator might have made bodies attract each other with forces varying according to the inverse *cube* of the distance, or any other *power*, instead of the inverse square. But we cannot conceive even Him to have made the properties of geometrical figures, or of numbers, to be other than they are. Some things we conceive that even Omnipotence cannot alter; and among such things, are such as this, that twice two is four, and not three, or five. And this being so, we cannot say that in the properties of space or number, we trace the operation of a Supreme Mind, who has willed them to be what they are;—that in discovering mathematical truths, we are making ourselves acquainted with the guiding Ideas of the Creative Intelligence.

26 These two objections are, to a certain extent, opposite to each other. The one is, that man cannot discover intention in the Universe, because he cannot enter into the intentions of the Creator:—the other is, that he cannot do this, in some things, because they are governed, not by intention, but by necessity. In our argument, we have said that the aspects of the Universe are the operations of the Divine Mind; and that these operations are revealed to us by the operations of our own minds. The latter of these two clauses of the argument, that Divine Mental Operations can be revealed to us by our own mental acts, is questioned in the former objection: the former clause of the argument, that such aspects *are* Divine Operations, is questioned in the latter objection. Or, to put the matter otherwise. In our argument, we have said that the Laws of Nature are the utterances of an Intelligent Spirit, whose language we can understand: but the former objection deems it too pre-

sumptuous a hope, to understand the language of God, as delivered in the Creation; and the latter objection sees in space, number, and the like, not a language, but a blank tablet, prepared only by being ruled into lines along which every possible inscription must be written; and therefore so far, disclosing nothing of the meaning of the Author of the world.

27 Now in reply to these objections, we shall make ourselves most intelligible by speaking briefly, using bold expressions, such as have been commonly employed by persons writing on the like subjects, and trusting that any appearance of presumption or irreverence, which these may at first sight seem to involve, will be removed, when the full import of our views is made apparent.

28 As to the former point, then, we do venture to say, that the ideas of man's mind, on the subjects on which he clearly apprehends general truths, must be supposed to be, in some measure, of the same nature as the Ideas of the Divine Mind*; and on the latter point, we say that the properties of space and number and the like, must be supposed to be what they are by an act of the Divine Mind.

29 And these suppositions combined, remove, we conceive, the difficulties which have just been stated. Taking the properties of space,—the truths of geometry,—as affording the clearest example of the matter in hand, we say that the Universe was created by its Author, in space; that is, that space was one of the Fundamental Ideas involved in its construction, so that all its parts have relations which flow from this Idea, and cannot exist without having such relations; and that further, the mind of man partakes of this

* Among the most recent expositors of this doctrine we may place M. Henri Martin, whose *Philosophie Spiritualiste de la Nature* is full of striking views of the universe in its relation to God. (Paris. 1849.)

Idea of space, and apprehends all objects according to this Idea; and cannot apprehend them otherwise. As in the Creative Mind, all things are regarded according to the Universal Creative Idea of Space; so in the created mind, having this Idea as part of its nature, all things are regarded according to the Universal Regulative Idea of Space. To man, the relations of space are seen as necessary relations, because they are a portion of the supreme and original act by which the Universe was made what it is; and because also the mind of man is made a sharer in the conditions of the creative act. To man, space is found to be a necessary condition of the existence of objects, because God has created all objects under this condition, and has made man capable of apprehending the universality of the condition. Man can apprehend no object which is not in space, because God has created no object which is not in space: the act of Creation involves a Universal Thought of the Divine Mind, which thought, the mind of man, can, according to its human powers, admit and entertain.

30 Space has been called, by some philosophers, a *Form* of human perception; as moulding and shaping necessarily all that is perceived, so as to impress upon it its own relations. If we adopt this phraseology, we must say also, that Space is a Form of Divine Creative Agency:—a universal form of such agency:—a form in the Divine Mind, so far as Creation is concerned; and a form of the human mind, as resembling, so far, the Divine Mind. Man apprehends the phenomena of the Universe, under certain universal Forms, and as exhibiting certain Ideas,—because God has created the Universe under these Forms—has interfused through its structure these Ideas; and because the mind of man is subjected to the Forms to which creation was subjected; because it has the power of detecting some of the Ideas which run through the work of Creation.

31 But man has no power of creation; and has not the power of discovering, except slowly and imperfectly, the Ideas which pervade the created world; at least, so as to acquire a clear and steady view of them, such as may lead to the apprehension of general truths. The Divine Mind contains Ideas according to which the work of creation is performed, and exercises Acts of Creative Power. But in the Divine Mind we must suppose the Ideas and the Acts, the operation of contemplation and action, to be identical. The Thoughts of the Creator are Creations. We are not to conceive space as made first, and objects put in it afterwards. Both exist necessarily together. As space is a condition of the existence of objects, so are objects a condition of the existence of space. The work of creation is, thus, subjected to conditions; and some of these conditions man apprehends as necessary truths. How many classes of such necessary truths there are, or can be, we need not here inquire. It may suffice to take the relations of space, as an obvious example of such truths. But these truths, being part of the Divine Thought which *is* creation, are part of the human thought which may be gathered out of creation. Man may, by contemplation of the Creation, ascend, gradually, as we have said, to these truths. Contemplation, with him, is so far from being commensurate with the creation which he contemplates, that he can, by the utmost effort of his powers, only obtain some glimpses of the Divine Ideas, of which the complex aggregate is the whole creation. Yet of some of these Ideas he can obtain glimpses; and these glimpses appear, in some cases at least, under the aspect of necessary truths;—truths which not only are true, but, by the nature of things, *must be* true. And so far, even as seen by man, contemplative truth and fact, possibility and actuality, present themselves as identical.

358| 32 Perhaps, as has been intimated, it may seem to

some persons too bold a view, to identify, so far as we thus do, certain truths as seen by man, and as seen by God:— to make the Divine Mind thus cognizant of the truths of geometry, for instance. If any one has such a scruple, we may remark that truth, when of so luminous and stable a kind as are the truths of geometry, must be alike *Truth* for all minds, even for the highest. The mode of arriving at the knowledge of such truths, may be very different, even for different human minds;—deduction for some;—intuition for others. But the intuitive apprehension of necessary truth is an act so purely intellectual, that even in the Supreme Intellect, we may suppose that it has its place. Can we conceive otherwise, than that God does contemplate the universe as existing in space, since it really does so;—and subject to the relations of space, since these are as real as space itself? We are well aware that the Supreme Being must contemplate the world under many other aspects than this;—even man does so. But that does not prevent the truths, which belong to the aspect of the world, contemplated as existing in space, from being truths, regarded as such, even by the Divine Mind.

33 If these reflections are well founded, as we trust they will, on consideration, be seen to be, we may adopt many of the expressions by which philosophers heretofore have attempted to convey similar views; for in fact, this view, in its general bearing at least, is by no means new. The Mind of Man is a partaker of the thoughts of the Divine Mind. The Intellect of Man is a spark of the Light by which the world was created. The Ideas according to which man builds up his knowledge, are emanations of the archetypal Ideas according to which the work of creation was planned and executed. These, and many the like expressions, have been often used: and we now see, we may trust, that there is a great philosophical truth, which they all tend |359

to convey; and this truth shows at the same time, how man may have some knowledge respecting the Laws of Nature, and how this knowledge may, in some cases, seem to be a knowledge of necessary relations, as in the case of space *.

34 The view which we have now given, appears to be strongly confirmed by the history of philosophy, and especially by the recent history of philosophy, both in this country and in Germany. In the course of that history, the distinction of necessary truths, and truths derived from experience, has given rise to much discussion. The immense success which crowned the pursuits of knowledge derived from experiment and observation of external facts, in the seventeenth century, produced a tendency to believe that all knowledge was, and must be, so derived. This philosophy especially obtained acceptance in England. It was the belief produced by the labours of Locke and his followers. But while such philosophers maintained that all knowledge must

* Most readers who have given any attention to speculations of this kind will recollect Newton's remarkable expressions concerning the Deity: 'Æternus est et infinitus, omnipotens et omnisciens; id est, durat ab æterno in æternum, et adest ab infinito in infinitum . . . Non est æternitas et infinitas, sed æternus et infinitus; non est duratio et spatium, sed durat et adest. Durat semper et adest ubique, et existendo semper et ubique durationem et spatium constituit.'

To say that God by existing always and everywhere *constitutes duration and space*, appears to be a form of expression better avoided. Besides that it approaches too near to the opinion, which the writer rejects, that He *is* duration and space, it assumes a knowledge of the nature of the Divine existence, beyond our means of knowing, and therefore rashly. It appears to be safer, and more in conformity with what we really know, to say, not that the existence of God constitutes time and space; but that God has constituted *man*, so that *he* can apprehend the works of creation, only as existing in time and space. That God has constituted time and space as conditions of man's knowledge of the creation, is certain: that God has constituted time and space as results of His own existence in any other way, *we* cannot know.

be obtained solely from observation of facts, another set of philosophers asserted, that there were some portions of knowledge which were not so derived; and gave, as an especial instance, the truths of geometry. These, they said, were not derived from experience, as being only actually and empirically true; they were seen to be true, independently of experience, necessarily and universally. And the difficulties, on each side in this controversy, were undoubtedly very great. For those who held that all truths, even those of geometry, were derived from experience, could not deny that those truths seemed to be necessary, notwithstanding. And they explained this seeming, by saying that the experience was so constant and universal, that men could not escape from its impression; and that we have the power of making in our own minds, by means of our ideas of lines, angles, and the like, experiments, which supply what observation may leave deficient. And on the other hand, those who maintained that the truths of geometry were independent of experience, were obliged to allow that man is at no period of his existence free from the impressions of experience; and that these, to which he is subjected from his earliest hours, and without cessation, do really, and must, agree with the truths which geometry asserts: so that it must be impossible to find a mind which possesses these truths, without having had them verified by experience.

35 Now perhaps we may be allowed to say, that the view which we have been endeavouring to present, goes far to reconcile these opposing doctrines, and to solve the difficulties which they object to each other. For we should say, in accordance with those views, that the truths of geometry are not derived from experience, but from the constitution of the human mind; but that, on the other hand, this constitution is not independent of the external world, but is created as part of the world, and endowed with the Idea

of space, that it may be in harmony with the world of space. In the human mind, is the Idea of space as a condition of apprehending the external world. In the world, is the Reality of space, as a condition of being apprehended. And the Idea and the Reality proceed alike from the Divine Mind, and there they coincide. The Reality is *here*, because the Idea is *there*. The two cannot exist separate. But in the human mind, we can speak of the reality and the idea separately; and though each supposes and co-exists with the other, we cannot, in our philosophy, derive the one from the other. Why does the constitution of the world, as an object, correspond with the constitution of the mind, as a mind? First; because they are works of the same Maker; but that is not the whole reason: further; because the constitution of the world is marked with the Thoughts of the Divine Mind, and the human mind is, in part, a sharer in the Thoughts of the Divine Mind. And thus, we not only see that objects do exist in space, and under *spatial* relations; but we feel that they must so exist, and that spatial relations are necessary truths.

36 The recent history of this portion of philosophy in Germany exhibits the steps of this controversy, and the attempt at its solution, on a larger scale; though apparently with little real addition to the argument on each side, and with an unnecessary and unwise boldness, in the way in which the solution is presented. When the doctrine of the English philosophers, that all knowledge is derived from mere experience, had excited attention, the question naturally arose; how, notwithstanding this doctrine, geometrical truths (for instance) could be necessarily true. To this, Kant gave a reply which produced a great impression upon the minds of his countrymen; and which, for our purpose, may be briefly expressed by saying, that there are certain *Forms of Perception* in the human mind, of which Space is one; and that

362|

all that is perceived, is necessarily perceived under these Forms; and the Matter, the Object perceived, has the attributes of the Form of Perception impressed upon it. This doctrine, that the origin of Truth and of Knowledge is in the Mind, and that the external world merely ministers material, to be shaped into knowledge by man's native Power of Knowing, was carried onwards still more resolutely by Fichte, the disciple of Kant.

37 But it soon began to be perceived that this account, though admired at the time as a vast source of new truths, was essentially incomplete. We have, the Kantians hold, in the mind, certain *Forms* of Perception, which give birth to Knowledge, by shaping the *Matter* supplied to our observation. But then, the question must occur, what is this Matter? Has it, of itself, no form? What the Forms of perception are, which belong to the constitution of our own minds, we can in some measure understand. They are Space, and Time, and it may be, others. But what is matter, what are objects, when they do not yet exist in space and time, because they are not yet perceived? We grant that to the term *Matter*, in the philosophical antithesis of Matter and Form, there must be allowed the widest possible range, as also there must, to the term *Form*. But granting any range, however liberal, of meaning, what can *Matter* be, so that there shall be such an antithesis as that of Matter and Form in this case? The Form is in the Mind: but where is the Matter? It is in the World; in the Universe. But in order that these two, Mind and Universe, may have any relation to each other, any power of affecting each other, they must have something in common. In order that Matter may be shaped, and Form may shape, they must impose conditions on each other. Each limits and modifies the other: for without the mutual operation of the two, we have Shapeless Matter and Empty Form. What is the origin of

this mutual operation, limitation, reciprocation of conditioning effect?

38 The German solution of this problem, (so far as we are here concerned with it,) was effected by the introduction of a term, which has since played a distinguished part in the philosophy of that country; though it may be doubted whether it really has much power to help us to clear views in this line of speculation. Since Matter and Form are thus both limited by external conditions, neither the one nor the other can be the Origin, or Supreme Principle of things. That Origin or Principle must itself be something free from, and superior to, conditions: imposing them, not receiving them. Taking, as a mode of expressing this, a term long established as the antithesis to that which is *Conditioned*, the German philosophers called it *The Absolute*. The Absolute, then, is the origin and principle of all being and of all conditions of being;—of Form and Matter;—of Knowledge and Existence;—of the Mind and of the Universe. And according to this doctrine, the identity of the conditions which affect both the members of the derivative results;—both Form and Matter;—both the Mind and the Universe;—is to be found in their derivation from this common origin—from this one Principle. The Universe and the Mind are different sides of The Absolute; therefore they must necessarily have the closest correspondence with each other. Form and Matter are both conditions, mutually affecting the emanations of The Absolute; therefore these conditions affect both, in a manner which keeps them in entire harmony, and makes, both independent existence, and independent knowledge, (that is separated, and so, independent in the lower steps of their derivation,) possible, in the world, and in the nature of man.

39 If, by thus assuming the *Absolute*, as a common 364| origin of the Mind and of the Universe, which, being com-

mon to the two, should explain, or rather, should render possible in our conception, the agreements and differences of the two, these speculators had only intended to express their conviction that there is such an origin; recognizing all its peculiar attributes, as being something quite out of the reach of our powers of conception, and confining themselves to the suggestion of the Idea of such an Origin, as the intelligible solution of the difficulties to which the speculations of their predecessors had led,—there seems to be no reason why we should not have accepted the suggestion, as one mode of expressing the solution of the problem with which we have been employed, so far as the coexistence and relation of Being and Knowing are concerned. But they have not been content with this. They have undertaken to go much farther. Having thus assumed the *Absolute*, as the source of the difficulties which affect our knowledge, even our most complete and comprehensive knowledge, they have, nevertheless, assumed that they comprehend this Absolute;—that they can fathom its nature, and draw their deductions from it. Having adopted the Absolute, as the source of the phenomena of the Universe, and the reason why we can know a little about them, they soon are drawn on to assume that they know all about them;—that they know far more than could have been collected from the phenomena. Having thus placed, in the Absolute, the Principle which reconciles Necessity and Fact, Possibility and Actuality, they come to assert that they can not only reconcile these rivals, but deduce the one from the other;—that they can show how fact flows from necessity;—how Possibility becomes Actuality.

40 We must exemplify, in some degree, the way in which they have done this. We, in our philosophy, are content to deduce our knowledge of the Laws of Nature, from multiplied observation and experiment: they, in their philosophy, consider all such proceeding as implying a want

|365

of the power of acquiring real knowledge. *We* trace, by laborious researches, the rules which govern the phenomena of gravitation, or of light: we think it much, if we can discover the general laws of the forces which act in the solar system, or which regulate the polarization of light: but they conceive themselves able to determine by *à priori* reasonings, (if they are to be called reasonings,) what the laws of gravitation and of polarization *must* necessarily be. *We* admire the persevering and arduous course of scientific toil, by which Kepler discovered his Laws of the Planetary Motions, and by which Newton from these, educed the Laws of Solar and Planetary Forces: but Hegel, though he allows some credit to Kepler, for discovering the Laws of the Motions, allows none to Newton, for discovering the causes of these motions*; and thinks that he can, in a few lines, give quite satisfactory reasons why the motions follow Kepler's Laws. The reasons which Hegel thus gives, as superseding Newton's reasonings, and showing them to be false and worthless, have been carefully and fairly considered in this country, and discussed, so far as they admit of discussion. And we do not see why persons who have thus weighed these speculations should not describe them plainly; and say that these, and many of the like speculations of other German philosophers, are ignorant, presumptuous, and illogical, to an extent to which no injustice is done by calling them childish. The same description applies, and has been shown to apply, to Hegel's attempt to prove *à priori* the laws of polarization: a fact at which no one, acquainted with those laws and their history, will be at all surprized.

41 Perhaps it may serve to illustrate to the reader unfamiliar with the history of celestial mechanics, or of modern optics, what we venture to regard, as the presumptuous and superficial tone of speculation of this kind of German

* Encyclopædie, §§ 269, 270.

writers, if we add, that not only in matters of physical science, have they thus pretended to prove that all that ages of laborious research have with difficulty discovered, might have been foreseen as necessary, and may be proved *à priori*: but that they have entered upon the like speculations, with the like boldness and the like claim of success, in other departments of human knowledge; in which the undertaking to deduce, from certain original principles, a long train of results, will probably be generally regarded as showing no ordinary presumption, and as necessarily empty of any philosophical value. Thus Hegel has derived the whole course of history from *à priori* principles; showing that nations possessing such characters as he ascribes to the Orientals, the Greeks, the Romans, and the Germans, must necessarily have succeeded each other as prominent figures on the stage of the world's history*. He has, in like manner, given an *à priori* history of Law, through all ages; and indeed, of most of the subjects of human knowledge†. Now we may readily grant that a person taking these vast ranges of history, as fields in which he seeks for resemblances and differences, for gradations and contrasts, may hit upon many striking views, many epigrammatic turns of thought, many poignant antitheses, many comprehensive progressions. Such are to be found in every lively and eloquent view of a subject. But that these parallels and contrasts, these antitheses and epigrams, these

* *Philosophie der Geschichte.*

† We may remark, that Hegel's views of the trial of Socrates have been shown, by Bishop Thirlwall, to be bold but feeble paradoxes. *History of Greece*, Vol. IV. p. 526, (2d Ed.) As an example of the mode in which Hegel gives to his views an air of novelty by the invention of a peculiar phraseology, we may adduce what he says of the opponents of Socrates: 'Sie fassten den Socrates als den Menschen auf, der das Nichabsolutseyn des Anundfürsichgeltenden zum Bewusstseyn brachte.' Of which this is the nearest translation which we can give: 'They apprehended Socrates as the man who brought to the consciousness [of men] the Not-being-absolute of the By-and-for-itself valid [truth].'

metaphors and phrases, are to pass for a Philosophy of History or of Law, or of Taste, or of any large subject, shows, we may almost say, not only great presumption in him who delivers such a so-called *philosophy*, but a certain abjectness of spirit in those who receive it as a profound philosophy. That views of this kind may be employed with great effect, for some special purpose, in which eloquence has its proper office, was well shown by an earlier writer of the German school—Fichte: when, in his *Orations to the German Nation*, delivered in 1808, at Berlin, he constructed hypothetically, on *à priori* principles, two Nations; one an original, unmixed, unconquered nation, speaking its original language; the other a mixed nation, the result of repeated conquests of the country, and speaking an adopted language; and pointed out the natural antipathy of spirit in the two, and the superiority of character in the former. This was felt to apply to the Germans and the French; nations then opposed to each other in a struggle for life and death: and no doubt, those Lectures, delivered at that time with a noble courage, did much to warm the national heart of the Germans, and to rouse their national spirit, at a moment when that was a worthy task for a patriotic man of letters. But no one would, we may suppose, now place this view as a great philosophical discovery: and yet it seems to be far worthier of such a character, than all the historical speculations of Hegel.

42 But we need not dwell further on this subject. What we mainly wished to illustrate, by this reference to recent German philosophy, was, that there, as in our own country, the contemplation of Knowledge, as we know it, and Existence as we see it, suggests the necessity of something which is the common source both of our knowledge and of the existence of things. We conceive that we find this common source in the Deity, and in the relation borne to Him, both 368| by our own Minds, and by the Universe: while at the same

time, regarding Him as Deity, we must look for and find in Him, much more than this. For though we, in this part of our speculations, regard Him only as the author of the physical Laws of the Universe, which it is the object of physical science to discover, He must also be the author of the Moral Law of our being, the ground of our religious hopes, and the object of our spiritual aspirations. These views are, however, to be reserved for other occasions.

43 In thus dwelling so long upon physical Laws, as the evidence of a Supreme Intelligence, we may seem to have deviated from our first subject, which was, to weigh the evidence for and against the Earth being the sole habitation of intelligent creatures. And yet, in truth, the views to which we have been led, bear very strongly upon that argument. For if man, when he attains to a knowledge of such Laws, is really admitted, in some degree, to the view with which the Creator himself beholds his creation;—if we can gather, from the conditions of such knowledge, that his intellect partakes of the Nature of the Supreme Intellect;—if his Mind, in its clearest and largest contemplations, harmonizes with the Divine Mind;—we have, in this, a reason which may well seem to us very powerful, why, even if the Earth alone be the habitation of intelligent beings, still, the great work of Creation is not wasted. If God have placed upon the earth a creature who can so far sympathize with Him, if we may venture upon the expression;—who can raise his intellect into some accordance with the Creative Intellect; and that, not once only, nor by few steps, but through an indefinite gradation of discoveries, more and more comprehensive, more and more profound; each, an advance, however slight, towards a Divine Insight;—then, so far as intellect alone (and we are here speaking of intellect alone) can make man a worthy object of all the vast magnificence of Creative Power, we can hardly shrink from believing that he is so.

CHAPTER XIII.

The Omnipresence of the Deity.

---

1 THE reflections presented in the chapter which we have just closed, may be carried somewhat further; and so pursued, may, perhaps, offer some additional views of the Universe, which, for many minds, will have a great interest. We have there regarded the Laws of Nature,—which involve so much of mathematical relation which we can unravel, so much of symmetry and order which we can admire, —as disclosing to us the intention of a Divine Mind, which works according to Laws of Geometry, or Plans of Symmetry and Order. We have said, for example, that the law of gravitation by which all the planets are constantly drawn towards the Sun, with a force of which the intensity varies inversely as the square of the distance from the center;—a law which produces an endless train of results, of which it requires the utmost powers of the highest mathematical minds among men, to trace only a few;—must be conceived, as being a law impressed by a Divine Mind, which foresaw, at once, all these results, and which comprehends at a glance all the mathematical relations which they involve. But we have said further, that these mathematical thoughts of the Creator, which are thus manifested to us in the Creation, must be conceived as identical with the act of Creation. The Ideas of the Divine Mind, which are, for us, embodied in the objects of the Universe, are the Laws of the Universe. By being His Ideas, they are its Laws. If we could conceive the Laws of the Universe abolished, there would be no Universe; for the Universe is only, as we have attempted

to show, a collection of what we call, Laws of Nature. If the Divine Ideas ceased, or could be for a moment conceived to cease, to govern the Universe, the phenomena of the Universe would cease to be governed, and would cease to be.

2 But how are we to conceive that the Divine Ideas govern the Universe? In what manner are we to suppose that the Laws, impressed by the Creator, control and direct the matter which is subject to them? In what manner, for instance, are we to suppose that the Law of Planetary Gravitation to the Sun, operates upon the planets? Are we to conceive this Law, as a kind of subordinate implement or minister, established between the Creator and his work;—set up, once for all, and appointed to perform its task, without further interference from Him? Can we conceive this? A Law is not an implement or a minister. It is a system of thought, which requires a Will to carry it into effect: and whose is this will? Manifestly, it cannot be any other, than the Will of the Creator himself. He must constantly execute what He himself planned. He must carry into effect the Law which is His Law. To assume any inferior Will, as the means by which the Law is enforced and maintained, would be an imagination for which we have no ground, and which is at variance with the mode in which we are compelled to conceive the relation of the Creator and his work. The Creator himself, then, is constantly maintaining the Laws of Nature, which He has appointed to be Laws. It is from His constant efficacy and agency, that they have their force. As He has decreed that the planets shall be urged to the Sun, by a force which is regulated by a certain Law; so by His perpetual act, they are so urged. His agency is universally at work, that His work may go on according to his plan. He has created the Universe which we behold; and this creation is real, because He is constantly

maintaining and regulating it. The phenomena according to their Laws, the courses of the planets, the order of nature, are the Act which He is constantly performing from the beginning to the end of the world.

3 That it is by the universal diffusion of the Divine Agency, that the Laws of the Planetary Forces, and in like manner, all other Laws of Nature, are realized and maintained, is a view which has often been put forwards, and which contains nothing which needs to startle or disturb those who have been accustomed to dwell upon the *Omnipresence* of the Deity, as one of the truths of Natural Religion. But it may be of use to offer some further remarks on this subject.

4 We may remark, in the first place, that this necessity of the Divine Agency, to carry into effect any of the Laws of Nature, for instance, the Law of Planetary Force, is not at all removed, by our imagining, or even by our discovering, any machinery by which the effect may be mechanically produced;—any train of intermediate mechanism which may connect the Sun and the Planets. Undoubtedly, at first sight, such intermediate mechanism does appear to make the operation more intelligible. We are accustomed to see matter operate upon matter by contact, and not otherwise: and when the discovery of the force of gravitation seemed to require that bodies should be supposed to act upon other bodies which are at a distance from them, the supposition was repelled by many, as repugnant to common sense. Even Newton, at one period of his speculations at least, declared that such a hypothesis was intolerable; and was greatly inclined to suppose an ether, or medium of some kind, which, extending from the Sun to the farthest planets, might establish a material connexion between them; and thus, make their mutual mechanical action possible. This suggestion has not been followed out, so as to lead to any theory of

372|

gravitation, which has obtained general acceptance among mathematicians. But what we have to remark is, that if such a theory had been contrived, and put in such a form that it should explain the phenomena, it would not remove the necessity which exists, as we are trying to show, for the Divine Agency, to realize and maintain the Laws of Nature. For this machinery, whatever it is, whether it operate by means of solid, fluid, or etherial links of connexion, must act according to the Laws of Nature; and we shall still have to ask, what it is that gives to those Laws their efficacy. Probably such a theory, if proposed, would be founded on mechanical principles, and upon the Laws of Motion. But what is it that upholds the Laws of Motion? What agency is it, for instance, which makes the first Law of Motion subsist in all parts of the Universe? This first Law is, as we have said, that a body put in motion, and not impeded, will go on for ever;—will move indefinitely onwards without changing its velocity. But, as we then asked, Who has determined what will happen for ever? Who has regulated what will come to pass through infinite time and infinite space?—so now we may further ask, Can we imagine that body has of itself a power to do this? Must we not suppose, not only that a Supreme Mind has determined what a body so circumstanced shall do, but that a Supreme Will compels the body to do this? The body is a mass held together by certain Laws of aggregation;—it is a crystalline mass, for instance, confused or clearly developed. What power shall we conceive, as fitted to carry such a mass onwards from place to place? to hold its particles together, while they thus move through indefinite regions of space, and continue to move for an indefinite period of time, so that it still continues the same body? Can the Laws of Crystallization, so combined with the Laws of Motion, be upheld, except by a continuation of the same act which made them to be Laws? And this, which is thus true of the Laws of motion, is equally true, and equally manifest, with regard to all other Laws, on which any mechanical explanation of the Law of gravitation can rest. Therefore we say, that no such explanation of the Planetary Forces, however persuasive as a theory, (and very persuasive theories on that subject have been propounded,) can make it conceivable, that such forces can continue constantly to operate, in all parts of the Solar System, and as we have reason to believe, in all parts of the Universe, without a Divine Agency, constantly acting, to realize the Law; as well as a Divine Thought, exercised at first to give form to the Law.

5 We may here call to mind what we have already ventured to say;—that, in regarding the world as the expression of the Thoughts of a Divine Mind, we must conceive the Act of Creation to be coincident with the Thought of Creation. We must conceive that when God makes a Law of Nature, it becomes a Law of Nature. The Idea and the Actuality are coincident in time. But if we conceive this to be the case as to Time, can we forbear from supposing it also to be the case as to Space? If *when* God connects certain results with certain conditions, they become forthwith connected by his act; must we not suppose that *where* he connects results with conditions, they are there connected by his act? The Laws of Nature are established, so far as we know, for infinite Time, and for infinite Space. But who or what can carry them into effect through infinite time and space, except a Being who is himself infinite? And if we must suppose the thought of the Divine Mind to be never, for a moment, barren and empty; so that no vacant moments or ages can intervene, between the appointment of the Law and its realization: must we not also suppose that no vacant space can intervene, the radius of a planet's orbit or the breadth of a hair, between the First Universal Agent and

the object acted upon? Must we not suppose God to be in every part of space, effecting what he has decreed shall take place in every part of space? As Nature, in its Laws, is the expression of His Ideas, is not Nature, in its Laws, also the expression of His Activity? If he is every where present, must he not be every where active? And must not his activity be the cause of all motion, and of all change?

6 If we thus suppose God, Infinite and Eternal, to be present in all parts of space, at all moments of time, and to carry into effect by his power the Laws of Nature, which he has established in his wisdom, many difficulties are removed, and many obscurities cleared. It, then, no longer becomes wonderful, that we should have Laws of Motion which determine the course and velocity of bodies through infinite space and for infinite time: for wherever and whenever they come, He is there to control and direct. It is, then, no longer inconceivable that a body should act on a body at a distance; for He is present at each body, and may urge them together according to any Law, which he may have appointed: while, at the same time, the Law may be determined by some special pre-eminence of simplicity, or produced mediately by some intermediate mechanism. On such a supposition, we must suppose also that the laws of light and heat, of air and moisture, are His Laws; not only in the sense of their being appointed by Him, but also, of their being carried into effect by Him. We must suppose also that all the phenomena of vegetable life are the manifestations of his constant and universal energy. Not only has He established the laws by which leaf and flower, and fruit and seed, constantly grow out of each other; but His active power is constantly at work, educing the flower from the stem, the fruit from the flower, the seed from the fruit, the germ of the new plant from the seed. Plants, thus going through their continual round of being, have

a kind of life: and the Divine Spirit which by its agency fills the universe, is the life of their life. And thus we can adopt the language of the poet, when he says that God

> Warms in the Sun, refreshes in the breeze,
> Glows in the stars, and blossoms in the trees;
> Lives through all life, extends through all extent,
> Spreads undivided, operates unspent.

and many other expressions, used by contemplative writers of all kinds*, in which God is spoken of in the like manner.

7 It is, however, proper to remind the reader that such expressions as these are conceived, by many pious and thoughtful theologians, to be likely, if not used with great caution, to lead to dangerous error. Many of the expressions which are often combined with these, are rightly judged to be unworthy and unfit representations of the relation between the Creator and the Creation. Such, for instance, is the couplet which occurs at the beginning of the passage which we have cited:

> All are but parts of one stupendous whole,
> Whose body Nature is, and God the soul.

This is a mode of speaking of the subject by all means to be avoided. God is not the Soul of the World:—He is the Maker, Upholder, and Lord of it. Nature, that is material nature, is not His Body, in opposition to the Soul; Nature is his Work. It is a work, not executed once for all, and left to itself: but a work which He is always working;—*Opus quod operat Deus a principio ad finem.* But though

* The philosophical reader will have, brought to his mind, the expression of Pascal, that God is a circle whose center is everywhere, and circumference nowhere. And still more to our purpose, we may recollect Newton's view, that the world 'can be the effect of nothing else than the wisdom and skill of a powerful, ever-living agent, who, being in all places, is more able by his will to move the bodies within his boundless, uniform *sensorium*, thereby to form and reform the parts of the universe, than we are by our will, to move the parts of our own bodies.'

He is always working, and present in every part of his work, He is not to be identified with his work. His work is not nature merely, but man also; man the intellectual, moral, religious, spiritual being. And man is not his work only, but his subject also. The souls of men are not only his creation, but also his empire. There, he not only works, but he governs:—gives Laws, judges, condemns, approves. To call God the Soul of the World, is a mode of expression which is fallacious, even as to His relation with material things; and is altogether at variance with his relation to spiritual creatures. And though we do not here dwell upon this latter relation, we must still warn the reader against delusive forms of speech, which may seem to be suggested by the views which we have ventured to present.

8 In offering such warnings, we only follow the example of Newton who, in the celebrated Scholium at the end of the *Principia*, says of God:—'Hæc omnia regit non ut Anima Mundi sed ut universorum dominus, et propter dominium, Dominus Deus vocatur:' 'He governs all this system of things, not as the soul of the world, but as the Lord of All Things; and on account of this Dominion, he is called the Lord God.'

9 Besides the error of speaking of God as the Soul of the World, into which men, as we have seen, may be led by dwelling upon the view of his animating presence in every part of the Universe, there is another kind of erroneous opinion, which has sometimes prevailed on this subject, and which is described by the name of *Pantheism*. The doctrine so described may be characterized briefly by the expression which its name almost involves;—that the Universe is God. Since God, present through all time and all space, does all that is done in the Universe;—since the Laws of Nature are, at the same time, His thoughts and His acts;—since the objects which make up Nature are merely the collection of

such Laws:—it may seem that we do not need to include any thing in our conception of God, beyond these Laws and these Acts. The general aggregate of Ideas and Activities which makes up the Universe, may also seem to be that which is God, so far as He is, or can be, known to us. But yet, a little attention will show us that, even with regard to Material Nature, this is a very different view from that which we have presented. According to our view, God is not a collection of Laws; he is a Lawgiver:—not a collection of Acts; he is an Agent. The objects which make up the Universe, do not make up Him: they are His works, not Himself. Behind and beyond all these material elements, all these cosmical laws, all this universe of things, we conceive a Divine Intelligence, a Conscious Spirit, a Mind involving Design and Will, in short, a Person, whose thoughts and works, the elements and their Laws, are. This, we say, is the doctrine that we have presented, even looking at the material world alone. Even there, we strongly and broadly distinguish between God and his World, which Pantheism confounds. But still more broadly and strongly is God distinguished from the Universe, when we take into account, as we have just said, the other and more important province of God's empire:—the minds and souls of men; with all the relations that may arise between God and them, not because they are identical with God, and parts of Him, but because they are, or may be, estranged from Him and opposed to Him.

10 But avoiding these fallacious and erroneous forms of expression:—that God is the Soul of the Universe; and that God is the Universe; the doctrine of the universal presence of God, and his universal activity, according to the laws of nature which he has established, may be dwelt upon by us with pleasure, as it has been dwelt upon by pious and thoughtful men in all ages. Every one will recollect the

numerous expressions to this effect which occur in the Book of Psalms. The 139th Psalm will at once occur to every reader, with its deep conviction of the presence of God in every part of the universe, and his operation in every province of nature. 'Lord, thou hast searched me out and known me; thou knowest my down-sitting and mine uprising; thou art about my path and my bed.' And again: 'Thou hast fashioned me behind and before, and laid thine hand upon me. My bones are not hid from thee. Thine eyes did see my substance yet being imperfect: and in thy book were all my members written, which day by day were fashioned, when as yet there was none of them.' The mystery of the developement of the living creature naturally turns the thoughts of the pious man to the pervading and universal agency of the Author of Life and of Organization. And again, in the 104th Psalm we have other expressions, implying the agency of God in every part of the creation, animate and inanimate. 'He laid the foundations of the earth. Thou coveredst it with the deep, as with a garment. At his word the waters go up as high as the hills, and there flow out in springs to the valley beneath.' And this is provided for the use of animals: 'All beasts of the field drink thereof: the wild asses quench their thirst: the fowls of the air have their habitation beside them: He bringeth forth grass for the cattle; and herbs, and wine, and oil, for the service of man.' Through His agency the trees are full of sap. The night does not interrupt His beneficent care: 'Then all the beasts of the forest move, and seek and find their food at his hand.' And in the same spirit, St Paul speaks of God's care for man, (Acts xiv.), 'Who left not himself without witness, in that he did good, and gave us rain from heaven, and fruitful seasons, filling our hearts with joy and gladness.' And in addressing the Athenians, he uses the expressions which imply that all being is involved in the being of God. 'For in Him we live, and move, and have our being.' And to the same effect is the teaching of Christ himself, (Luke xii. 6.) 'Are not five sparrows sold for two farthings? and not one of them is forgotten before God.'

11 This view of all the phenomena and objects of the world, as the work of God; not a work made, and laid out of hand, but a field of his present activity and energy; cannot fail to give an aspect of dignity to all that is great in creation, and of beauty to all that is symmetrical, which otherwise they could not have. And accordingly, it is by calling to their thoughts the presence of God as suggested by scenes of grandeur or splendour, that poets often reach the sympathies of their readers. And this dignity and sublimity appear especially to belong to the larger objects, which are destitute of conscious life; as the mountain, the glacier, the pine-forest, the ocean; since in these, we are, as it were alone with God, and the only present witnesses of His mysterious working.

12 Now if this reflection be true, the vast bodies which hang in the sky, at such immense distances from us, and roll on their courses, and spin round their axles with such exceeding rapidity; Jupiter and his array of Moons, Saturn with his still larger host of Satellites, and with his wonderful Ring, and the other large and distant Planets, will lose nothing of their majesty, in our eyes, by being uninhabited; any more than the summer-clouds, which perhaps are formed of the same materials, lose their dignity from the same cause;—any more than our Moon, one of the tribe of satellites, loses her soft and tender beauty, when we have ascertained that she is more barren of inhabitants than the top of Mount Blanc. However destitute the planets and moons and rings may be of inhabitants, they are at least vast scenes of God's presence, and of the activity with which he carries into effect, everywhere, the laws of nature. The light which

380|

comes to us from them is transmitted according to laws which He has established, by an energy which He maintains. The remotest planet is not devoid of life, for God lives there. At each stage which we make, from planet to planet, from star to star, into the regions of infinity, we may say, with the patriarch, 'Surely God is here, and I knew it not.' And when those who question the habitability of the remote planets and stars are reproached as presenting a view of the universe, which takes something from the magnificence hitherto ascribed to it, as the scene of God's glory, shown in the things which he has created; they may reply, that they do not at all disturb the glory of the creation which arises from its being, not only the product, but the constant field of God's activity and thought, wisdom and power; and they may perhaps ask, in return, whether the dignity of the Moon would be greatly augmented if its surface were ascertained to be abundantly peopled with lizards; or whether Mount Blanc would be more sublime, if millions of frogs were known to live in the crevasses of its glaciers.

CHAPTER XIV.

Man's Intellectual Task.

---

1 WE must not overlook very important limitations of what has been said, respecting the acts which take place in the creation, as being all manifestations of the Divine Agency. Such a view may be conceived to be true, with regard to all the acts, if they are to be so called, of insentient matter;—with regard to all the laws affecting mere material things, which have no power of thought or will. All such objects are passive results of the Divine Activity:—they are collections of laws originally made, and continually executed, by God: But there are acts of another kind, performed by creatures:—acts which are really acts; acts of thought, and acts of will. Are these, also, only parts of the Divine activity? Does God think and will *in us*, as he moves matter, or developes plants? Are our thoughts His thoughts, in the same sense in which the motions of the planets, the play of chemical affinities, the efflux of light, the operation of warmth and moisture, the rise of vegetable juices, the successions of vegetable races, are His acts?

2 To answer in the affirmative, would plainly involve us in opinions concerning the thoughts and acts of the Deity, in the highest degree incongruous, and at variance with the ideas which we cannot but form of his perfection and wisdom. That our dark, imperfect, struggling efforts to obtain a sight of truth, can be, directly, the acts of a supreme, unerring, luminous, all-comprehending Intelligence, is contrary to all that we can conceive of the nature of intelligence. Our perception of truth is constantly mixed with

error; and is, in fact, only a few gleams of light, in the midst of obscurity and darkness. Our knowledge is a knowledge of a very few Laws, which operate in a universe which is pervaded by an infinite number of Laws. Our explanations of Nature are scanty and limited views of a scene which, as we can discern, extends, in every direction, and in every point of view, infinitely beyond our knowledge. Our thoughts cannot be as His thoughts, except by a remote similitude, and to an infinitesimal extent.

3 And in fact, our speculations give us no ground, and no encouragement, to make such a supposition. On the contrary, they give us clear and broad reasons, why we should say that, though the events which take place in the *material* world, according to the Laws which the Author of the Universe has, in his wisdom, ordained, are, in a certain sense, also His acts, the emanation of His power; yet that the events which take place in our *intellectual* world, the thoughts which we think, cannot be said, at all in the same sense, to be His thoughts, or the direct result of His agency. For, in the material world, there is no error, no falsehood. The objects which make up that world, always go right;—always proceed according to the Laws laid down for them;—are always true to their principles. The planets never deviate from the courses appointed for them, according to the Laws of motion. If they ever wander from the orbits which we had calculated for them, it is because we had calculated wrong, not because they go wrong. If they seem to transgress the obvious Laws of Motion, it is because they obey the Laws of Motion, in a higher and truer sense. If we cannot calculate what they ought to do, according to such Laws, they nevertheless do exactly what, according to such Laws, they ought to do. The mathematician, in attempting to calculate their motions, is stopped by difficulties in his own art; difficulties of elimination, of integration, and the like. He finds the problem

insoluble, or is compelled to content himself with an approximate solution. But the bodies themselves solve the problem; or rather, the Divine Presence and Agency solve the problem for them; and that, not approximately, but exactly, rigorously, unerringly. The Laws of Nature are always perfectly obeyed, completely realized. And the same is the case in other departments of nature. Every species of plant constantly grows according to the laws of its kind. Its leaves, its flowers, its fruit, its seeds, are always precisely what they ought to be. It may happen that, under peculiar circumstances, there are abnormal growths, monstrous productions, seeming deviations. But these seeming deviations from Law, are really governed by Law. They are due to the modifying circumstances; and the effect of these circumstances, (light, heat, moisture, close proximity of several germs, and the like,) act by Laws as fixed, as the Laws of ordinary growth: and thus, these Laws are as faithfully followed as the ordinary laws. What seems most irregular, is still rigorously governed by Rule. The Laws of Nature are always exactly carried into effect: and therefore, we are never tempted, by ascribing to God the execution of those Laws, to say that God is the author of error, failure, disorder, confusion, or falsehood.

4 But, what takes place in the Intellectual World, in the region of human thought and knowledge, is entirely of a different kind. The Law of man's belief and conviction is Truth; for he cannot believe, or be convinced of, anything without regarding it as true. And that there is a Truth, which man may attain to the knowledge and belief of, we cannot doubt. The whole structure of Science, so far as it is clearly discovered and firmly established, consists of such Truth. The human mind, when its intellectual powers and habits are unfolded, is constantly and irresistibly impelled in the pursuit of such Truth. To seek to know what is true, and

384|

to believe it when found, is the Law of man's intellectual nature. But how scantily, how imperfectly, to how small an extent, with how little of fulness or clearness in the results, is he able to obey this Law! How perpetually is he misled, by some tempting form of error, to mistake falsehood for truth;—to believe it and cling to it, as the very truth he was in search of! *Misled*, we do not hesitate to say; because a little while later, he discovers that his supposed truth was falsehood. He finds a new, and more solidly proved truth, which supersedes his former favourite belief. And with what labours and struggles, what bewilderment and doubt, what conjectures and hypotheses, what recurrence backwards and forwards, from the Thoughts which he thinks to the Facts which he sees, and back again, from Facts to Thoughts,—does he attain to true thoughts! through what oscillations between Phenomena and Ideas, does he arrive, in any department of nature, at a stable Theory! All these wandering and erroneous movements, all these deviations from the line which points to Truth, all these transgressions of the Law which draws him to Truth, cannot be regarded as being, in their general assemblage, acts of a Divine Intellect, which cannot, for a moment, acquiesce in anything that is not true. It cannot be, that the Divine Mind manifests itself alike, in all these struggles and wanderings, all these errors and contradictions, all these oscillations between wrong and right; or more frequently, between two opinions both wrong. This blindness and confusion, this mistake and inconsistency, this adoption of the false, and dulness to the true, belongs to man, not to God.

5 And yet, as we have already said, there is in the very power of discovering truth, something divine. That man can, at all, and to the extent to which he has done it, learn the Laws by which the Universe is governed, is an evidence

that he has a share of that Intelligence by which the Laws of the Universe were devised. That he can attain to any truth, shows that he has in him a principle derived from the Source of all Truth. That he can find, in the depths of his mind, and with whatever labour, Ideas which give congruity to the Phenomena of the Universe, proves that his Ideas have some resemblance to the Ideas by which the Phenomena of the Universe are regulated in the Divine Mind. He has in his soul a light which has emanated from that Source of Light, by which the Creation was illuminated into an orderly scheme; however this light may be dimmed and obscured, distorted and perverted, by the other elements of his nature. He has the Divine Treasure, though in earthen vessels.

6 This, then, is the view to which we are led, by combining what we have said of man's darkness, and of his light; of his love of truth, and his proneness to error. He has, implanted in him, a germ of pure Intellect, or of insight of truth, by which he is enabled to discover and know the truth; and at the same time, this germ can only be developed slowly, laboriously, and by the aid of favourable circumstances; and he has also, along with this germ of truth, an abundant store of germs of error, which may easily check, or quite overpower, the more genuine and precious growth. He has an intellectual light, but the light has to struggle with clouds and shadows; and may be dimmed, or even, (so far as the higher kinds of knowledge are concerned,) extinguished. He has a spark of celestial flame in his soul; but the flame needs fuel, to be gathered from the material world; and needs, too, shelter from the adverse blasts which blow across it from other portions of his human nature. His power of knowing truth may be regarded as the element of a divine nature; but man is not merely intellect; he has many other elements in his human nature; and intellect,

which has its office in conjunction with these, may be drawn aside, or overmastered, by the other principles with which man is endowed;—endowed for purposes which may be regarded as even more important than the mere right exercise of intellect;—as higher than the discovery, even of the largest and loftiest scheme of mere scientific truth.

7 If we consider the office and duty of man, regarding him merely as an intellectual creature, employed in the contemplation of the creation, and called upon, by the law of his nature, to aim at the attainment of a true view of the Universe, in which he exists; we may say, that his great business consists in controlling and subduing the influences within him, which may prevent or retard his discovery of truth, and in unfolding and perfecting the powers which he has in him tending to such discovery. It is his appointed task to foster the divine germ of pure intelligence; to clear his spark of the original light from cloud and dimness; to cherish the sacred flame of divine insight; to unfold, and exercise, and carry onwards, towards their perfect work, the powers which he possesses of seeing as God sees. However imperfect his progress in this task must, after all is done, be, still it is his task; and his progress, if slow, is still, possibly, indefinite: if he cannot do all, he may still do much. And this task must be forwarded, both by a study of the Universe without, and of the Light within:—both by supplying material to the fire, and by arranging the fuel so that the flame may burn brightly. He must discover the Laws of the Universe, both by seeing clearly the Universe, and by seeing clearly what is Law. By studying the Phenomena of Nature, and by unfolding the Ideas of his mind, he must bring the Phenomena and the Ideas into coincidence in his mind, as they coincide in the Mind of God.

8 This is the Intellectual Task of man:—a task involving a series of efforts and struggles;—capable of failure, or

of a success, more or less complete. It is therefore a Trial: —a Trial imposed upon him by the Author of his being, who gave him his place in the Universe, and his capacity and desire of knowledge. And since the Author of man's being has thus placed him here, and imposed upon him such a task, and such a trial, we now see, perhaps, why He has, in a great degree, left him to think for himself, and to seek truth by his own acts. In these matters, in the business of the intellectual world, God does not think man's thoughts for him; because man is placed here, that he may exercise and apply his own intellectual powers, and show, to the eye of his Creator, Governor and Judge, what use he can and will make of the gifts which have been bestowed upon him. The planets and the stars, the elements and the vegetable creation, are not left to do anything for themselves. God does all for them; for they are subjected to no trial. They have no task, no business, no duty in this higher sense. They always perform, they cannot but perform, their task, their office; which is, to obey the Laws of Nature, and so, to carry on the business of the material world. Nature never errs: but man errs; and is made liable to error, that he may be capable of discovering truth. If he saw all truths by intuition, without the possibility of error, he would be already divine. If he had no liability to error, but no apprehension of truth, he would be a stock or a stone, an elementary or vegetable production, and not an intellectual creature. By being capable of error, and yet capable of truth;—perpetually drawn into error, and yet never willingly acquiescing in it as error;—endued with a power of going right, which yet does not secure him from going wrong;—he is capable of being placed in a condition of intellectual trial;—of showing what use he can make of divine endowments and privileges;—of trying how far he can, *on this road*, advance towards the Supreme Intelligence

388|

from which his love of truth and capacity of truth are derived.

9 We may once more observe, that these reflections are not without a very strong bearing upon the question with which we were at first principally concerned;—the possibility of the earth's being the only seat of intelligent life. For if, on the earth, the Creator have placed a race who are not only endowed with a portion of the Divine Intellect, but who are placed there in order, (at least among other purposes,) that they may cultivate and develope this gift, and thus, rise nearer and nearer to the condition of the Divine Intellect, and be fitted, so far, for an immortal existence; we cannot have any ground to think that the scheme of creation is too narrow; or that it needs, in order to give it sufficient dignity and value, and a worthy object in our eyes, that other worlds should be stocked with races of creatures, with regard to whom we cannot assume such a condition, without the most wanton and baseless exercise of fancy: and who, even if they were in a like condition, could only add the value of number to that aspect of the intellectual world, of which the least valuable feature is its number. For how few of mankind realize the intellectual destiny thus placed before them! And if, by any act of the Divine Government, the number of those who rightly aspire to approach to the Divine Nature, were greatly increased upon earth, would not this be a far more suitable, and a far more intelligible method, of extending the Divine Kingdom of Intellect, than any multiplication of worlds could be, supposing the worlds to be such as this world of earth was, before that act of the Divine Government by which man was created, in his intellect at least, after the Image of God!

CHAPTER XV.

Man's Moral Trial.

---

1 THERE is one other line of speculation, suggested by some of the reflections which we have had to make, which we must not decline to enter upon briefly; because the reader would probably feel the question to have been imperfectly treated, if our discussion were to stop here: though, in truth, the subject now to be considered, is not very closely connected with such philosophical views as we have had to contemplate; and is very difficult to deal with in a way at all satisfactory, within the limits of the present Essay. We have hitherto spoken of man as in a state of *intellectual* trial only; and have attended to the other principles of his nature, besides his intellect, only as they affect and disturb the clear and sure working of his intellect;—as they make truth difficult for him to discover, or mislead him to falsehood. But man is, we have the strongest reason to believe, in a state of moral, as well as intellectual trial; indeed, far more; for all persons are called upon to obey moral rules, and are condemned, if they transgress such rules: while the cultivation of the intellect, especially in its higher forms, and as leading to a comprehensive and scientific knowledge of the universe, is the Task of few; and success in this Task, is the privilege of still fewer. It will, then, be well, if we can see how far this view of man, as a creature placed on earth with a view to a moral trial, harmonizes, in any degree, with the other speculations which we have been pursuing.

2 As the intellectual distinctions of opinions are ex-

pressed by speaking of opinions as *true* and *false*, so the moral differences of actions are most simply expressed by speaking of actions as *right* and *wrong*. As it is a Law of man's nature that he seeks truth, and cannot acquiesce in any opinion except by believing it to be true; so it is a Law of man's nature that he must do what is right; and he cannot conceive any reason for doing an act, superior to this, that it *is* right. It may be painful; it may be a loss; it may be evil spoken of; but if it be right, he knows that he *must* do it; he *should* do it; he *ought* to do it. For these are the phrases by which the direction or command of the Moral Law is expressed. To describe an action from which the Law of his nature excludes him, there is no term more complete and final than to say, that it is wrong to do it; he ought not to do it. *Right* and *wrong*, the adjectives, *ought* and *ought not*, the verbs, express the supreme moral differences of actions. Connected with these, and sharing, in different ways, their signification, are other terms, as *Virtue*, *Duty*, and the like; which also serve to express the moral Law; and others, as *Vice*, *Crime*, *Sin*, which express transgressions of the Law under different points of view.

3 Man has certain impulses to act, as he has an intellectual impulse to know; or rather, he has many impulses to actions of various kinds; of which the impulse to know, is only one. The other impulses to act, or, as we may call them, *Springs of Action*, are generally more vigorous and effective than the desire of knowledge. Such are the rude animal Springs of Action; hunger and thirst and lust, the desire of shelter, and warmth, and favourite food. Along with these, come the love of offspring, and other kinds of love; and again, anger, which gives force and energy to the acts prompted by other impulses. But in man, all these Springs of Action take a mental and reflective character, in consequence of his rational nature. He desires Security;

he desires Property; abstractions which include, in a more general form, many objects of mere animal desire. But further than this, he has a desire for Social Existence; a tendency to live as man with men: to be a member of a civil community, bound to the other members of it by the use of language, and by the various ties which human society involves. These desires, bodily, mental, and social, which form the principal Springs of Action in man, must be so regulated and controlled, that his actions, prompted by his desires, conform to the moral rule;—that they are *right* actions.

4 The desires which have been enumerated are regulated and controlled by certain institutions which form the conditions of man's social existence;—the institutions of Property, Family, Contract, Government. It is as acting under these institutions, that man has to be tried, whether he will do what is right, or what is wrong. His trial is, whether he will conform to the rules which man's social existence thus involves: first, whether he will conform his actions, so as not to transgress the instituted human Laws by which Property, Family, and the like, are upheld; next, whether he will conform his internal springs of action, his Affections, Desires, Habits and Will, to these conditions; and finally whether, not content with avoiding, even in thought and desire, what is wrong, he will aim at higher developements of his moral being;—at acts of Duty and habits of Virtue, in the highest form. This is the trial which man has to undergo upon earth.

5 The Law—the Moral Law to which man is required to conform, and in conforming to which his trial consists;—is a Law of which God is the Legislator. It is a Law which man has the power of discovering, by the use of the faculties which God has given him. By considering the nature and consequences of actions, man is able to discern, in a

392|

great measure, what is right and what is wrong;—what he ought and what he ought not to do;—what is duty and virtue, what is crime and vice. Man has a Law on such subjects, written on his heart, as the Apostle Paul says. He has a conscience which accuses or excuses him; and thus, recognizes his acts as worthy of condemnation or approval. And thus, man is, and knows himself to be, the subject of Divine Law, commanding and prohibiting; and is here, in a state of probation, as to how far he will obey or disobey this Law. He has impulses, springs of action, which urge him to the violation of this Law. Appetite, Desire, Anger, Lust, Greediness, Envy, Malice, impel him to courses which are vicious. But these impulses he is capable of resisting and controlling;—of avoiding the vices and practising the opposite virtues;—and of rising from one stage of Virtue to another, by a gradual and successive purification and elevation of the desires, affections and habits, in a degree, so far as we know, without limit.

6 Now in considering the bearing of this view upon our original subject, we have, in the first place, to make a remark similar to that which we made with reference to man's condition on earth, as a state of intellectual trial: namely, that the existence of a body of creatures, capable of such a Law, of such a trial, and of such an elevation as this, is, according to all that we can conceive, an object infinitely more worthy of the exertion of the Divine Power and Wisdom, in the Creation of the universe, than any number of planets occupied by creatures having no such lot, no such law, no such capacities, and no such responsibilities. However imperfectly the moral law may be obeyed; however ill the greater part of mankind may respond to the appointment which places them here in a state of moral probation; however few those may be who use the capacities and means of their moral purification and elevation;—still, that there is

such a plan in the creation, and that any respond to its appointments,—is really a view of the Universe which we can conceive to be suitable to the nature of God, because we can approve it, in virtue of the moral nature which He has given us. One school of moral discipline, one theatre of moral action, one arena of moral contests for the highest prizes, is a sufficient center for innumerable hosts of stars and planets, globes of fire and earth, water and air, whether or not tenanted by corals and madrepores, fishes and creeping things. So great and majestic are those names of *Right* and *Good, Duty* and *Virtue*, that all mere material or animal existence is worthless in the comparison.

7 But further: let us consider what is this moral progress of which we have spoken;—this purification and elevation of man's inner being. Man's intellectual progress, his advance in the knowledge of the general laws of the Universe, we found reason to believe that we were not describing unfitly, when we spoke of it as bringing us nearer to God;—as making our thoughts, in some degree, resemble His thoughts;—as enabling us to see things as He sees them. And on that account, we held that the placing man, with his intellectual powers, in a condition in which he was impelled, and enabled, to seek such knowledge, was of itself a great thing, and tended much to give to the Creation a worthy end. Now the moral elevation of man's being is the elevation of his sentiments and affections towards a standard or idea, which God, by his Law, has indicated as that point towards which man ought to tend. We do not ascribe *Virtue* to God, adapting to Him our notions taken from man's attributes, as we do when we ascribe Knowledge to God: for Virtue implies the control and direction of human springs of action;—implies human efforts and human habits. But we ascribe to God infinite Goodness, Justice, and Truth, as well as infinite Wisdom and Power; and

Goodness, Justice, Truth, form elements of the character at which man also is, by the Moral Law, directed to aim. So far, therefore, man's moral progress is a progress towards a likeness with God; and such a progress, even more than a progress towards an intellectual likeness with God, may be conceived as making the soul of man fit to endure for ever with God; and therefore, as making this earth a prefatory stage of human souls, to fit them for eternity;—a nursery of plants which are to be fully unfolded in a celestial garden.

8 And to this, we must add that, on other accounts also, as well as on account of the capacity of the human soul for moral and intellectual progress, thoughtful men have always been disposed, on grounds supplied by the light of nature, to believe in the existence of human souls after this present earthly life is past. Such a belief has been cherished in all ages and nations, as the mode in which we naturally conceive that which is apparently imperfect and deficient in the moral government of the world, to be completed and perfected. And if this mortal life be thus really only the beginning of an infinite divine plan, beginning upon earth and destined to endure for endless ages after our earthly life, we need no array of other worlds in the universe to give sufficient dignity and majesty to the scheme of the Creation.

9 We may also here repeat, with reference to the moral scheme of the world, a remark which we have already made with reference to man's intellectual condition. If by any act of the Divine Government the number of those should be much increased, who raise themselves towards the moral standard which God has appointed, and thus, towards a likeness to God, and a prospect of a future eternal union with him;—such an act of Divine Government would do far more towards making the Universe a scene in which God's goodness and greatness were largely displayed, than could be done by any amount of peopling of planets with creatures

who were incapable of moral agency; or with creatures whose capacity for the developement of their moral faculties was small, and would continue to be small till such an act of Divine Government were performed. The Interposition of God, in the history of man, to remedy man's feebleness in moral and spiritual tasks, and to enable those who profit by the Interposition, to ascend towards a union with God, is an event entirely out of the range of those natural courses of events which belong to our subject: and to such an Interposition, therefore, we must refer with great reserve; and using great caution, that we do not mix up speculations and conjectures of our own, with what has been revealed to man concerning such an Interposition. But this, it would seem, we may say:—that such a Divine Interposition for the moral and spiritual elevation of the human race, and for the encouragement and aid of those who seek the purification and elevation of their nature, and an eternal union with God, is far more suitable to the Idea of a God of Infinite Goodness, Purity, and Greatness, than any supposed multiplication of a population, (on our planet or on any other,) not provided with such means of moral and spiritual progress.

10 And if we were, instead of such a supposition, to imagine to ourselves, in other regions of the Universe, a moral population purified and elevated without the aid or need of any such Divine Interposition; the supposed possibility of such a moral race would make the sin and misery, which deform and sadden the aspect of our earth, appear more dark and dismal still. We should therefore, it would seem, find no theological congruity, and no religious consolation, in the assumption of a Plurality of Worlds of Moral Beings: while, to place the seats of such worlds in the Stars and the Planets, would be, as we have already shown, a step discountenanced by physical reasons; and discountenanced the more, the more the light of science is thrown upon it.

CHAPTER XVI.

The Design of Animal Springs of Action.

---

1 PERHAPS, in urging these opinions, as those which principally give us a satisfactory and adequate view of the object of the scheme of God's Creation, we may appear to some persons to overlook, or unreasonably to reject, a doctrine which is often put forwards:—that the happiness of man, and the enjoyment of all sentient creatures, is the pervading purpose of the Creation. It may be urged that this purpose is shown to be really pursued, by the endless number and variety of the contrivances which have the enjoyment and comfort of animals for their object; and that it agrees entirely with what we are led to conceive of the nature of God, as infinitely benevolent; and as using his wisdom and his power to further the designs of his benevolence. And it may be held, further, that the aspect of the world really supports and justifies this view. The cheerfully tempered Natural Theologian, looking at the indications of the vast diffusion of animal enjoyment, says, 'It is a happy world after all.' And if this be so, it may appear that the happiness which exists in a world, is a sufficient reason for the existence of that world itself. It may be urged, that since God has taken so much care to provide for the enjoyments of animals, *that* must be held by us to be a worthy object of Creative Power and Care.

2 And even with regard to man, it may be said, that there is no Law of his Nature, more general and pervasive than this;—that he seeks happiness. Happiness is his being's end and aim; that which gives the attractiveness to

everything which he desires and pursues; that which expresses the charm of everything which he enjoys. Indeed others will be disposed to go further, and to hold Morality itself to be merely a system of rules, the end of which is to promote human happiness, and which derive their force and their value from their subservience to that end. And thus, it would seem, that the scheme of Creation ought to be discussed, as to its bearing upon this object:—the happiness and enjoyment of its inhabitants; and its provisions, its extent and its plan, ought to be judged of with reference to that, as its dominant purpose.

3 Let us speak of this view, first, as it concerns the condition of man. That happiness is the object, more or less latent, of all rational human action, we readily grant:—so far even, as to assent to the common doctrine that acts of duty and virtue must be performed, by a rational being, with a conviction of their suitableness to this object: and that they would not be suitable to man's nature, except he were persuaded that either in the present life, or in some other state of his being, they will promote his happiness. Nor shall we here deny, that from the purpose of promoting our own happiness and that of all mankind, rules of duty may be drawn, which agree, in the main, with the rules of duty which we obtain by regarding man's whole nature, and the necessary distinction of right and wrong, which he, in virtue of his nature, applies to all actions and impulses to action. The morality founded upon the promotion of human happiness, agrees, in the main, with the morality founded upon man's moral nature. But yet, except our morality, however founded, be such as to point out the moral elevation of man's nature, as a great and worthy object in itself, and as the source of his highest destinies in this life, and in a future life, so far as morality deals with that question; it will be a morality which will not suffice to give us an adequate view of

man's peculiar place in the creation, and of the character of the scheme of creation, so far as they depend upon man's peculiar nature and endowments. If we consider man's terrestrial happiness and enjoyment as the ultimate end for which he was placed on earth, and therefore, as the object and measure of morality, we do not give to man any place in the creation superior to animals. On that supposition, it may be that his pleasures are promoted by larger and more subtle contrivances than those of other animals; by the rules of morality, among others: but still, the value of the human part, as well as of every other part of creation, must, on this supposition, be estimated by the amount of pleasure produced. And in this view, it would certainly seem that there are intelligible reasons why other planets, as well as the earth, should be stocked with sentient creatures, capable, like the inhabitants of the earth, of pleasure and enjoyment.

4 But let us somewhat further consider the opinion thus propounded;—that the animals besides man, and, as *we* say, inferior to man, are created and placed on the earth with a view to their enjoyment; and that the amount of enjoyment so produced may be regarded as a sufficient and satisfactory reason why they are so created and placed, by the hand of an infinitely benevolent Creator. That among animals there is a great amount of enjoyment, we by all means grant. All the appetites and needs of their nature are provided for, by means which are accompanied with pleasure, when the appetites are indulged and the needs supplied: and besides the mere gratification of appetite, and the enjoyment of bodily comfort, there appear to be other pleasures which animals enjoy—a kind of gaiety, as shown in the songs of birds, and the gambols of brutes; and also, it would seem, in some, a kind of pleasure in companionship. But our estimate of these latter kinds of pleasure must be doubtful and obscure. The gambols of animals, as

of kittens and puppies, seem often to be a sort of rehearsal of the future struggles of the chase or of the fight, which their condition will require. The songs of birds may be the expression of pain or need, as well as of pleasure and complacency. But let all these indications of pleasure be fully estimated. Still we must, in such a discussion, estimate the pain also, which undoubtedly prevails, very strongly and extensively, in the creation. We cannot fail to recollect how many tribes of animals derive their food exclusively from the wounds, mutilation, and death of other tribes. And the pain of wounds is the same to brutes as to men; and even the paroxysms of terror and agony which the fear of death produces, do not seem to require human reason to give them their pang. The more timid and helpless animals would seem to live a life of constant terror and painful watching, of anxiety and distress: while creatures of prey pass through alternate periods of forced and painful abstinence, and fierce and greedy indulgence. The animal creation appears to be an almost universal war of race against race; and this war is accompanied throughout by extreme suffering on the one side, and gluttony and cruelty on the other. It is hardly to our purpose at present to take into our account the enjoyments and sufferings of the animals living in a state of domestication, under the control of man: but we may observe, that though the animals on which he feeds, are kept generally in a state of comfort, till they are put to death, the animals which work for him, the horse, the ass, the ox, the dog, are often treated with a degree of cruelty, which as it is frequently brought under our eyes, cannot fail to shake our belief that the animal world contains a vast excess of enjoyment over pain. Taking into account all the means which we have, of judging of the pleasures and pains of animal life, we may often be led to doubt, whether, merely as a means of producing pleasure, a perfectly benevolent Being would

400|

have thought it worth while to create a world of brute animals: especially considering that the pleasures can at most be only the pleasures of appetite and anger, satiety and repose.

5 But, it will be asked, on what account, then, were animals created? What reason can we assign for their existence, if their own enjoyments are not reason enough? It would be very presumptuous to pretend to answer such a question as this, in a dogmatical manner: and, indeed, we may most willingly leave to their own belief, those who can see in the animal creation a sufficient amount and balance of pleasure to make *that* the ground of its having been created. But to many persons, the animal creation, and its fortunes, as they are most commonly seen, offer a far less cheerful spectacle than this. It appears to them, to be full of suffering and pain, unmerited, unredressed, uncompensated. And this aspect is all the more distressing and uncheered, because we can see, in this pain and suffering, no process of discipline, no means of educing and developing higher qualities; no preparation for another state, in which the hardships of the present lot shall be softened, and the inequalities and seeming injustice of the present good and evil remedied by a future good and evil of a higher order. To such spectators, the scene of the being and actions of animals, if it be regarded as including the whole scheme of creation, cannot but appear, as in a very high degree, unsatisfactory, incomplete, and melancholy.

6 And this view by no means applies only to the races of animals which now exist upon the earth, as the contemporaries of man. The past ages of animal creation, as revealed to us by the discoveries of geologists, exhibit, so far as we can judge, exactly a similar series of actions and sufferings. The Earth has, through all periods, during which animals have lived upon it, been the scene of constant

war and slaughter, pursuit and fear, between different races of creatures. At all times, and every where, the stronger have persecuted, and preyed upon the weaker. Pain and death, and the fear of pain and death, have, throughout, been the sources of life, and of the means of life and enjoyment.

7 And however we may say that death ends the pains of animals, and that the life which is destroyed is compensated by the life that is supported and perpetuated, it is not easy to make, out of this, a condition on which we can look with satisfaction and complacency;—on which we can dwell as a scheme of pure benevolence. At least we may be excused if, in order to obtain some additional satisfaction, we apply, to this part of the order of creation, some of the views which we have already been led to entertain with regard to the insentient provinces of nature. We may be allowed to ask, whether any analogies, borrowed from the structure of animals, may also apply to their habits; so as to explain, or to suggest some explanation of, those features in animal life which do not so readily appear to be explained by that notion of creative design to which we have referred:—the notion of a creative design, that animals, with their habits of mutual violence and injury, should exist on their own account alone, and in order that there might be, in the world, the full amount of animal enjoyment.

8 A notion which we were led to entertain, with regard to the structure of animals, was this:—that though many parts of that structure exhibit wonderful contrivances, in each species, for the special good of the species, and are, thus, evident instances of a design which we can understand; yet, that other parts of the structure exhibit the manifest indications of general laws, which are not seen by us to be necessarily or primarily subservient to such special design;
402| but are there, *as* general laws; which exist, so far as we

can see, merely as general laws, in addition to the means required by the special design in each case. Thus to take, once more, the class of vertebrated animals, as that class in which, both special design, and general law, are most clearly manifested; and which has, for us, a peculiar interest, as involving, among its forms, the structure of man;—we see that there is a general plan of such animals; however different, to a common eye, may be their appearance, mode of action, circumstances, and habits. All have a back-bone, a skull, fore and hind extremities: the vertebræ in every case consist of the same rudimentary elements, variously modified and developed:—so do the limbs;—so does the skull. In man, this developement is more complete; and in his structure, many of the elements receive a meaning which could not be understood, when they were merely contemplated in a ruder and more undeveloped form of animal being. The radius and ulna of the fore arm, the bones of the wrist and fingers, of the foot and toes, are discoverable in the structure of animals which lived thousands, and it may be hundreds of thousands of years, before (so far as we can judge) there was upon the earth any creature possessed of a fore arm, with the power of turning the extremity upwards and downwards; or before there were in the animal creation, fingers, or toes like fingers. Such a vast scheme of foresight in the structure of animals, may well excite our admiration, if we regard the general law by which such rudiments exist in the skeleton, ages before they are called into action, as a general law, to which this part of the animal creation was subordinated from the first; with a prevision of its ultimate developement, and as a means of giving the basis and means of such developement. But the general law by which such rudiments of human limbs were prepared so long before, existed, as it must appear to us at least, not for the sake of those extinct animals, but because there was, in the Creator's mind, some reason for proceeding in the work of the animal creation, by such a law. The general law is there, as a general law, applicable to the most complete animal structures; and any attempt to explain the existence of the rudimentary parts of the skeleton, by regarding the structure and habits of the inferior animals only, must necessarily have been unsatisfactory; because the general law was devised and introduced with a purpose, which went far beyond the being and life of the inferior animals.

9 Now may we not apply a view of the same kind to the springs of action of animals, as well as to the limbs and bodily instruments by which those springs of action operate? As, in the scheme of creation, it was thought good by the Creator, that an immense class of animals should have fore-legs, and hind-legs, jaws, and eyes, and noses;—though of very different forms and structures, yet all with a certain community of plan;—so it was thought good by the Creator, that all these animals should have hunger, and thirst, and the appetite of generation; that all should have the impulse of taking and keeping what their needs require; that all, or almost all, should have the love of their offspring; that many should be gregarious; that some should live in pairs, some make themselves habitations; some live in orderly communities with distinctions of classes and offices among the individuals of the community. No doubt, in each case, these impulses, these desires, these springs of action, have a special accordance with and subservience to the animal's mode of life. They are means of securing the sustenance, continuance, and enjoyment of the species. But these springs of action, the appetites and bodily desires, *in their general form,* exist so universally, that they are parts of a general plan, as well as examples of special adaptation. And not the appetites and bodily desires only, are thus diffused among animals, but certain affections or sentiments; and

404|

especially anger, which co-operates with the desires, giving an almost supernatural energy and vivacity to the actions done under its prompting, repelling injury when threatened, and punishing it when inflicted.

10 Now these general impulses, which are the springs of action in animals, are also the springs of action in man. The Appetites and Bodily Desires exist in man, no less than in animals. They exist in man, however they are included, summed up, and transformed into more general and abstract shapes, by the operation of man's rational and reflective powers. The gratification of the appetites and bodily cravings, the desire of freedom from pain and fear, assume, collectively, the form of a Desire of Security and Well-Being. The mere animal impulse to obtain and keep objects of desire, is developed into the tenacity of our Property. The mere animal love of partner and of offspring, is unfolded into the human Love of Family. The gregarious instinct takes the higher form of a rational tendency to live in an organized Society. Yet still, the germs of all these leading springs of *human* action, are to be found in the mere *animal* springs of action. And even Anger, the impulse which appears, at first sight, most entirely animal and brutal, and which, even in man, retains some strong traces of its ferine affinities, is sublimed, under the influence of human capacities and human relations, into the Indignation and Wrong; and thus gives energy to all the most genuine impulses of human, and even of moral action. And thus the springs of action in animals, may all be regarded as *rudiments* of the proper and universal springs of human action. And the general law of the animal creation, that such springs of action should exist, may be regarded as instituted, not for the sake of animals alone, but as a great feature of the animal creation, with man at its head, who is far more than a mere animal; and in whom, the animal springs of

action assume a form and a purpose, which could not have been conjectured, from their operation in animals merely: just as the meaning of the rudimentary bones of the human hand could not have been conjectured, from merely considering the paw of the lion, or the foot of the crocodile.

11 If this view of the subject be admissible, we may then venture to say, that the life of animals, and the appetites, desires, and affections by which that life is supported, do indeed exist for their own sakes, and are provided with a regard to the life of each species; but we may say also, that these appetites, desires, and affections, are part of a general plan of the immaterial constitution of animals:—part of a universal law of sentient nature; which does not exist for the sake of its direct results merely, the animal qualifications to which it leads: but as a general law, which is there on its own account; which may lead to other results of a higher kind; as we see that, in the human species, it does: and which does not lead to failure, when it produces pain as well as pleasure, suffering as well as gratification; since these, in the operation of such a law, are necessarily correlative with each other. There cannot be the operation of these animal instincts without animal suffering. The Laws of the animal creation are, for the most part, Laws of War; or rather, a Law of War, without the limit which *Law*, in its human sense, implies. Man only is capable of *Laws of War;* and hence, of Laws of Peace.

12 If, then, we have to contemplate an animal world, in which there is no creature of the nature of man, in whom the animal impulses are rationalized, controlled, balanced, subjected to Law, and made the means of moral actions; we have to contemplate a world, in which the aspect of the habits of living creatures is incomplete, rudimentary, and undeveloped; and is, therefore, naturally unsatisfactory, unintelligible, and painful. And no multiplication of such

406|

worlds, to any amount, would make the creation a spectacle on which we can dwell, with the same kind of satisfaction, and conviction of its consistency, worthiness, and intelligibility, which we experience when we regard the world as the seat of man; in whom the animal impulses are unfolded and combined with his rational nature; so that there is presented to us a scene of moral Law and order, of moral discipline and control, of moral progress and elevation.

13 Perhaps it may be said, that all which we have urged to show that other animals, in comparison with man, are less worthy objects of creative design, may be used as an argument that other planets are tenanted by men, or by moral and intellectual creatures like men; since, if the creation of *one* world of such creatures exalts so highly our views of the dignity and importance of the plan of creation, the belief in *many* such worlds must elevate still more our sentiments of admiration and reverence of the greatness and goodness of the Creator; and must be a belief, on that account, to be accepted and cherished by pious minds.

14 To this we reply, that we cannot think ourselves authorized to assert cosmological doctrines, selected arbitrarily by ourselves, on the ground of their exalting our sentiments of admiration and reverence for the Deity, *when the weight of all the evidence which we can obtain respecting the constitution of the universe is against them.* It appears to us, that to discern one great scheme of moral and religious government, which is the spiritual center of the universe, may well suffice for the religious sentiments of men in the present age; as in former ages such a view of creation was sufficient to overwhelm men with feelings of awe, and gratitude, and love; and to make them confess, in the most emphatic language, that all such feelings were an inadequate response to the view of the scheme of Providence which was revealed to them. The thousands of millions of inhabitants of the Earth,

to whom the effects of the Divine Plan extend, will not seem, to most religious persons, to need the addition of more, to fill our minds with sufficiently vast and affecting contemplations, so far as we are capable of pursuing such contemplations. The possible extension of God's spiritual kingdom upon the earth will probably appear to them a far more interesting field of devout meditation, than the possible addition to it of the inhabitants of distant stars, connected in some inscrutable manner with the Divine Plan.

15 To justify our saying that the weight of the evidence is against such cosmological doctrines, we must recall to the reader's recollection the whole course of the argument which we have been pursuing.

It is a possible conjecture, at first, that there may be other worlds, having, as this has, their moral and intellectual attributes, and their relations to the Creator. It is also a possible conjecture, that this world, having such attributes, and such relations, may, on that account, be necessarily unique and incapable of repetition, peculiar, and spiritually central. These two opposite possibilities may be placed, at first, front to front, as balancing each other. We must then weigh such evidence and such analogies as we can find on the one side or on the other. We see much in the intellectual and moral nature of man, and in his history, to confirm the opinion that the human race is thus unique, peculiar and central. In the views which Religion presents, we find much more, tending the same way, and involving the opposite supposition in great difficulties. We find, in our knowledge of what we ourselves are, reasons to believe that if there be, in any other planet, intellectual and moral beings, they must not only be *like* men, but must *be* men, in all the attributes which we can conceive as belonging to such beings. And yet to suppose other groups of the human species, in

408| other parts of the universe, must be allowed to be a very bold hypothesis, to be justified only by some positive evidence in its favour. When from these views, drawn from the attributes and relations of man, we turn to the evidence drawn from physical conditions, we find very strong reason to believe that, so far as the Solar System is concerned, the Earth *is*, with regard to the conditions of life, in a peculiar and central position; so that the conditions of any life approaching at all to human life, exist on the Earth alone. As to other systems which may circle other suns, the possibility of their being inhabited by men, remains, as at first, a mere conjecture, without any trace of confirmatory evidence. It was suggested at first by the supposed analogy of other stars to our sun; but this analogy has not been verified in any instance, and has been, we conceive, shown in many cases, to vanish altogether. And that there may be such a plan of creation,—one in which the moral and intelligent race of man is the climax and central point to which innumerable races of mere unintelligent species tend,—we have the most striking evidence, in the history of our own earth, as disclosed by geology. We are left therefore, we conceive, with nothing to cling to, on one side, but the bare possibility that some of the stars are the centers of systems like the Solar System; an opinion founded upon the single fact, shown to be highly ambiguous, of those stars being self-luminous: and to this possibility, we oppose all the considerations, flowing from moral, historical, and religious views, which represent the human race as unique and peculiar. The force of these considerations will, of course, be different in different minds, according to the importance which each person attaches to such moral, historical, and religious views: but whatever the weight of them may be deemed, it is to be recollected that we have on the other side a bare possibility, a mere conjecture; which, though suggested at first by astronomical discoveries, all more recent astronomical researches have failed to confirm in the smallest degree. In

|409

this state of our knowledge, and with such grounds of belief, to dwell upon the Plurality of Worlds of intellectual and moral creatures, as a highly probable doctrine, must, we think, be held to be eminently rash and unphilosophical.

16 On such a subject, where the evidences are so imperfect, and our power of estimating analogies so small, far be it from us to speak positively and dogmatically. And if any one holds the opinion, on whatever evidence, that there are other spheres of the Divine Government than this earth, —other regions in which God has subjects and servants,—other beings who do his will, and who, it may be, are connected with the moral and religious interests of man;—we do not breathe a syllable against such a belief; but, on the contrary, regard it with a ready and respectful sympathy. It is a belief which finds an echo in pious and reverent hearts*; and it is, of itself, an evidence of that religious and spiritual character in man, which is one of the points of our argument. But the discussion of such a belief does not belong to the present occasion, any further than to observe, that it would be very rash and unadvised,—a proceeding unwarranted, we think, by religion, and certainly at variance with all that science teaches,—to place those other, extra-human spheres of Divine Government, in the Planets and in the Stars. With regard to the planets and the stars, if we reason at all, we must reason on physical grounds; we must suppose, as to a great extent we can prove, that the laws and properties of terrestrial matter and motion apply to them also. On such grounds, it is as improbable that visitants from Jupiter or from Sirius can come to the Earth, as that men can pass to those stars: as unlikely that inhabitants of those stars know and take an inter-

* "For doubt not that in other worlds above
There must be other offices of love,
That other tasks and ministries there are,
Since it is promised that His servants, there,
Shall serve Him still."—TRENCH.

est in human affairs, as that we can learn what they are doing. A belief in the Divine Government of other races of spiritual creatures besides the human race, and in Divine Ministrations committed to such beings, cannot be connected with our physical and astronomical views of the nature of the stars and the planets, without making a mixture altogether incongruous and incoherent; a mixture of what is material and what is spiritual, adverse alike to sound religion and to sound philosophy.

17 Perhaps again, it may be said, that in speaking of the shortness of the time during which man has occupied the earth, in comparison with the previous ages of irrational life, and of blank matter, we are taking man at his present period of existence on the earth:—that we do not know that the race may not be destined to continue upon the earth for as many ages as preceded the creation of man. And to this we reply, that in reasoning, as we must do, at the present period, we can only proceed upon that which has happened up to the present period. If we do not know how long man will continue to inhabit the earth, we cannot reason as if we did know that he will inhabit it longer than any other species has done. We may not dwell upon a mere possibility, which, it is assumed, may at some indefinitely future period, alter the aspect of the facts now before us. For it would be as easy to assume possibilities which may come hereafter to alter the aspect of the facts, in favour of the one side, as of the other*. What the future destinies of our race, and of the earth, may be, is a subject which is, for us, shrouded in deep darkness. It would be very rash to assume that they will be such as to alter the impression

* For instance, we may assume that in two or three hundred years, by the improvement of telescopes, or by other means, it may be ascertained that the other planets of the Solar System are not inhabited, and that the other stars are not the centers of regular systems.

derived from what we now know, and to alter it in a certain preconceived manner. But yet it is natural to form conjectures on this subject; and perhaps we may be allowed to consider for a moment what kind of conjectures the existing state of our knowledge suggests, when we allow ourselves the licence of conjecturing. The next Chapter contains some such conjectures.

CHAPTER XVII.

The Future.

1 WE proceed then to a few reflections to which we cannot but feel ourselves invited by the views which we have already presented in these pages. What will be the future history of the human race, and what the future destination of each individual, most persons will, and most wisely, judge on far other grounds than the analogies which physical science can supply. Analogies derived from such a quarter can throw little light on those grave and lofty questions. Yet perhaps a few thoughts on this subject, even if they serve only to show how little the light thus attainable really is, may not be an unfit conclusion to what has been said; and the more so, if these analogies of science, so far as they have any specific tendency, tend to confirm some of the convictions, with regard to those weighty and solemn points,—the destiny of man, and of mankind,—which we derive from other and higher sources of knowledge.

2 Man is capable of looking back upon the past history of himself, his Race, the Earth, and the Universe. So far as he has the means of doing so, and so far as his reflective powers are unfolded, he cannot refrain from such a retrospect. As we have seen, he has occupied his thoughts with such contemplations, and has been led to convictions thereupon, of the most remarkable and striking kind. Man is also capable of looking forwards to the future probable or possible history of himself, his race, the earth, and the universe. He is irresistibly tempted to do this, and to endeavour to shape his conjectures on the Future, by what he

knows of the Past. He attempts to discern what future change and progress may be imagined or expected, by the analogy of past change and progress, which have been ascertained. Such analogies may be necessarily very vague and loose; but they are the peculiar ground of speculation, with which we have here to deal. Perhaps man cannot discover with certainty any fixed and permanent Laws which have regulated those past changes which have modified the surface and population of the earth; still less, any laws which have produced a visible progression in the constitution of the rest of the universe. He cannot, therefore, avail himself of any close analogies, to help him to conjecture the future course of events, on the earth or in the universe; still less can he apply any known Laws, which may enable him to predict the future configurations of the elements of the world; as he can predict the future configurations of the planets for indefinite periods. He can foresee the astronomical revolutions of the heavens, so long as the known Laws subsist. He cannot foresee the future geological revolutions of the earth, even if they are to be produced by the same causes which have produced the past revolutions, of which he has learnt the series and order. Still less can he foresee the future revolutions which may take place in the condition of man, of society, of philosophy, of religion; still less, again, the course which the Divine Government of the world will take, or the state of things to which, even as now conducted, it will lead.

3 All these subjects are covered with a veil of mystery, which science and philosophy can do little in raising. Yet these are subjects to which the mind turns, with a far more eager curiosity, than that which it feels with regard to mere geological or astronomical revolutions. Man is naturally, and reasonably, the greatest object of interest to man. What shall happen to the human race, after thousands of 414| years, is a far dearer concern to him, than what shall happen

to Jupiter or Sirius: and even, than what shall happen to the continents and oceans of the globe on which he lives, except so far as the changes of his domicile affect himself. If all our knowledge of the earth and of the heavens, of animals and of man, of the past condition and present laws of the world, is quite barren of all suggestion of what may or may not hereafter be the lot of man, our knowledge will lose the charm which would have made it most precious and attractive in the eyes of mankind in general. And if, on such subjects, any conjectures, however dubious;—any analogies, however loose, can be collected from what we know, they will probably be received as acceptable, in spite of their insecurity; and will be deemed a fit offering from the scientific faculty, to those hopes and expectations,—to that curiosity and desire of all knowledge,—which gladly receive their nutriment and gratification from every province of man's being.

4 Now if we ask, what is likely to be the future condition of the population of the earth as compared with the present; we are naturally led to recollect, what has been the past condition of that population as compared with the present. And here, our thoughts are at once struck by that great fact, to which we have so often referred; which we conceive to be established by irrefragable geological evidence, and of which the importance cannot be overrated:—namely, the fact that the existence of man upon the earth has been for only a few thousand years:—that for thousands, and myriads, and it may be for millions of years, previous to that period, the earth was tenanted, entirely and solely, by brute creatures, destitute of reason, incapable of progress, and guided merely by animal instincts, in the preservation and continuation of their races. After this period of mere brute existence, in innumerable forms, had endured for a vast series of cycles, there appeared upon the earth a creature,

even in his organization, superior far to all; but still more superior, in his possession of peculiar endowments;—reason, language, the power of indefinite progress, and of raising his thoughts towards his Creator and Governor: in short, to use terms already employed, an intellectual, moral, religious, and spiritual creature. After the ages of intellectual darkness, there took place this creation of intellectual light. After the long continued play of mere appetite and sensual life, there came the operation of thought, reflection, invention, art, science, moral sentiments, religious belief and hope; and thus, life and being, in a far higher sense than had ever existed, even in the slightest degree, in the long ages of the earth's previous existence.

5 Now this great and capital fact cannot fail to excite in us many reflections, which, however vaguely and dimly, carry us to the prospect of the future. The present being *so* related to the past, how may we suppose that the future will be related to the present?

In the first place, *this* is a natural reflection. The terrestrial world having made this advance from brute to human life, can we think it at all likely, that the present condition of the earth's inhabitants is a final condition? Has the great step from animal to human life, exhausted the progressive powers of nature? or to speak more reverently and justly, has it completed the progressive plan of the Creator? After the great step by which man became what he is, can and will nothing be done, to bring into being something better than man now is; however that future creature may be related to man? We leave out of consideration any supposed progression, which may have taken place in the animal creation previous to man's existence; any progression by which the animal organization was made to approximate, gradually or by sudden steps, to the human

416| organization; partly, because such successive approximation is questioned by some geologists; and is, at any rate, obscure and perplexed: but much more, because it is not really to our purpose. Similarity of organization is not the point in question. The endowments and capacities of man, by which he is man, are the great distinction, which places all other animals at an immeasurable distance below him. The closest approximation, of form or organs, does nothing to obliterate this distinction. It does not bring the monkey nearer to man, that his tongue has the same muscular apparatus as man's, so long as he cannot talk; and so long as he has not the thought and ideas which language implies, and which are unfolded indefinitely in the use of language. The step, then, by which the earth became a *human* habitation, was an immeasurable advance on all that existed before; and therefore there is a question which we are, it seems, irresistibly prompted to ask, Is this the last such step? Is there nothing beyond it? Man is the head of creation, in his present condition; but is that condition the final result and ultimate goal of the progress of creation in the plan of the Creator? As there was found and produced something so far beyond animals, as man is, may there not also, in some course of the revolutions of the world, be produced something far beyond what man is? The question is put, as implying a difficulty in believing that it should be so; and this difficulty must be very generally felt. Considering how vast the resources of creative power have been shown to be, it is difficult to suppose that they are exhausted. Considering how great things have been done, in the progress of the work of creation, we naturally think that even greater things than these, still remain to be done.

6 But then, on the other hand, there is an immense difficulty in supposing, even in imagining, any further step, at all commensurate in kind and degree, with the step which carried the world from a mere brute population, to a human

population. In a proportion in which the two first terms are *brute* and *man*, what can be the third term? In the progress from mere Instinct to Reason, we have a progress from blindness to sight: and what can we do more than see? When pure Intellect is evolved in man, he approaches to the nature of the Supreme Mind: how can a creature rise higher? When mere impulse, appetite, and passion are placed under the control and direction of duty and virtue, man is put under a Divine Government: what greater lot can any created being have?

7 And the difficulty of conceiving any ulterior step at all analogous to the vast and most wonderful of the revolutions which have taken place in the condition of the earth's inhabitants, will be found to grow upon us, as it is more closely examined. For it may truly be said, the change which occurred when man was placed on the earth, was not one which could have been imagined and constructed beforehand, by a speculator merely looking at the endowments and capacities of the creatures which were previously living. Even in the way of organization, could any intelligent spectator, contemplating anything which then existed in the animal world, have guessed the wonderful new and powerful purposes to which it was to be made subservient in man? Could such a spectator, from seeing the *rudiments of a hand*, in the horse or the cow, or even from seeing the hand of a quadrumanous animal, have conjectured, that the hand was, in man, to be made an instrument by which infinite numbers of new instruments were to be constructed, subduing the elements to man's uses, giving him a command over nature which might seem supernatural, taming or conquering all other animals, enabling him to scrutinize the farthest regions of the universe, and the subtlest combinations of material things?

418| 8 Or again; could such a spectator, by dissecting the

tongues of animals, have divined that the tongue, in man, was to be the means of communicating the finest movements of thought and feeling; of giving one man, weak and feeble, an unbounded ascendancy over robust and angry multitudes; and, assisted by the (writing) hand, of influencing the intimate thoughts, laws, and habits of the most remote posterity?

9 And again, could such a spectator, seeing animals entirely occupied by their appetites and desires, and the objects subservient to their individual gratification, have ever dreamt that there should appear on earth a creature who should desire to know, and should know, the distances and motions of the stars, future as well as present; the causes of their motions, the history of the earth, and his own history; and even should know truths by which all possible objects and events not only are, but must be regulated?

10 And yet again, could such a spectator, seeing that animals obeyed their appetites with no restraint but external fear, and knew of no difference of good and bad except the sensual difference, ever have imagined that there should be a creature acknowledging a difference of right and wrong, as a difference supreme over what was good or bad to the sense; and a rule of duty which might forbid gratification by an internal prohibition?

11 And finally, could such a spectator, seeing nothing but animals thus with all their faculties entirely immersed in the elements of their bodily being, have supposed that a creature should come, who should raise his thoughts to his Creator, acknowledge Him as his Master and Governor, look to His Judgment, and aspire to live eternally in His presence?

12 If it would have been impossible for a spectator of the præhuman creation, however intelligent, imaginative, bold and inventive, to have conjectured beforehand the endowments of such a creature as man, taking only those which we have thus indicated; it may well be thought that if there

is to be a creature which is to succeed man, as man has succeeded the animals, it must be equally impossible for us, to conjecture beforehand, what kind of creature *that* must be, and what will be *his* endowments and privileges.

13 Thus a spectator who should thus have studied the præhuman creation, and who should have had nothing else to help him in his conjectures and conceptions, (of course, by the supposition of a præhuman period, not any consciousness of intelligence, though most active intelligence would be necessary for such speculations.) would not have been able to divine the future appearance of a creature, so excellent as man; or to guess at his endowments and privileges, or his relation to the previous animal creation; and just as little able may we be, even if there is to exist at some time, a creature more excellent and glorious than man, to divine what kind of creature he will be, and how related to man. And here, therefore, it would perhaps be best, that we should quit the subject; and not offer conjectures which we thus acknowledge to have no value. Perhaps, however, the few brief remarks which we have still to make, put forwards, as they are, merely as suggestions to be weighed by others, cannot reasonably give offense, or trouble even the most reverent thinker.

14 To suppose a higher developement of endowments which already exist in man, is a natural mode of rising to the imagination of a being nobler than man is: but we shall find that such hypotheses do not lead us to any satisfactory result. Looking at the first of those features of the superiority of man over brutes, which we have just pointed out, the human hand, we can imagine this superiority carried further. Indeed, in the course of human progress, and especially in recent times, and in our own country, man employs instead of or in addition to the hand, innumerable instruments to make nature serve his needs and do his will. He

works by Tools and Machinery, derivative hands, which increase a hundred-fold the power of the natural hand. Shall we try to ascend to a New Period, to imagine a New Creature, by supposing this power increased hundreds and thousands of times more, so that nature should obey man, and minister to his needs, in an incomparably greater degree than she now does? We may imagine this carried so far that all need for manual labour shall be superseded; and thus, abundant time shall be left to the creature thus gifted, for developing the intellectual and moral powers which must be the higher part of its nature. But still, that higher nature of the creature itself, and not its command over external material nature, must be the quarter in which we are to find anything which shall elevate the creature above man, as man is elevated above brutes.

15 Or, looking at the second of the features of human superiority, shall we suppose that the means of Communication of their thoughts to each other, which exist for the human race, are to be immensely increased, and that this is to be the leading feature of a New Period? Already, in addition to the use of the tongue, other means of communication have vastly multiplied man's original means of intercourse of thought:—writing, employed in epistles, books, newspapers; horses and posting establishments; ships; railways; and, as the last and most notable step, made in our time, electric telegraphs, extending across continents and even oceans. We can imagine this facility and activity of communication, in which man so immeasurably exceeds all animals, still further increased, and more widely extended. But yet so long as what is thus communicated is nothing greater or better than what is now communicated among men;—such news, such thoughts, such questions and answers, as now dart along our roads;—we could hardly think that the creature, whatever wonderful means of intercourse with

its fellow-creatures it might possess, was elevated above man, so as to be of a higher nature than man is.

16 Thus, such improved endowments as we have now spoken of, increased power over materials, and increased means of motion and communication, arising from improved mechanism, do little, and, we may say, nothing, to satisfy our idea of a more excellent condition than that of man. For such extensions of Man's present powers are consistent with the absence of all intellectual and moral improvement. Men might be able to dart from place to place, and even from planet to planet, and from star to star, on wings, such as we ascribe to angels in our imagination: they might be able to make the elements obey them at a beck: and yet they might not be better, nor even wiser, than they are. It is not found generally, that the improvement of machinery, and means of locomotion among men, produces an improvement in morality, nor even an improvement in intelligence, except as to particular points. We must therefore look somewhat further in order to find possible characters, which may enable us to imagine a creature more excellent than man.

17 Among the distinctions which elevate man above brutes, there is one which we have not mentioned, but which is really one of the most eminent. We mean, his faculty and habit of forming himself into Societies, united by laws and language for some common object, the furtherance of which requires such union. The most general and primary kind of such societies, is that Civil Society which is bound together by Law and Government, and which secures to men the Rights of property, person, family, external peace, and the like. That this kind of society may be conceived, as taking a more excellent character than it now possesses, we can easily see: for not only does it often very imperfectly attain its direct object, the preservation of Rights, but it

422|

becomes the means and source of wrong. Not only does it often fail to secure peace with strangers, but it acts as if its main object were to enable men to make wars with strangers. If we conceive a Universal and Perpetual Peace to be established among the nations of the earth, (for instance by some general agreement for that purpose,) and if we suppose, further, that they employ all their powers and means in fully unfolding the intellectual and moral capacities of their members, by early education, constant teaching, and ready help in all ways; we might then, perhaps, look forwards to a state of the earth in which it should be inhabited, not indeed by a being exalted above Man, but by Man exalted above himself as he now is.

18 That by such combinations of communities of men, even with their present powers, results may be obtained, which at present appear impossible, or inconceivable, we may find good reason to believe; looking at what has already been done, or planned, as attainable by such means, in the promotion of knowledge, and the extension of man's intellectual empire. The greatest discovery ever made, the discovery by Newton, of the laws which regulate the motions of the cosmical system, has been carried to its present state of completeness, only by the united labours of all the most intellectual nations upon earth; in addition to vast labours of individuals, and of smaller societies, voluntarily associated for the purpose. Astronomical observatories have been established in every land, scientific voyages and expeditions for the purpose of observation, wherever they could throw light upon the theory, have been sent forth; costly instruments have been constructed, achievements of discovery have been rewarded; and all nations have shown a ready sympathy with every attempt to forward this part of knowledge. Yet the largest and wisest plans for the extension of human knowledge in other provinces of science by the like means, have remained

hitherto almost entirely unexecuted, and have been treated as mere dreams. The exhortations of Francis Bacon to men, to seek, by such means, an elevation of their intellectual condition, have been assented to in words; but his plans of a methodical and organized combination of society for this purpose, it has never been even attempted to realize. If the nations of the earth were to employ, for the promotion of human knowledge, a small fraction only of the means, the wealth, the ingenuity, the energy, the combination, which they have employed in every age, for the destruction of human life and of human means of enjoyment; we might soon find that what we hitherto knew, is little compared with what man has the power of knowing.

19 But there is another kind of Society, or another object of Society among men, which in a still more important manner aims at the elevation of their nature. Man sympathizes with man, not only in his intellectual aspirations, but in his moral sentiments, in his religious beliefs and hopes, in his efforts after spiritual life. Society, even Civil Society, has generally recognized this sympathy, in a greater or less degree; and has included morality and religion, among the objects which it endeavoured to uphold and promote. But any one who has any deep and comprehensive perception of man's capacities and aspirations, on such subjects, must feel that what has commonly, or indeed ever, been done by nations for such a purpose, has been far below what the full developement of man's moral, religious, and spiritual nature requires. Can we not conceive a Society among men, which should have for its purpose, to promote this developement, far more than any human society has yet done?—a body selected from all nations, or rather, including all nations, the purpose of which should be to bind men together by a universal feeling of kindness and mutual regard, to associate them in the acknowledgement of a common Divine Lawgiver, Governor, and

424|

Father;—to unite them in their efforts to divest themselves of the evil of their human nature, and to bring themselves nearer and nearer to a conformity with the Divine Idea; and finally, a Society which should unite them in the hope of a union with God and with each other, in which the parts of their nature which seem to claim immortality, the Mind, the Soul, and the Spirit, should endure for ever in a state of happiness arising from their perfected condition? And if we can suppose such a Society, fully established and fully operative, would not this be a condition, as far elevated above the ordinary earthly condition of man, as that of man is elevated above the beasts that perish?

20 Yet one more question; though we hesitate to mix such suggestions from analogy, with trains of thought and belief, which have their proper nutriment from other quarters. We know, even from the evidence of natural science, that God *has* interposed in the history of this Earth, in order to place Man upon it. In that case, there was a clear, and, in the strongest sense of the term, a *supernatural interposition* of the Divine Creative Power. God interposed to place upon the earth, Man, the social and rational being. God thus directly instituted Human Society; gave man his privileges and his prospects in such society; placed him far above the previously existing creation; and, endowed him with the means of an elevation of nature entirely unlike any thing which had previously appeared. Would it then be a violation of analogy, if God were to interpose again, to institute a Divine Society, such as we have attempted to describe; to give to *its* members their privileges; to assure to them their prospects; to supply to them his aid in pursuing the objects of such a union with each other; and thus, to draw them, as they aspire to be drawn, to a spiritual union with Him?

It would seem that those who believe, as the records of

the earth's history seem to show, that the establishment of Man, and of Human Society, or of the germ of human society, upon the earth, was an interposition of Creative Power beyond the ordinary course of nature; may also readily believe that another supernatural interposition of Divine Power might take place, in order to plant upon the earth the Germ of a more Divine Society; and to introduce a period in which the earth should be tenanted by a more excellent creature than at present.

21 But though we may thus prepare ourselves to assent to the possibility, or even probability, of such an interposition; there are considerations which warn us that we must carefully abstain from associating such a belief with any anticipation of a new and future Period of the Earth's History. Those persons who most firmly believe that such an Interposition is likely, are, in general, those who also believe that it has already taken place. It is true that such persons believe also, that though this Divine Interposition in the History of the Earth is already past, the effects of the Interposition will hereafter be most important in their bearing on the dealings of God with man; when man shall cease to inhabit the earth; or at least shall cease to inhabit it in his present condition. But such persons would commonly, and most wisely, reject all attempts to confirm or illustrate this belief, by any analogy, drawn from geology, or any other material science.

22 We find no encouragement, then, for any attempt to obtain, by the light of the analogy of the past, any definite view of a future condition of the Creation. And that this is so, we cannot, for reasons which have been given, feel any surprise. Yet the reasonings which we have in various parts of this Essay, pursued, will not have been without profit, even in their influence upon our religious thoughts, 426| if they have left upon our minds these convictions:—That

if the analogy of science proves any thing, it proves that the Creator of man can make a Creature as far superior to Man, as Man, when most intellectual, moral, religious, and spiritual, is superior to the brutes:—and again, that Man's Intellect is of a divine, and therefore of an immortal nature. Those persons who can, on any grounds of belief, combine these two convictions, so as to feel that they have a personal interest in both of them;—who have grounds to hope that their imperishable element may, hereafter, be clothed with a new and more glorious apparel by the hand of its Almighty Maker;—may be well content to acknowledge that science and philosophy could not give them this combined conviction, in any manner in which it could minister that consolation, and that trust in the Divine Power and Goodness, which human nature, in its present condition, requires.

THE END.

PART THREE

“A Dialogue on the Plurality of Worlds” from the second edition

THE DIALOGUE

In the preface to the second edition of the *Essay,* in introducing the Dialogue which was now to be a feature of the book, Whewell lamented that he had "very unwillingly been led so far into controversy." I hope that Whewell's God has forgiven him for this statement, for no greater untruth could be found throughout the universe, inhabited or otherwise. Whewell loved a good punch-up and, like a great many vigorous, overbearing people, could only truly respect someone who stood right up to him. Whewell hoarded clippings of reviews of his book and there were few critical comments which were not to be thrown back at his critics and tormentors. Fortunately, Whewell is a fair debater, and in dealing with his opponents, he takes care to give the case made against him before he turns to criticism. We can therefore judge the full nature of the disagreement and the strength of the case made by Whewell, on the one hand, and of any specific opponent, on the other. It is therefore not necessary to review each interchange in the *Dialogue,* merely noting that every part of Whewell's thesis—the science, the natural religion, the revealed theology—came under attack.

I will not presume to judge the merits of the various arguments or to declare a victor. I will say, however, that Whewell hit a raw nerve of the mid-Victorian consciousness. He may or may not have been right in his conclusions. He was surely right, however, in jabbing at the complacency of his fellows, and in showing them that their easy and comforting assumptions needed more thought, review, and evidence than had hitherto been supplied. Whewell's great works on the history and philosophy of science showed him to be an important synthetic philosopher. Now he was showing himself to be an important critical philosopher.

Using the notes left by Whewell himself in his papers, the detective work of Isaac Todhunter when he came to write the scientific biography, and further efforts by Michael Crowe, the following is a fairly complete list of the critics referred to in Whewell's Dialogue. Note that sometimes Whewell introduces a new letter when the topic changes, even though the critic does not change. At other times, when the topic does not really change, Whewell runs his critics together (sequentially). Usually Whewell lets the critic speak for himself, but sometimes (as with Brewster) Whewell lets someone report on the critic's arguments. Whewell himself is "Z."

A. Sir Roderick Murchison (geologist)
B. Augustus de Morgan (logician)
C. Sir James Stephen (historian)
D. Sir James Stephen
E. Sir James Stephen
F. "Chorus" (everyone!)
G. Sir James Stephen (p. 7)
G. Sir Henry Holland (physician) (pp. 10–13)
H. Sir Henry Holland
I. Sir Henry Holland
K. Adam Sedgwick (geologist)
L. "Mr H" (as marked by Whewell on his own copy of the *Dialogue,* otherwise unknown); James Garth Marshall; Sir John F. W. Herschel (astronomer and philosopher)
M. Sir Henry Holland
N. J. D. Forbes (geologist)
O. J. D. Forbes
P. J. D. Forbes

Q. J. D. Forbes
R. J. D. Forbes
S. J. D. Forbes
T. T. H. Huxley (morphologist)
U. Sir Henry Holland
V. Richard Jones (political economist); J. D. Forbes
W. Sir James Stephen; Sir John F. W. Herschel
X. "R.M.M." (Todhunter remarks intriguingly that these initials denote "one who still survives but bears another name")
Y. J. D. Forbes
Z. William Whewell
AA. Sir David Brewster (physicist)
AB. Sir David Brewster (pp. 70–71)
AB. Richard Owen (morphologist) (p. 37)
BB. Hugh Miller (geologist and essayist)

A DIALOGUE

ON

THE PLURALITY OF WORLDS;

BEING

A SUPPLEMENT

TO THE ESSAY ON THAT SUBJECT.

The arguments here assigned to the various speakers, have been urged in conversation, in writing, or in print, since the Essay was first printed.

The remarks assigned to "*AB*," and the replies to them, have been added in the present edition.

A DIALOGUE

ON

THE PLURALITY OF WORLDS.

A. You have published a book which you entitle *Of the Plurality of Worlds;* but the main tendency of what you say (although your assertions do not go quite so far,) is to *disprove* the plurality of inhabited worlds, and to leave the impression that this earth on which we live is the only world; at any rate, the only world of intellectual and moral creatures. You ought therefore rather to have entitled your book *Of the Unity of the World.*

Z. I do not think such a title would have conveyed my meaning, so well as the one which I have employed. When *the Unity of the World* has once been put forward distinctly as the antithesis to *the Plurality of Worlds,* (as I have ventured to put it in the course of the Essay,) the former phrase may be understood as suggesting this antithesis; but without such a preparation, the term *unity* would have been understood as implying a connexion of all the parts, rather than an exclusion of a plurality of similar parts. The Plurality of Worlds is a current and celebrated question; and whichever side I take,

I may adopt for my title the phrase by which the question has commonly been described.

B. Your Essay goes to prove, not the Plurality of Worlds, but the '*singularity* of the World.' You might have taken that title.

Z. Certainly, if I wished to attract attention by the 'singularity' of my title-page. But you know as well as I, that the term *singular* is not opposed to *plural* in common language, though it is in our grammars: so far from it, indeed, that to express *singular* in that sense, we use *unique*, or some similar term.

A. Well; we will allow you your title; but you will find it difficult to prove to the world in general the singularity, or uniqueness, or whatever you call it, of the world in which we live. The plurality of worlds is a notion which has taken possession of the popular mind, and will be difficult to eject.

Z. Of that I am quite aware. I only proposed to shew that this notion has been taken up on insufficient grounds; and that the most recent astronomical discoveries point the opposite way.

A. But you are aware that many of your readers, for instance, my friends *C*, *D*, &c. urge what they consider very strong arguments, against the opinions of the Essay. Do you think that you can answer all these objections to the satisfaction of impartial judges, in a plain and simple manner?

Z. I am quite ready to make the attempt: and the more so, inasmuch as I think I can add something to some of the arguments which I have used. But I must make one stipulation.

A. What is that?

Z. If any objections are urged which have been fully discussed in the Essay, I must be allowed to refer the objectors to the discussion as there given. There would be no use in going over the same ground again, when nothing new is alleged. And now you may begin your objections.

C. How can you think that all the stars which we see in the vault of heaven on a fine winter's night are either made for nothing, or made to give a scanty gleam of light to the earth?

Z. I do not pretend to know for what purpose the stars were made, any more than the flowers, or the crystalline gems, or other innumerable beautiful objects; as I have said Chapter XI. of the Essay, Articles 1—11. But I do not at all doubt that the stars are intended and fitted to draw our thoughts to God, as I have said Chap. XI. Article 24, and Chap. XII. Art. 9: and in other places. And this they never failed to do, during the long centuries when, as yet, men never dreamt of their having inhabitants. A pious man, looking at the flowers of the field, said, 'These are the *smiles* of the Creator.' A person of like habits of mind would see, in the stars, still brighter flashes of the Divine

Eye; though he might no more think of peopling them, than of peopling the flowers with inhabitants.

D. But why should the planets and stars not be inhabited? How can you pretend to limit the power of the Creator? Might he not make creatures fitted to live in the stars?

Z. This I have answered Chap. XI. Art. 15, &c. No doubt the Creator might make creatures fitted to live in the stars, or in the small planetoids, or in the clouds, or on meteoric stones; but we can not believe that he has done this, without further evidence.

D. But we have the evidence of analogy. The planets are like the earth. The stars are like the sun.

Z. In Chap. VIII. Art. 9, &c., I have considered this likeness, as regards the stars; and as to the evidence of analogy, I have remarked, Art. 29, that the question is whether there *is* an analogy. It appears to me probable that there is not. The likeness of the planets to the earth, I have considered at length in Chapters IX. and X. We are still among the arguments, which, as I conceive, I have answered.

E. I find the strongest argument of all against your doctrine in the extraordinary improbability that *one subordinate world*, in the totality of creation as we see it around us, should be thus *specialized.*

Z. I have replied to this argument Ch. VIII. Art. 27, 28. Of improbabilities, in such a case as this, we cannot know anything. It is improbable that we shall have unlike results in like cases; but we do not know that the cases *are* like: or rather, we know that the cases are in many respects unlike. As to the inhabited world being a 'subordinate' body; that fact is so far from making it more improbable that it should be *the* inhabited body, that most persons, even who hold the other subordinate bodies, the planets, to be inhabited, do not believe that the primary body, the sun, is inhabited.

F. You may reason as you will, but you shall never persuade me that the planets and the stars are not inhabited.

Z. That, I confess, is an argument which I have not answered; nor can.

G. But let us come more to points of detail. You, Mr Essayist, say, (Chap. VII. Art. 12) that the resolution of the Nebulæ into stars, by Lord Rosse's and other powerful telescopes, is merely distinguishing the nebulous mass into *lumps*, and that we have no right to call these lumps, *stars*. But surely this is bold phraseology. We have the same right to call these bright objects *stars*, as we have to call any others so. You might just as well call Arcturus or Sirius *lumps*.

Z. Not exactly. I call Arcturus and Sirius *stars*, because they *are* stars. The name has always been used to describe such objects. I avoid calling the dots into which Nebulæ are resolved, *stars*, because I do not at all know that they are objects of the same kind as Arcturus and Sirius.

F. Why not? they differ only in being smaller, and requiring a powerful telescope to see them: and that arises only from their being much more distant.

Z. They differ also in being elements of Nebulæ, which is not the case with Arcturus or Sirius: they differ, very likely, as a cloud of dust differs from a rock. The dust may be resolvable into microscopic masses of stone; but it may also consist of vegetable and animal fragments: and it is possible that its small portions may not be of stony consistence. I would not call a cloud of dust a host of rocks, merely because a small speck of stone may possibly appear, in the microscope, as a rock.

And then as to the Nebulæ being much more distant than the Fixed Stars:—are you sure of that? How do you know it?

F. Why, are not all intelligent persons agreed that it is so?

Z. Not all astronomers, certainly. I have commenced my speculations on this subject, not with an opinion only, but with a proof, supplied by Sir John Herschel, that the fact is not so:—that the Nebulæ, as a class of objects, are *not* more distant than the Fixed Stars. The proof is drawn from the Magellanic clouds;—see the Essay, VII. 10; and is, I think, a very solid starting point for such speculations.

I have also proved, I think, (VII. 14) that the dim parts of nebulæ cannot be much farther distant than the bright parts; and that therefore nebular dimness does not prove a greater distance than stellar brightness.

But I can also add Lord Rosse's authority for the opinion that the nebulæ are not more distant than many of the fixed stars. In his address to the Royal Society, which he delivered as President last November, he says, 'There are double stars known to be physically double from their motions, which are probably as distant as some of the nebulæ. In certain nebulæ, stars are so peculiarly situated that we can scarcely doubt their connexion with the nebular system in which we see them, and some of these stars are as bright as some of the stars known to be physically double: as bright even as some of the stars which the latest Pulkova observations have shewn to have sensible parallax, and whose distance therefore is approximately known.' Here, you see, authority the highest, observations the best and most recent, countenance me in placing the nebulæ at distances of the same order as fixed stars:—even as the fixed stars which have parallax, and therefore are the nearest to us.

F. Would you then make the irregular nebulæ to be mere floating clouds of luminous matter, at a little distance from us?

Z. Not at a little distance; for they are at any rate as far off as the nearest fixed stars; which is, you know, 200,000 times the distance of the sun. But as for their being luminous clouds, I do not think their consistency is unreasonably disparaged by calling them so. A great astronomer once told me that he doubted whether one of these nebulæ contained luminous matter enough to light a good sized room. I do not say this. But I think if we were in the middle of one of them without a sun, it is pretty certain we should not be able to read small print by the light.

G. In your zeal for reducing the Nebulæ to a state of thin luminous vapour, you overlook the close agreement between the Nebulæ and the Galaxy. Both contain nebular light of the same aspect; and we cannot reasonably assume that they are substances of different kind.

Z. And what inference do you draw from this similarity?

G. This. The nebulous portions of the Galaxy are resolvable, by telescopic aid, into stars. These stars, we have reason to believe, are of the same nature as bright Fixed Stars; for there appears to be a gradation from the bright stars lying in the Galaxy, to the smallest stars of the Galaxy; and from these smallest obvious stars, to the telescopic stars into which the nebular parts of the Galaxy are resolved. And therefore the telescopic stars into which the other Nebulæ are resolved, are of the same nature as the bright Fixed Stars. You see I have four Terms compared, and I assert that these Terms are of the same kind: Bright stars; small stars in the Galaxy; resolved stars in the Galaxy; resolved stars in Nebulæ. And I say that in the transition from any one of these Terms to any other, there is nothing which authorizes us to suppose that we are passing from an object of one kind to an object of another kind.

Z. An excellent argument; and one which it is very important for me to examine well. If all the four links of your chain hold, you will go near to pull down the doctrine of the Essay about Nebulæ.

G. So I think. And yet, which of the links has any flaw in it?

Z. You speak of them much more strongly than the case will justify, when you talk of them as links without flaw. The arguments are all of the *pourquoi non* form; which, as I have said, generally admits of being retorted. Three or four *pourquoi nons* do not make a very strong chain, even if none of the questions be answered.

But let us take the special case. You say, Why should not the smallest stars of the Galaxy be of the same nature as the larger? And why should not the nebular parts of the Galaxy be of the same nature as the star-powder? And why should not the other Nebulæ be of the same nature as the nebular parts of the Galaxy?

To this I might reply, by reversing the series.

The Nebulæ are not of the same nature as the Fixed Stars, as Herschel proves from the Magellanic clouds; as Lord Rosse holds on the ground of his own observations; and as appears by the straggling nebulæ. Therefore either the other nebulæ are unlike the galactic nebulous matter, or the galactic nebulæ are different from the galactic stars; or the galactic stars are different from the fixed stars.

And in truth, is it not plain that in this, as in other instances, it is not identity, but gradation of cases, which will explain the phenomena? The largest definite fixed stars are different, in consistency at least, from the filaments and tails of straggling nebulæ. But they may be of the same substance; and in that case, there may be intermediate kinds of bodies, smaller stars, smaller than the largest stars, more definite than the nebulæ. And while the parts of the universe nearest to us are gathered into stellar masses, of which our sun is a large and definite portion (*why not* the largest?), the farther portions of the galactic stratum may consist of smaller stars; and the farthest portion, of nebular matter not gathered into stars at all.

G. But this is a mere arbitrary conjecture.

Z. But the equality or approximate equality of all stars, obvious and telescopic, merely because they are stars, is no less a mere arbitrary conjecture. My conjecture has the advantage of explaining the difference of stars and nebulæ; which is so strongly supported by the best astronomers, as I have shewn you.

G. But your scheme makes the solar system the center of the Universe.

Z. Not quite that. But if it did, are you prepared to add, *which is absurd?*

G. It is at least contrary to the general current of modern speculation; which, ever since the discovery of the telescope, has been running strongly against the ascription of any such peculiar character to the solar system.

Z. But it is a very common, almost universal occurrence, that such strong currents of speculation overrun the point of equilibrium where truth resides.

H. You would have us believe, then, that the stars of the Milky Way are smaller and smaller, as they are farther from the sun.

Z. If you reject such a supposition, what is the alternative hypothesis?

H. That the stars in the Milky Way are, in a general way, of the same size as our Sun. And *why not?*

Z. Perhaps we may soon see a possible reason why not. But as I now understand you, this collection of stars, of the order of our Sun, forms a stratum in which we are placed, and see its mass edgeways, and thus we see the Milky Way. And this stratum seen afar off is a nebula. Is not this your doctrine?

H. Sir William Herschel's, as well as mine.

Z. But this stratum of stars, is it finite or infinite in extent?

H. Finite, I suppose; for otherwise our Nebula would differ from all others; to say nothing of other difficulties in the way of an infinite stratum of stars.

Z. Then being finite, it has a boundary, has it not?

H. Of course.

Z. And the stars next this boundary are as large as those near the middle of the stratum?

H. In a general way; but probably with great irregularities. And still, Why not?

Z. Have we any examples in the distribution of bodies in the Universe, of a great number of co-ordinate masses, of about the same size, with no subordination to a principal body, or gradation from great to small? You will understand the meaning of my question, if you consider that we have many examples of the opposite arrangement;—a principal body with many inferior ones; a large body with many small ones. Thus we have the vast Sun, and the comparatively small masses of the planets: the great mass of Jupiter attended by his minute satellites: the bulk of Saturn accompanied by his satellites: of Uranus, by his. But we have no where, so far as I know, a collection of satellites without a primary planet; or of planets without a sun. Why, then, a collection of suns without any body of superior dignity to keep them in order?

H. But *that* analogy would lead us to suppose that there *is* a body of superior dignity;—a primary sun about which our sun revolves as a secondary.

Z. To suppose such a sun, when we do not see it, would be very wild. The analogy is not strong enough to bear the weight of such an inference, directly in the teeth of all facts. If such a primary sun were big enough for our system to revolve round, it must be big enough for us to see. Or if you suppose the solar system to revolve round an *invisible* center, you invent a mechanism as arbitrary as the crystalline spheres of the ancients, without anything like the same amount of excuse. But if you are bent upon having a central body, why should you not make *our* Sun the principal body in the universe, as it is the greatest which we know of?

H. But the stars do not revolve round our Sun.

Z. I believe it will be as easy to explain the minute apparent proper motions of the stars by supposing them to revolve round our Sun, in due directions, and at due distances, as in any other way. But even if the Sun were the largest body in the universe, it by no means follows that all the other bodies must revolve about him. They might all have independent motions, slightly modified by the natural attraction of all:—slightly, I say, on account of the immense distances of all, and the probable extreme tenuity of substance of many. My own opinion is, too, that these motions are far too minute and uncertain to draw from them, as yet, any conclusion, as to cosmical arrangements; therefore I avoid this obscure part of astronomy.

I. You argue too much from the spiral form of some of the Nebulæ. It belongs to very few of them, and is not a general feature in the aspect of the Nebulæ.

Z. I argue from such features in the Nebulæ as I can find noticed by the best observers. The features from which any inference can be drawn are very few, and we must make the most of those which we have. But with regard to the spiral form of the Nebulæ, I reason as Lord Rosse reasons, in the Presidential Address which I have already quoted: 'It is probable that in the Nebular systems motion exists. If we see a system with a distinct spiral arrangement, all analogy leads us to conclude that there has been motion, and that if there has been motion, it still continues.' I add to this, that if the motion be very slow, the mass must be very much attenuated. I pray you, attend for a moment to this part of my reasoning.

The radius of the Earth's orbit is about 100 times the Sun's diameter, or 200 times the Sun's radius. If the sun's mass were diffused in patches of pretty uniform density through the sphere which the earth's orbit girdles, the density of the solar matter would be 8,000,000 times less than its present density, which is that of water. But in this case, each patch would tend to make a complete revolution round the center in one year; and if by the resistance of the other parts it were made to describe a spiral, it would still move at nearly the same rate. But we have the strongest reason to believe that the motion of the patches of the spiral nebulæ, and of the nebulæ of all forms, is very much slower than this;—slower by hundreds of times, for no such motion has yet been detected. And therefore the density of the nebulæ must be very much smaller than that extremely small density which I have already mentioned. It must be millions of times, probably, less than the density of air.

And then, as to there being *few* spiral nebulæ, listen again to Lord Rosse, who says that probably there are only a few normal forms of these objects, and that their great apparent variety probably arises from the different positions in which we look at them. He would have them carefully measured and sketched, in the hope that out of the apparent confusion we may succeed in extracting their normal forms. We cannot doubt that one, at least, of the principal normal forms is a spiral, or group of spirals. And if we take any other tolerably definite form of Nebula, we may reason upon it in the same manner. It must be kept in its definite form by a motion of revolution; which will be slow, in proportion to the tenuity of the mass. But all these motions of revolution are hitherto imperceptibly slow: therefore the mass of each nebula is incalculably attenuated.

If I wished for new evidence of the tenuity of nebular matter from its general aspect, I might again quote this Address of Lord Rosse. He says that among the Nebulæ, 'there are vast numbers much too faint to be sketched or measured with any prospect of advantage; the most powerful instruments we possess shewing in them nothing of an organized structure, but merely a confused mass of nebulosity of varying brightness.' Do you still hold that this confused mass of nebulosity consists of innumerable suns, each the center of a planetary system?

I. But why should the substances in this our world be so much more dense than those in distant parts of the universe?

Z. If I prove to you that they *are* so, I shall hardly need to give you reasons why they *should be* so. But yet I might say that they should be so, in order that this our world might be a world. We are, and live upon, the solid and substantial part of the universe; the nebulæ are tracts in which no creatures could live, except such as could make the zodiacal light their dwelling-place.

And so it appears, I think, that I am supported by the observations and opinions of the best astronomers, when I say that I have cleared away the inhabitants of the Nebulæ.

K. But you will find the clearing away the inhabitants of the Fixed Stars, or their planetary systems, a harder task. Why should you not allow us to believe that there are such inhabitants? You cannot deny that our Sun, seen from one of the fixed stars, would look as Sirius looks to us, and that all the Planets would be imperceptible. Is not this good reason for supposing that there is a Sirian system as there is a Solar system?

Z. Let me ask in reply, whether it is not a strange way of finding good reason for supposing the existence of a system, to go so far off that you cannot see whether it is there or not? You may go so far from the Sun that he will look like Sirius, as you may go so far from a house that it will look like a rock; but this does not prove, nor make it probable, that a class of objects which look like rocks are really houses.

K. This proves that such objects *may be* houses; and so, the stars may be inhabited.

Z. May be, so far as we can judge from looking at them in this way. But perhaps we can obtain also other means of judging.

K. Yes; but here again you take the exceptional cases to reason from:—Double Stars, Changeable Stars, and the like. And yet, how few stars of this kind are there, compared with the whole collection of stars!

Z. I reason from such cases, because such cases give me something to reason from. But here again, as with the nebulæ, I think you will find that the most recent views of the best astronomers do not represent as so very rare, the cases from which I reason. In the great work which M. Struve has lately published, containing the record of his labours on Double Stars at Dorpat, he gives, as the result of his careful examination and comparison of the whole body of facts in stellar astronomy, some conclusions which may perhaps startle you and others who regard Double and Multiple Stars as cases so completely exceptional. He examines especially the brighter stars,—those comprized between the first and fourth magnitudes,— and arrives at the conclusion that *every fourth star* of such stars in the heavens is physically double. He even ventures to assert that when we have acquired a more complete knowledge of double stars, it will be found that *every third* bright star is physically double. Applying these considerations to the stars of inferior orders of magnitude, he finally arrives at the following conclusion, which he admits to be of an unexpected character:—that the number of insulated stars is indeed greater than the number of compound systems; but only three times, perhaps only twice as great. Now if we recollect that the double and multiple stars cannot be supposed to have systems of planets except by the making most arbitrary hypotheses, in defiance of all analogy; and that, of the single stars, a large proportion may be irregular masses, or limited nebular patches, very unlike the sun, as we know that some are; it would seem to be a very bold and baseless assumption, that they all, or the great majority of them, have attendant planets, as our Sun has.

K. But surely it would be very bold to maintain that of all the innumerable stars which spangle the sky, and which astronomers have hitherto held to be bodies of the same nature as our sun, not one is really like the sun, in having planets revolving round him.

Z. It would be very bold to maintain that all have not; and no less bold, it seems to me, to maintain that many have, or even that one has, without further evidence. Of the nature of the stars, we know scarcely anything, except that they are seemingly self-shining, very nearly fixed, (as to our sense,) and exceedingly distant from us. In this state of ignorance, to assert that they have, and that they have not, attendant planets, would be alike rash. To assert that if they have attendant planets, these planets have inhabitants, would be more rash; and still much more, if it appear probable, as I think it will, that the other planets of our system have not inhabitants. If we fix our thoughts on any 'bright particular star,' we may easily *imagine* it to have inhabitants; as we may easily imagine a central cavity in the earth tenanted by inhabitants. But the only way of passing judgment on such imaginations with regard to the

Fixed Stars, is to look at the Fixed Stars as a class of objects. Now as a class of objects which we may imagine to be inhabited, the Fixed Stars stand between the Planets and the Nebulæ; more remote than the Planets, but limited in form like them; distant and luminous like the Nebulæ. I think I have proved that the Nebulæ are not habitations; I think I make it probable that the Planets in general are not. This being so, it seems contrary to the scheme of gradation on which nature generally proceeds, to say that the intermediate class are habitations:—or rather, centers of systems of inhabited masses, of which masses the very existence is quite unproved.

K. But their being the centers of such systems is proved by analogy from their resemblance to the Sun.

Z. But the evidence of this resemblance was always very loose; and becomes the feebler, the more it is examined, as I have been telling you on Struve's authority. It appears, by his researches, that one third of the Fixed Stars differ from our Sun, in the broad fact of not being single masses, but systems of two or more luminous masses lying near each other, or revolving about one another; which entirely puts an end to any probability of a planetary system on the ground of analogy. To these cases of differences, which exist between the stars and the sun, you must add the other *known* kinds of difference, as the cases of variable stars:

and you will hardly doubt, that when there exist several known kinds of difference, in a class of objects of which we know so little, there must be other kinds of difference, which are *unknown.* So that the assumption of stellar planetary systems, on the ground of the close resemblance between the stars and the sun, appears to dwindle away to nothing, when closely scrutinized.

L. I allow that if you disprove the existence of inhabitants in the Planets of our system, I shall not feel much real interest in the possible inhabitants of the Sirian system. Neighbourhood has its influence upon our feelings of regard, even neighbourhood on a scale of millions of miles. And this being so, I think you are very hard upon our neighbours in Jupiter, when you will not allow them to be anything better than 'boneless, watery, pulpy creatures'. (Chap. IX. Art. 19).

Z. I had no disposition to be hard upon them, when I entered upon these speculations. I drew what seemed to me probable conclusions from all the facts of the case. If the laws of attraction, of light, of heat, and the like, be the same there as they are here, (which we believe to be certain,) the laws of life must also be the same; and if so, I can draw no conclusions other than those which I have stated.

L. One might extract many pleasant imaginations out of this doctrine of the fluidity of Jupiter.

Who can conceive the configuration of the creatures that dwell there? They may exist as immense alga-like or medusa-like creatures, 'floating many a rood,' and feeling, observing, thinking, and operating at each of their infinitely multiplied extremities. Then what crystal palaces they may build on the Tabasheer nucleus of that huge aquatic globe! What water-organs in the nature of Sirenes they may construct! To what a perfection may they have brought the science of Hydropathy! There is no limit to the ejaculatory questions of this kind which might be proposed.

Z. Absolutely none. And I should hope that these imaginations may supply the void made by dispeopling Jupiter of the imaginary creatures which have hitherto occupied him. You must suppose Micromagas to be a huge leviathan; or if you choose, to be Hobbes's Leviathan, which was a kind of polyp, as appears by his frontispiece.

M. But why do you make Venus so different from the Earth? She seems to resemble it much in physical condition.

Z. I have given you in the Essay my reasons for what I have done. It is not I, but Herschel, who tells you that she is not marked with permanent spots, such as denote land and water. But when he tells us this, I cannot think that her physical condition is like that of the earth. And as to the amount of heat which must prevail there and which makes it difficult to suppose inhabitants in her, according to the known laws of life;—*that* is a matter of calculation, if the difference of heat in different planets depends on the sun. And that something, apparently the nearness of the sun, makes her surface different from that of the exterior planets, Herschel also tells us. If, notwithstanding this, you are resolved to believe that Venus is inhabited, I have only to say that I see no grounds for such a conclusion.

N. But if Venus and other planets have atmospheres, is not that a good ground for believing them to be inhabited? You argue that the Moon is not inhabited, because she has no atmosphere. Supposing that I assent to this. If it be good as a negative argument, it is also good as a positive one. If we find other planets where water and air, evaporation and clouds, do appear to exist, we have a contrary argument provided (as seems to me) of at least equal force as regards the general question.

Z. You will hardly say so, I think, on deliberate consideration. For the existence of life, several conditions must concur; and any of these failing, life, so far as we know anything of it, is impossible. Not air only, and moisture, but a certain temperature, neither too hot nor too cold; and a certain consistence, on which the living frame can rest. Without the other conditions, an atmosphere alone does not make life possible; still less, prove its existence. A globe of red hot metal, or of solid ice, however well provided with an atmosphere, could not be inhabited, so far as we can conceive. The old maxim of the logicians is true; that it requires all the conditions to establish the affirmative, but that the negation of any one proves the negative.

O. In assuming Jupiter to be a mass of water on account of his low density, you seem to leave out of account the fact that Saturn is much less dense still. If you argue that because Jupiter is of the density of water, he *is* water, you should argue also that because Saturn is of the density of cork, he *is* cork.

Z. Yes, if I held that cork is one of the universally diffused elements of the universe. You must recollect that I try to explain how the water got there.

O. But in the case of Saturn, it seems most natural and consistent to assume a porous or cavernous structure. The only alternative is to suppose that he consists of a fluid lighter than water: and such a fluid is nowhere found, except in minute quantities; and then, I believe, always derived from organic bodies on the earth's surface. Do you think it likely that Saturn is composed of alcohol or ether?

Z. Certainly not. What I have suggested is a combination of the same well-known elements, water, (including aqueous vapour), and air. His outer parts are probably air and vapour, his inner parts liquid water. Your porous or cavernous structure appears to me a most arbitrary assumption;—indeed, another form of the cork hypothesis.

P. You appear to have involved yourself in a difficulty in your theory of Jupiter: after presenting it in the text you add a defence in a note. In a temperature such as Jupiter must have, the water of which he is composed must become ice; and then what becomes of your polyps and marine monsters?

Z. Even let them go. I have no special love for them. I spoke of them mainly for the purpose of gratifying (as far as I could) those who wish to have the planets inhabited. I said that, in my opinion, the choice lay between such a population and none; I have no wish to *defend* my theory, as you call it, of Jupiter's inhabitants.

Q. But you must allow that most of your objections do not apply to Mars. Why should not Mars be inhabited?

Z. Fontenelle's favourite figure *Pourquoy non!* But, as I have said, the figure is two-edged, and cuts one way as well as the other. Why should not Mars be *un*inhabited? Or, to put it otherwise, why should not Vesta, why should not Ceres, why should not Juno be inhabited?

Q. O, they are so very small! Vesta is only 250 miles in diameter; Ceres, only 163; Juno, only 79. The whole surface of Juno is not so large as Newfoundland, and who knows how much of it is solid?

Z. There is then, it seems, a degree of smallness which makes you reject the supposition of inhabitants. But where does that degree of smallness begin? The surface of Mars is only one fourth that of the Earth. Moreover if you allow all the small planets between Mars and Jupiter to be uninhabited, the planetary bodies which you acknowledge to be probably uninhabited, far outnumber those with regard to which even the most resolute pluralists of worlds hold to be inhabited. The majority swells every year. Since the publication of the Essay three have been added. The Planetoids are now twenty-nine. The fact of a planet being inhabited, then, is, at any rate, rather the exception than the rule; and therefore must be proved in each case by special evidence. And of such evidence, I know not a single trace.

R. By the way, are you not rash when you suggest that these small planets—planetoids, you call them—have been produced by a collapse of sidereal matter which has failed to make one large planet? (Chap. IX. Art. 28).

Z. It seems to me, that however rash I may be in such a procedure, I am much less adventurous than the astronomers who discovered these planets, and who held that they were fragments of a larger planet which had been blown to pieces by some unknown cause. This, as you know, was the hypothesis of Olbers; and in the *Connaissance des Temps* for 1814, Lagrange investigated the explosive force which was necessary to produce the result: and found that the velocity which the explosion produced need only be 20 times the velocity of a cannon-ball*. Is it a more rash speculation, to form a planet by condensation of sidereal matter, than to shatter one to pieces by an unknown explosive force?

S. In your speculations, you adopt the Nebular hypothesis: and adopt it with the most thorough-going adhesion, (Chap. x. Art. 6). You not only accept it as applied by the elder Herschel and Laplace, to the starry world; but as it is applied by Laplace to the evolution of our solar system out of a revolving cloud of solar matter, gradually shrinking by cooling, and throwing off in the process rings or shreds of matter, which ultimately became planets with their revolving train of satellites; and which finally, in the last age of the world, condensed into the substantial sun, which serenely governs the progeny born of his own body. This strange cosmogony—the wildest imagining

* Grant, Hist. Ast. p. 241.

which ever emerged from the brain of a mathematician—appears to me an unaccountable delusion. How can a grave man like you give currency to it?

Z. My excellent friend, I am sorry that my dealing with this hypothesis *as a hypothesis* disturbs you. But pray consider for a moment: what cosmogony is there, or can there be, which is not strange, which is not wild? It may be the wisest and best course, not to have, or to seek, any cosmogony at all:—not to ask how, by what steps and through what changes, the world came to be what it is. It may be the most reverential habit, to take the natural world as it is, and to make no inquiries about its probable history. But the whole tendency of modern science is entirely opposed to this habit. When we occupy ourselves with geology, for instance, our express and principal aim is to learn the history of the earth; to know what changes it has undergone; to trace the stages of its past existence backwards, through all the variations of organic life, till we come to a condition in which there was as yet no organic life. Are we bound, by some paramount obligation, to stop there? Is it wrong to carry our speculations further? to ask how the mineral mass of the earth came to be what it is? through what changes that also may have passed? But if we may make such inquiries at all, what is simpler than to ask whether these mineral masses have passed through a fluid, or a gaseous condition, before they became solid? We know that most materials, probably all, are capable of these three conditions. May we not make a hypothesis as to the mode in which the three conditions have succeeded each other, and see what comes of the hypothesis? I grant you that our hypothetical speculations of this kind are likely to be very insecure; our conclusions very precarious. We have here very few facts which point to any conclusions. The case is very different from what it is in geology, where we have a great multitude of facts, pointing to conclusions in a great degree definite. But we can speculate, bearing in mind this insecurity. We can take care that our confidence in our conclusions does not go beyond our evidence. To forbid us to speculate, even in this spirit;—to arrest our inquiries at a certain point, and beyond that, not to allow us even to guess;—is a very arbitrary proceeding; nor is it at all likely that men's minds, in these our times, will submit to it. But at any rate, men cannot submit to such prohibitions, and at the same time talk about the Plurality of Inhabited Worlds. That doctrine, so far as scientific grounds are concerned, is as strange as any cosmogony;—is as much a conjecture as the primitive nebulosity of worlds;—is a hypothesis supported by fewer facts and analogies than the nebular hypothesis.

Indeed the doctrine of the Habitation of many Worlds is, as I conceive, an essential part of the

Nebular Hypothesis. The rule that worlds shall be inhabited, is the last part of the rule, according to which, on that hypothesis, worlds are formed. Nebulæ condense to suns, and revolving suns throw off planets, and planets become seats of life, it is held, by certain universal laws of nature. I dispute, I deny these laws of nature. I think it quite unproved that nebulæ condense to suns; that suns throw off planets; that planets in general, (if there are any systems of planets besides ours,) become seats of life. I know that the planet on which we live *is* a seat of life. I believe it has become so by acts out of the common order of the laws of nature; therefore I do not think it reasonable to assume, without any evidence, that other planets are also seats of life. I think it probable that the solid materials of our globe were formerly fluid; possible, that they were formerly gaseous and nebular. But I do not therefore hold that all nebular objects will produce suns and planets like ours. Even when I employ, hypothetically, the hypothesis of the gradual condensation of nebular into solid matter, I never give it the general application which belongs to Sir William Herschel's nebular hypothesis. Even if I suppose for a moment the solar system to have been formed from a nebula, I never suppose that all the nebulæ are in the course of forming planetary systems. On the contrary, that is precisely the opinion which I reject.

S. But this doctrine of the formation of our solar system from a nebula has been combined with irreverent opinions, which you yourself would shrink from, by popular but superficial authors. I am sorry therefore that you should give the doctrine additional currency among us.

Z. If you mean, as I think you do, that this doctrine of the natural derivation of planets from nebulæ has been combined with the natural derivation of man from the most elementary animals, I am aware that such a combination has been made. But I must say, I think the doctrine of the natural derivation of species one from another, has too little real connexion with any conjectures about nebulæ, to make these dangerous on that account. If you are moved by such dangers, geology is the part of science which is really dangerous, as supplying some plausible evidence of the derivation of higher species from lower. For my part, I have not only rejected, but, as seems to me, given good reasons for rejecting, any natural derivation of organic life from brute matter, and any derivation of human from animal existence. On the latter point I have laboured (in Chapter VI.), to an extent which I feared might be deemed prolix and superfluous; but which was naturally occasioned by the importance of that point in my reasoning.

S. I grant that, in you, the nebular hypothesis, if it be an extravagance, is a mere *physical* extravagance; and is associated with solid and reverential views of the Divine Providence, from which it has generally been, in a marked degree, dissociated.

T. In your chapter on the 'Unity of the World' (Chap. XII. Art. 3), you quote a passage from Professor Owen's 'Nature of Limbs,' which you suppose to favour your views; but I have rarely met with an instance of more excessive carelessness than you here betray.

Z. Pray shew me this. If it be so, I will not only correct the error, but thank you very sincerely for enabling me to do it.

T. Look here. You quote this passage as from Mr Owen:

'If the world were made by any antecedent mind or understanding, that is, by a Deity, there must needs be an Idea and Exemplar of the whole world before it was made, and consequently actual knowledge, both in the order of time and nature, before things. But conceiving of knowledge as it was got by their own finite minds, and ignorant of any evidence of an ideal archetype for the world or any part of it, they [the Democritic philosophers who denied a Divine creative mind] affirmed that there was none, and concluded that there could be no knowledge before the world was as its cause.'

Now this passage which you thus quote, stands in Mr Owen thus:

'*The learned Cudworth tells us that—The Democritic Atheists* reason thus. If the world,' &c.

So that you absolutely attribute to Professor Owen the *words* of Cudworth, and the *opinion* of the Democritic Atheists, and finally garble the Hunterian Professor's commentary to make the whole fit.

Z. A grave charge indeed! But surely upon the face of it you go too far. Do I not introduce these opinions of the Democritic Atheists for the very purpose of shewing that Prof. Owen refutes them? Would it not have made the matter clearer if you had gone on to my next sentence, still a quotation from Mr Owen: '*Now* however the recognition of an ideal exemplar for the vertebrated animals proves that the knowledge of such a being as man must have existed before man appeared;' and therefore that the Democritic argument is worthless? Do you not think the 'excessive carelessness' here, is on your side?

T. But you cannot deny quoting Cudworth's words as Owen's.

Z. Let us see them. How much of the passage which you have read is Cudworth's?

T. The whole, of course.

Z. Indeed! You have looked at Mr Owen's book, and therefore you ought to have seen that the double commas " " which mark the quotations, include only the first sentence; and that at any rate, the second sentence is Mr Owen's own. And as the first sentence merely serves to introduce the second, it did not seem too great a liberty to omit the reference to Cudworth, which is quite unessential.

T. But if Mr Owen thought the reference necessary in quoting Cudworth's words, you ought to have thought it necessary too.

Z. Excuse me. It was proper for Mr Owen to give his authority for his assertion of a fact in philosophical history, but not at all necessary for me, who had only to do with his argument.

But you must allow me to say, that though you have referred to Mr Owen's book, you do not appear to have referred to Cudworth's. The fact is, that Mr Owen has *not* quoted Cudworth's *words*, even in that single sentence. He has altered them so much, as to make that sentence also his own. The passage referred to occurs in the 'Intellectual System of the Universe,' Book I. Chap. 5, page 847; and stands thus:

"But the Democritick and Epicurean Atheists, will make yet a further assault, from the Nature of Knowledge, Understanding, after this manner; If the world were made by a God, or an Antecedent Mind and Understanding, having in itself an Exemplar or Platform thereof before it was made, then must there be Actual Knowledge, both in order of Nature and Time, before Things; whereas Things," &c.

Now Mr Owen, in employing this passage, has not only omitted the superfluous phrases, but altered some of the terms; putting 'Idea and Exemplar,' for 'Exemplar or Platform,' and leaving out the developement of the argument. Mr Owen acted very judiciously in doing this; but by doing it, he made the passage his own:—very properly; since it was to form part of his argument; or rather, contained the reasoning which he undertook to refute; and which he has so strikingly refuted.

And now, where is the garbling?

AB. In addition to your explanation, I am now able to say, as a matter of fact, that Mr Owen himself is of opinion that you have quoted and represented him fairly.

T. But what does all this prove against the Plurality of Worlds?

Z. I do not pretend to disprove a Plurality of Worlds. But I ask in vain for any argument which makes the doctrine probable. And as I conceive the Unity of the World to be the result of its being the work of one Divine Mind, exercising creative power according to its own Ideas; so it seems to me not unreasonable to suppose, that man, the being which can apprehend in some degree those Ideas, is a creature unique in the creation.

T. But Mr Owen is in favour of the Plurality of Worlds; he says, only two pages before that which you quote from:

'The naturalist and anatomist, in digesting the knowledge which the astronomer has been able to furnish, regarding the planets and the mechanism of the satellites for illuminating the night season of the distant orbs that revolve round one common sun, can hardly avoid speculating on the organic mechanism that may exist, to profit by such sources of light, *and which must exist if the only conceivable purpose of these beneficent arrangements is to be fulfilled.*'

Mark the last clause.

Z. I mark it; and mark you too the sequel. Mr Owen goes on to argue, very ingeniously, that the moons of Jupiter exist to give light;—that light being governed, there, by the same laws as here,—the animals must have eyes organized on the same dioptric principles as on the earth; and that as the bony orbits in which eyes are placed, in this planet, are constructed of modified vertebræ, it will not appear so hazardous, to infer that the animals on other planets may also be vertebrate.

T. And what have you to say against this?

Z. Nothing. *If* there be anywhere animals which are bony, the simplest supposition is, that they are also vertebrate. I do not know how any one can reason better, as to the structure of the animals which inhabit other planets, *if there are such animals.*

T. But you see, Mr Owen says that such animals must exist, if the only conceivable purpose of the beneficent planetary arrangement is to be fulfilled.

Z. I have learnt much from Mr Owen. I

have learnt from him, in many most striking cases, to admire purpose in organic arrangements, where purpose is apparent. But I have learnt from him also, that to infer facts from 'an only conceivable purpose,' is a very hazardous process. I have employed Mr Owen's views in order to illustrate the insecurity of this process, (Chap. xi. Art. 4), and have done so with the most entire conviction that I was following the truest and largest lessons of his noble philosophical progress. In this very case of vertebrated animals, we have, as I conceive, one of these lessons. Why, I have already asked, (Chap. xi. Art. 3) should so many animals have a skeleton of the same plan—this vertebrate plan—bone for bone? In many instances, there are multitudes of bones in the skeleton of a species, which answer no known purpose: why then may there not be planets and satellites which answer no purpose? none, that is, which we have yet discovered? And does not every body allow that there are such? Do you hold that all the planetoids between Mars and Jupiter, now numbering twenty-nine, are inhabited? Some of these perhaps are not bigger than Mount Blanc, and perhaps as irregular in form. Can you not conceive such a mass revolving round the Sun, separate from the Earth, void of inhabitants?

T. Certainly I should not expect such planetoidal masses to be inhabited.

Z. No, of course you would not. And therefore you must give up, for them, 'the only conceivable purpose.' And what difference can it make in the conceivableness of the thing, if they are a myriad times as large?

T. But in the case of the skeleton, the bones which have no known purpose are there in virtue of a general law, namely, the vertebrate plan.

Z. And doubtless in the case of the planets, they, inhabited or not, are there in virtue of general laws, whether we know the laws or not. But moreover, I have tried to point out what these laws may be, in the Chapter on the Theory of the Solar System.

U. You assume that every intelligent being must be human, and subject to the conditions of man's physical existence (Chap. IV. Arts. 6—10), because we cannot conceive intelligence unaccompanied by human attributes. But Mr Mill, in his System of Logic, has exploded the notion that what is inconceivable cannot be true.

Z. Some writers, especially writers who have dealt with theological subjects, have asserted that not only what is inconceivable may be true, but that all the highest truths must be inconceivable. But it would surprise me very much if Mr Mill were to allow you to say that *that* of which you know nothing but that it is inconceivable, is probable.

U. But we do know more of the inhabitants

of other planets. We know that their existence is agreeable to analogy.

Z. To the 'analogy' of one case out of thirty-six planets, of which planets there are twenty-nine to which you dare not apply the analogy. But I have discussed this argument in the Essay (Chap. VII. Art. 29), and shall not dwell further upon it.

U. Even supposing that no being possessing what we call intelligence can have any other corporeal garb and limitations than those of man, the arguments against the possibility of his existence in other planets are merely negative. Considering what we know of the enormous influence of atmospheric conditions and physical conformation upon climate on the earth, and that we possess only the roughest notions with respect to either, (atmosphere or conformation,) in any planet but our own,—a cautious mind, I think, would hesitate long before drawing any conclusion as to climatal impossibilities, for even human existence, in Mercury or Jupiter.

Z. That is, Mercury may be as cool as the earth, and Jupiter as hot. Certainly I cannot prove that it may not be so, except by supposing that propinquity to the sun is the main cause of heat. But I think 'a cautious mind' would hesitate still longer, before concluding that the planets, differing in amount of heat, so far as solar propinquity determines heat, from 7 times more than the earth, to 900 times less, and possibly, with atmospheric conditions and physical configuration that increase their unfitness for habitation, are nevertheless all inhabited.

U. But your arguments are merely negative. You only prove that we do not know the planets to be inhabited.

Z. If, when I have proved that point, men were to cease to talk as if they knew that the planets *are* inhabited, I should have produced a great effect.

U. Your basis is too narrow for so vast a superstruture, as that all the rest of the universe besides the Earth is uninhabited.

Z. Perhaps: for my philosophical basis is only the earth, the only known habitation. But on this same narrow basis, the earth, you build up a structure that other bodies *are* inhabited. What I do is, to shew that each part of your structure is void of tenacity, and cannot stand.

It is probable that when we have reduced to their real value all the presumptions, drawn from physical reasoning, for the opinion of planets and stars being either inhabited or uninhabited, the force of these will be perceived to be so small, that the belief of all thoughtful persons on this subject will be determined by moral, metaphysical, and theological considerations. But on these I will not now venture further than I have done in the Essay.

V. Your readers perpetually come back to the

argument, that you suppose that an intelligent being cannot exist in other physical conditions than those of the earth; and that this is an arbitrary assumption. The Creator may adapt an intelligent creature to conditions the most varied, as we see by the adaptation of animals to different conditions on the earth.

Z. If my readers perpetually come back to this argument, I may be allowed to come back to the answer which I have given to it (Chap. XI. Art. 15). But if you please, we will return to the argument, in another form. You say, then, that there may possibly be intelligent creatures, which are not subject to the same physical conditions as we are. How far do you go in this liberal line of speculation? Do you believe in Ariel, and the other Sylphs, the spirits which have their place in the air, and have frames and qualities adapted to their sphere of action? Do you believe in the Naiads, or Nixies, or Water-sprites, or Kelpies, or whatever their due appellation is, which inhabit the water? Do you believe in the Gnomes, the inhabitants of the dark masses of earth, and the Salamanders which are at home in the fire?

V. These are Rosicrucian inventions, or poetical imaginations. You have no real intention of comparing them to scientific conclusions.

Z. To such scientific conclusions as the dwellers in Mercury or Saturn, it seems to me they may very fairly be compared. Those who hold them as realities, have only to use your argument, that we must not limit the possible conditions of intelligent existence, by the empirical conditions of man's existence.

V. There is no evidence of their real existence.

Z. The Rosicrucian philosophers would contest this with you; and would use, in defence of the habitability of air, water, earth and fire, many of the arguments which you employ to uphold the habitability of the planets.

But if you want a class of intelligent beings for which you have more abundant evidence, what think you of Ghosts that walk, and Departed Spirits that return? No doubt these are not bound by the laws of our living bodily existence; they are seen, but cannot be felt. They mock the grasping hand, and glide through the bolted door. But what of that? Why should they not? Grasps and bolts have hold only on a gross bodily frame like that of man. With the testimony which is current of such things, and with your ready rejection of ordinary physical laws, surely you ought to believe in these, much rather than in the inhabitants of Mercury or Saturn, of whom you have no testimony at all.

V. But I do not suppose the inhabitants of other planets to be exempt from the ordinary laws of matter, as ghosts are supposed to be.

Z. But then your limitation of the possible deviation from the ordinary physical laws of man's being is quite arbitrary; and I do not see what right you have to draw the line in one place rather than another. Let us see where you are disposed to draw it. Do you think it probable that the Sun is inhabited?

V. I know that Sir William Herschel held that it might be so;—that the inhabitants who live below the cloudy canopy of light and heat which forms the surface, might be skreened from its glare and fire. And Arago, his commentator and biographer, evidently leans to the same opinion, which he considers to be almost generally adopted. But I rather hold to the public opinion, which prevailed in 1787, when a certain Dr Elliot being charged at the Old Bailey with the murder of a young lady, his advocate urged, as a proof of his *insanity*, that he believed the sun to be inhabited.

But this notion very strongly tends to shew the instinctive belief of man in the diffusion, if not of his species, at least of beings analogous to it.

Z. But the belief of the advocate, and the public opinion to which he appealed,—what does that prove? the instinctive belief that such a notion is mad, I should think. And the operation of this latter instinct is very old. Lactantius laughs at some of the philosophers of the Stoic sect, who having ascribed inhabitants to the moon, doubted whether they should also allow them in the sun. 'Why not?' says he. 'What would they cost you, when you have gone so far? But I suppose you were afraid they would be burnt to cinders, and the mischief laid at your door.'

V. But you see, in this very passage, how general a belief of this kind has always been. The preponderance of belief in all ages has been in favour of the Plurality of Worlds. This involuntary prepossession of mankind at large may almost be considered in the light of an argument of some weight. You have done wisely in not going into the history of opinions on this subject.

Z. I hope I have done wisely in not entering upon that history; for I was much tempted to go into it; and refrained, only because I was afraid of confusing the argument, and introducing questions of erudition.

But you astonish me much by saying that there has always been a preponderance of the belief in the plurality of worlds; meaning by that, of course, the belief that the planets, moon, and sun, are inhabited.

V. But is it not so?

Z. As far as my reading goes, quite the contrary. This belief has always been regarded as extravagant, and has been maintained only by a few men, who have become by-words on that account. Moreover it has generally been combined with other opinions, as fantastical as the Rosicrucian plurality of worlds, which you so unhesitatingly reject.

V. What records of such opinions do you find?

Z. Xenophanes held that there were inhabitants in the moon; but then, they were somehow on an earth which was inside the moon's orb. So Lactantius tells us. Plutarch, in his curious Dialogue 'On the Face which appears in the Moon's Orb,' gives the arguments for and against the doctrine of the moon being inhabited. But the doctrine that it is, is combined with divers mythological fancies, and was evidently regarded as having only the same kind of reality which they had: (in the time of the Emperor Hadrian, recollect.) Kepler translated this Dialogue; and he himself indulged in speculations on the same subject, in which he called the moon *Levana*, and spoke of her inhabitants; but as seems to me, with much the same amount of conviction as Plutarch. The narrative is put in the mouth of a Lapland witch. In the same way, Lucian makes the moon an island floating aloft and carrying inhabitants, much as Swift represents Laputa. And even when you come to the modern assertors of the Copernican system, who, as I allow, were naturally led to *guess* that there might be inhabitants in the planets, their guesses are so vague and fantastical, that I do not see why we should ascribe to them more weight than to the Rosicrucian doctrine of Sylphs and Gnomes. Thus Nicolas of Cus, who asserted the possibility of the Heliocentric system before Copernicus, in his book 'De Doctâ Ignorantia,' says, suitably to the design of the book, which is to shew the imperfection of human knowledge;—that we cannot know whether the inhabitants of other spheres are of a higher nature than man;—although, as he adds, it does not appear how any nature can be more noble, or more perfect, than the intellectual nature of man. He suspects however that in the region of the Sun, the intellectual inhabitants are more clear and illuminated, and more spiritual, than they are in the Moon, where they are 'magis lunatici;' as in the Earth they are 'magis materiales et grossi.'

Giordano Bruno was one of the early assertors of the Copernican system in England. He published a book (in prose and very bad Latin verse), entitled 'De Immenso et Innumerabilibus,' in which undoubtedly he asserts the Plurality of Worlds; but he goes much further than the moderns would follow him. He declares that the earth is inhabited in its interior, as well as at its surface: and that animals in the shape of spheres of fire fly through the air, one of which he himself saw.

I do not know of any writers whose speculations on this subject can be treated more seriously, except perhaps Huyghens, in his *Cosmotheoros*, a posthumous work; which however Humboldt, with great reason, as it seems to me, thinks unworthy of its author:—and by far the most thoughtful and earnest of such speculators, Wilkins, afterwards Master of Trinity College, Cambridge. He laboured at considerable length to prove, not only that the moon is inhabited, but that we need not despair of being able to make visits to the inhabitants.

480|

V. You treat the subject with levity; but undoubtedly it is a subject on which many persons bestow very serious thoughts and feelings.

Z. I am aware of that. I too have my serious thoughts and feelings on the subject; as I hope appears in what I have written. Nor would I treat with levity any serious feelings on this subject. But if the serious thoughts and feelings which flow from a belief in the Plurality of Planetary Worlds have really been produced by a guess, lightly made at first, quite unsupported by subsequent discoveries, and discountenanced by the most recent observations, though too remote from knowledge to be either proved or disproved:—what is there to prevent my speaking of such thoughts and feelings as mere results of imagination; and comparing the views on which they rest, with those which I think to be more probable, and more consistent with real seriousness?

W. But is there not something very elevating and consoling in the thought there may be in the stars innumerable worlds, far happier and better than ours? For my part, I can scarcely bear to think that it is not so. If the stars be seats of happy conscious beings, I can reconcile myself to the misery and guilt of earth: but if it be not so, the aspect of the universe is dark and desolate. In such a universe, I do not know how it can be said that 'God is love.'

Z. But surely the disciple who so repeatedly tells us that *God is love*, never dreamt that the stars were inhabited; nor did any of those who derived joy and hope from the words for fifteen hundred years afterwards, ever find such a belief necessary to their comfort. The need of this view is quite a modern feeling.

W. But to imagine worlds of intelligent creatures, purer and happier than man, is a consolation and relief to a mind oppressed with the prospect of the guilt and misery which are so prevalent on the earth.

Z. Be it so. But why place those worlds in the planets, or in the stars, with which we have no reason whatever to connect them? If the ground for believing, or wishing to believe, such better worlds, be the badness of this human world, the better worlds should surely be in some way connected with the state of man; so that their goodness in some way relieves or compensates his vice and misery. If other worlds, quite removed from this terrestrial world, are free from the terrestrial measure of vice and misery, that difference surely makes our terrestrial condition more sad, and not more consolatory. If we might have been so much better and happier, we may well lament the lot which has made us what we are.

W. But perhaps it is our own fault which has made us what we are.

Z. And if we amend our faults, or are in any way freed from them, we may be transferred to these better planetary abodes. Is that the doctrine?

482| *W.* Why not?

Z. I know no reason why not, except that to place in planets and stars these abodes of our purified and elevated nature, appears to me an assumption without either ground or object. Moreover, in that case, the planets exist for the sake of man, as his future abode. And therefore that supposition makes man the center of the providential scheme of the universe; which is precisely what is objected to the opinion that the earth alone is inhabited. I have no objection to this supposition. I can readily believe that God has done that for man, which he has not done for any other creature, and that man is the center of the providential scheme according to which the world has been conducted from the beginning. But I thought that you could not feel satisfied without other fields of moral agency.

W. Surely you will not deny that there is in this world so much vice and misery, that it is oppressive to have to think of it as the best world in existence.

Z. To think of man, being what he is, as the ultimate work of Providence, would indeed be oppressive. But if man, as he is, be the germ of something better, which he is to be, there is, in that thought, more consolation, I should think, than in millions of inhabited planets which have no connection with man.

W. But how difficult is it to discover how man is to be made better, and to make him so!

Z. Difficult indeed! And it is, I believe, the difficulty of this task, which makes men so willing to take refuge from the painful effort in the easy imagination of planetary elysiums. But yet this task is our real business; however many worlds there may be in the planets, or however few, our business is to make this world of earth better than it is. If we turn aside from that undertaking in indifference or despair, what shall we be the better for the mild light of the host of moons which shine on Jupiter's inhabitants, or the splendour of the skies where planets revolve round a double star, of which the two components wear the colours of the ruby and the emerald?

X. When you say that our business is to improve the world in which we live, you must recollect that it is our business quite as much to improve ourselves. This is every body's business: to improve the world may be the office of few only.

Z. No doubt: we ought to try to mend both; the *Microcosm*, as well as the *Macrocosm:* the little world within us, as well as the great *Cosmos* without.

X. Yes: but have you a right to call the universe a *Cosmos?* Do you not, by divesting the stars of their use, make *Cosmos* into *Chaos?*

Z. Nay, surely a world characterized by unity is as orderly as a world of plurality. That man is the center of creation, and that all the rest of the universe exists on his account, is, whether true or not, at any rate a simple and intelligible scheme. The accumulation of innumerable worlds of moral and intellectual creatures, without connection or subordination, may rather be termed a chaos; 'A mighty maze, and *all* without a plan,' so far as we can see, or even guess. When the term *Cosmos* was invented and applied, those who used it never dreamt of other inhabited worlds besides the earth, and other rational beings besides man.

Y. But in your inclination to make man the center of creation, and the object of all the rest of the universe, are you not forgetting the admonitions of those who warn us against this tendency to self-glorification. You will recollect how much of this warning there is in the Essay on Man:

> Ask for what end the heavenly bodies shine—
> Earth for whose use? Pride answers, 'tis for mine:

And the rest of the passage. To imagine ourselves of so much consequence in the eyes of the Creator, is natural to us, selfoccupied as we are, till philosophy rebukes such conceit.

Z. It is quite right to attend to such warnings. But warnings may also be useful on the other side: warnings against self-disparagement; against the belief that man is *not* an important object in the eyes of the Creator. I do not know what philosophy represents man as insignificant in the eyes of the Deity. Kant, you will recollect, said that two things impressed him with awe; the Starry Heaven without him, and the Moral Principle within.

And still less does Religious Philosophy favour the belief of man's insignificance in the eyes of God. What great things, according to the views which Religion teaches, has He done for mankind, and for each man!

Y. Yes, you build the philosophy of your Essay on a religious basis. You take for granted the truths of Revealed Religion, and reason from them. But that is not the way to arrive at true views of the physical universe.

Z. You must excuse me for saying that I do not reason in the way which you ascribe to me. I obtain my views of the physical universe from the acknowledged genuine sources; observation and calculation. The doctrine of inhabited planets and stars rests in a very small degree on physical grounds: as far as I can see any grounds of physical reasoning on that subject, I reason physically. But the doctrine is defended upon theological grounds also. I do not attempt to disprove the Plurality of Worlds by taking for granted the truths of Revealed Religion; but I say that the teaching of Religion, may, to a candid inquirer, suggest the wisdom of not taking for granted the Plurality of Worlds. Religion seems, at first sight at least, to represent Man's history and position as unique. Astronomy, some think, suggests the contrary. I examine the force of this latter suggestion, and it seems to me to amount to little or nothing.

As illustrating the general bearing of such speculations, I may perhaps be allowed to say that the views contained in the Essay have in several instances, as I have learnt, been received as very consolatory and satisfactory, by persons who are in the habit of deriving their opinions from religious principles.

Y. And yet many such persons had found great satisfaction and pleasure in the speculations of Chalmers, which proceed so far in the opposite direction.

Z. That may be. But I may say, on the other hand, that I have found religious persons, who, returning to the perusal of Chalmers's Astronomical Discourses, now that the first burst of their popularity arising from their great eloquence, is past, have found his suppositions and pictures painfully bold; and his views very far from consolatory. They see little comfort in the suggestion that the inhabitants of earth are more wicked and more miserable than those of any other part of the creation.

Y. But surely the contemplation of the possible existence of creatures purer and more elevated than ourselves, has a tendency to soothe and raise our minds. And such a belief is strongly encouraged by Religion.

Z. A belief in Ministering Angels, and Blessed Spirits, you mean. I grant you: but once more, how do you connect this belief with the planets and the stars? According to the views of religious men, such Ministers have their offices, in a great measure, among men:

> Myriads of spiritual creatures walk *the earth*,
> Unseen, both when we wake and when we sleep.

And if you would have your thoughts raised and solemnized by a view of a company of spirits, purer and higher than man in his mortal condition, I will quote you a passage from a modern Divine, of a soberer eloquence than the great Preacherof Scotland, and I think, of a sounder philosophy.

'How does the zealous Christian, alive to the honour of God, and troubled, like Elijah, to see his followers few, rejoice in every speculation by which he can persuade himself that the borders of his dominion are extending; that the cords of his tent are gaining length. How does he hail the Christian settlements amongst the horde of savages in the solitary islands of the Southern main! How does he anticipate in his glowing thoughts the day when the knowledge of the Lord is to cover the earth as the waters cover the sea! How does he even indulge in fanciful but innocent conjectures on the relation the Incarnation and Atonement of Christ may bear to the *inhabitants of other planets* besides our own! So vast does he naturally think must be the effect of the Sacrifice of Him who was with God, and who was God, and by whom all things were made. He can scarcely persuade himself to contemplate that great mystery merely in respect to this little earth. He is reluctant to circumscribe its virtues to the limited compass of a plot of ground like this. Behold then *we shew unto him a more excellent way!* Let him turn his meditations to this doctrine of the Communion of Saints, and his heart becomes *lawfully* enlarged. In that direction he may give his thoughts leave to wander in the full assurance of faith. There he finds a vast population of souls, some in the body, some out of the body, wherever dwelling, wherever the Paradise of God may be; of which the earth however is but the ante-chamber, and death the door; and he beholds troops of spirits in unceasing succession in the act of emigrating to that ample colony, never to be overpeopled, till the day when God shall have made up the number of his elect!'

This, and more in the same admirable strain, you will find in a Sermon preached at Cambridge by Professor Blunt, in November 1849. And we cannot end our discussion better than so.

AA. No, my friend, you are not to be allowed to end the discussion so. Here are forty pages of objections to your Essay, from a Critic who writes in a very vehement and angry strain.

Z. If he writes in a vehement and angry strain, I will have nothing to say to him. I have presented, gravely and calmly, the views and arguments which occurred to my mind, on a question which many persons think an interesting one; and if any one will introduce any other temper into the discussion of this question, with him I will hold no argument. But how does it appear that he is vehement and angry?

AA. I will read to you a few of the expressions which he uses, and you shall judge. He declares that to conceive such a universe as he says that you conceive, 'indicates a mind dead to feeling and shorn of reason.' He maintains that 'minds of the highest caste' view the world as he views it; but that 'ordinary minds' have the difficulties which he ascribes to you; though he can hardly conceive that persons 'of the smallest mental capacity' have such difficulties. He says, of your assertion, that the earth and its human inhabitants are, as far as we yet know, in an especial manner the subject of God's care and government, 'that the sentiment is groundless and the reason futile.' He speaks of your 'very silly attempt at argument.' He says that you 'travel, *con amore*, to our moon, a fit study and a suitable residence for one who holds the opinions and invents the theories of our author.'

Z. Severe and original sarcasm!

AA. He speaks of doctrines which he opposes, as 'an inconceivable absurdity which no sane mind can cherish, but one panting for notoriety:' and as 'a supposition too ridiculous even for a writer of romance.' Are you satisfied? He ascribes to you a bold attempt to 'concuss' the reader into your views 'by repeated applications of the *argumentum baculinum*.'

Z. The *argumentum baculinum!* That used to mean a sound beating with a staff. It is plain that he is very sore; but I am quite at a loss to understand what I have done to hurt him. But, as you say, it is evident enough that he is very angry. I cannot conceive why he should be so. He believes, I suppose, that the planets and the stars are inhabited. I have no objection to his believing this. But does he give reasons for his belief? Does he attempt to solve the difficulties, which I have noticed as attendant upon the belief? Does he answer the arguments which I have offered, to shew that the belief is ill-founded? If he does this, he may be worth notice, angry though he be.

AA. He says that 'we deduce from this language of Scripture,' ('When I consider the heavens...what is man?') 'a positive argument for the plurality of worlds:' for *that* view makes the Hebrew poet's wonder intelligible.

Z. That the Hebrew poet knew or thought about the plurality of worlds, is a fact hitherto unnoticed by the Historians of Astronomy. To their consideration I leave it.

AA. He says that the similarity of the conditions of the different planets, when it was discovered, led men to believe that they were occupied by life: and that thus the belief that the Earth alone was inhabited became more and more presumptuous.

Z. Then he holds that all planets are inhabited?

AA. So it seems, by his argument.

Z. Does he hold this of the twenty-nine small planets between Mars and Jupiter?

AA. He does not, I think, say so: but to be consistent, he must so hold it.

Z. Does he believe that meteoric stones are of the same nature as the small planets?

AA. Not only so; but he claims this 'theory of meteoric stones' as his own, published forty years ago.

Z. Then I must ask him why he does not believe the meteoric stones to have been inhabited?

AA. I suppose he will say, because they have not been found to be so, when brought within the reach of examination.

Z. Then is it not likely that Juno, Irene, Lutetia, and the rest of the twenty-nine smaller planets, if brought within the reach of examination, would be found to be uninhabited also?

AA. That may be; and yet the larger planets, Mars, Jupiter, and the like, may be inhabited.

Z. But the argument that they *are* inhabited, because they are planets, fails him. Has he any other reasons?

AA. Yes; he has other reasons. He says that, if not inhabited, the planets would be wasted.

Z. This question of wasted means, I have fully considered Chap. XI, Art. 11, &c. and must leave what I have there said to be considered by the reader.

AA. The Critic is shocked, he says, at your speaking of planets which have 'failed in the making.'

Z. The expression is intended to remind the reader (as any intelligent reader will see) in how many instances natural objects seem to us to fail in the making. This appears in millions of undeveloped or uncompleted vegetable and animal embryos, which have marks of a fitness for ulterior uses far more clearly than the planets, and yet never are used. Is there any offence in saying that a seed which is killed, *fails* to produce the plant which it is fitted to produce? Doubtless such seeming *failure* is part of the Divine Plan. It is failure to our eyes only. But we are not to make a Divine Plan for ourselves, by excluding, as irreverent, the notion of such failures.

However, I have discussed this matter in Chapter XI, with reference to various provinces of creation.

AA. Your Critic calls that a 'Chapter of omniscience.'

Z. That is an odd expression; for the Chapter contains only a few facts familiarly known to all persons who take a comprehensive view of nature. If the Critic thinks such knowledge either extraordinary or inapplicable, he is very unfit to speculate on such questions as he deals with.

But what does he say of the difficulties attendant upon the doctrine of the plurality of worlds, such as I have stated in Chapter IV?

AA. He says 'after a careful perusal of this Chapter, we must acknowledge our inability either fully to understand its meaning, or to see its bearing on the real question of a plurality of worlds. It is a mere display of ingenuity, obliterating metaphysically the brightness of our perceptions, and coming over our minds like an Eastern fog on a spring morning, or like the tail of a comet over a cluster of stars.'

Z. If he cannot see the difficulties which the question involves, he is not one of those for whom I wrote. I am sorry that he has allowed me to obliterate metaphysically the brightness of his perceptions.

AA. He does, however, speak of the religious difficulty; and says, that 'Dr Chalmers has rather cut than untied the knot.' His own solution of the question concerning the redemption of other worlds appears to be this: that the provision made for the redemption of man by what took place upon earth eighteen hundred years ago, may have extended its influence to other worlds.

Z. In reply to which astronomico-theological hypothesis, three remarks offer themselves. In the first place, the hypothesis is entirely without warrant or countenance in the revelation from which all our knowledge of the scheme of redemption is derived. In the second place, the events which took place upon earth eighteen hundred years ago, were connected with a train of events in the history of man, which had begun at the creation of man, and extended through all the intervening ages: and the bearing of this whole series of events upon the condition of the inhabitants of other worlds must be so different from its bearing on the condition of man, that the hypothesis needs a dozen other auxiliary hypotheses to make it intelligible. And in the third place, this hypothesis, making the earth, insignificant as it seems to be in the astronomical scheme, the center of the theological scheme, ascribes to the earth a peculiar distinction, quite as much at variance with the analogies of the planets to one another, as the supposition that the earth alone is inhabited; to say nothing of the bearing of the Critic's hypothesis on the other systems that encircle other suns.

AA. Your Critic does not allow, as I have already told you, that man and his history are an especial care of God. He says 'the lily that neither toils nor spins,—the ravens that neither sow nor reap,—the sparrows, though of less value than man,—the crushed insect,—the broken planet,—the twinkling star,—are all as special objects of God's care, as the wise elephant,—the illiterate biped,—the presumptuous philosopher, and the great globe itself, and all that it inherits.'

Z. But surely a scheme for the redemption of man, bound up with the whole history of man and establishing a special relation between man and God, shows a care more special than the Divine operation which clothes the lily and feeds the sparrow; and this was implied, when it was said, 'Fear not; ye are of more value than many sparrows.'

Can you not find something more to the purpose in the criticism?

AA. Perhaps if we go into the detail of the physical argument we may. In opposition to your doctrine, that Jupiter is a sphere of water, he says. that if it were so, 'The light reflected from his surface, when he is in his quadratures, must contain, what it does not, a visible portion of polarized light: and if his crust be composed of mountains of ice, some of whose faces may reflect the light at nearly the polarizing angle, that fact would be distinctly indicated, which it is not, by a very large quantity of polarized light.'

Z. That is a very good remark; and would be of considerable importance, if the observations of which the result is asserted, have been made carefully; which I by no means deny. But I doubt whether the remark is applicable: for Jupiter's watery or icy mass must be clothed in a thick stratum of air, and aqueous vapour, and clouds.

AB. But even if the planet were free from clouds, the parts of the planet's surface from which polarized light would be reflected would only be as points compared with the whole surface; and the common light reflected from the whole surface would quite overwhelm and obliterate the polarized light.

Pray let me hear some more of his arguments.

AA. He says that your argument from Geology (Chap. v.) is erroneous, because you make the age of the earth in its present organic condition to be brief compared with preceding periods; whereas it will be infinite.

Z. I reason at present from the present state of things; as I have said, Chapter XII. Art. 20. But how does he know that the duration of the human race on the earth will be infinite?

AA. He says that 'neither reason nor religion leads us to believe that the Solar System will ever be destroyed, or that living beings will ever cease to occupy the earth.' He says that 'if this is not the case, we are involved in that inconceivable absurdity,' &c. &c., 'that the Almighty took millions of years to make a world which was to be occupied only for a few thousands.'

Z. It seems to me that he must be a very bold man to speak so confidently on such subjects.

AA. He is offended at your using that expression. You have said (Chap. VII. Art. 12) that he must be a very bold man, who says, of the specks into which Lord Rosse's telescope resolves the Nebulæ, 'These lumps, O man, are suns,' &c. The Critic says that you might in the same way call a man a very bold man, who says of geological strata, 'These are the fragments of strata, O man, which though seen only in patches, surround the whole globe,' &c.

Z. The illustration would have some propriety, if all that we knew of the patches of strata was, that they were patches of seemingly earthy matter, without any stratified structure or organic contents to identify them. In that case he *would* be a very bold man.

AA. He says that your doctrine of Nebulæ revolving in spirals is extravagant; and exclaims *Risum teneatis Astronomi!*

Z. As I have already said, such astronomers as Lord Rosse have no occasion for his warning, but consider this spiral motion very gravely. And so will any astronomer who looks at the representations of Nebulæ, which I have borrowed from Lord Rosse, and made my frontispiece.

AA. He denies that Encke's Comet revolves in a spiral.

Z. Very good. Let Encke decide between us.

AA. He says that Biela's Comet was never separated into two; that the luminous portions are connected by a transparent portion.

Z. You recollect, probably, that when Galileo discovered the inequalities of the Moon's surface, one of his opponents, who held that the heavenly bodies must be exactly spherical, said that the apparent cavities were filled with transparent crystal. Whereupon Galileo threatened to place upon the surface transparent crystal mountains, ten times as high as the cavities were deep. I must reduce the Critic to reason, if necessary, by proceeding to extremities in the same way. I may hold it probable that Biela's Comet is divided into several invisible masses, besides the two visible

498| ones.

AA. He has an illustration to shew how little is proved by the changing of colour, or extinction, of a few among the Fixed Stars. He supposes that London is lighted by a single electric light on the top of St Paul's; and that an aeronaut, knowing this, flying over France, sees many lights. He will suppose that they illumine cities and villages, though a few go out, or burn red or blue.

Z. Yes, if he *knows* that he is moving over France, an inhabited country: but in our case, that is precisely the thing unknown. That being the fact, each of these lights may be an active volcanic fire, or a patch of phosphorescent matter.

AA. He says that your modification of Fontenelle's illustration (Chap. VIII. Art. 26) is unjust, because the stars are more like each other than islands in the same sea are.

Z. The opposite opinion, that neighbouring islands, in the same sea, are much more like each other than the stars are, appears to me much more probable; and I have given reasons for thinking so.

Does he follow Fontenelle in his conjectures as to the character and condition of the inhabitants of different planets?

AA. Oh, he is much bolder than the old Cartesian, in his conjectures as to the ways in which the planetary races may differ from man. He says: 'The being of another mould [world?] may have his home in subterranean cities, warmed by central fires,—or in crystal caves, cooled by ocean tides,—or he may float with the Nereids upon the deep,—or mount upon wings as eagles, or have the pinions of the dove that he may flee away and be at rest.'

Z. Very pretty: and withal very much to the purpose; for such writing is a fair specimen of the mode in which the notion of creatures, governed by physical laws different from those which regulate the existence of man, may be embodied and realized. But then I return to my question (Chap. XI. Art. 17): Why should these creatures be confined to inhabit other planets? Why should they not dwell on, and move in and about, the earth? They are precisely the Gnomes, Naiads, and Sylphs of the Rosicrucians, about which I was talking a little while ago.

AA. There are some other objections made by the Critic in a worse spirit. He tries repeatedly to connect your speculations with those of the *Vestiges of the Creation.*

Z. If he were to try to connect me with an answer to that book, which went through two editions, under the title of *Indications of the Creator*, he would be nearer the mark. At least I adopt the sentiments of this latter book; and they agree with those of the Essay, as you may satisfy yourself by looking. In both works, the placing of man on the earth is regarded as an event out of the ordinary course of nature; directly the opposite of the doctrine of the *Vestiges.* The Essay is mainly employed, first, in proving this doctrine, and then in reasoning from it.

AA. Yes; the person who, after reading the Essay, so misrepresents it, using (to adopt his own language) 'the boldness of an anonymous writer,' is deserving of 'severe censure' and 'grave reproof.'

There is another passage which will give you some pain. The Critic says that he looks to a future existence with hopes which the study of the heavens has enlarged; and that 'in the infinity of space, and amid infinity of life, we descry the home and the companions of the future. That home and those companions have been denied us by the author of the work before us.'

Z. I am certainly pained at hearing such a remark. If any man have connected his hopes and images of an existence beyond this life with his astronomical opinions, I should be much grieved at being the instrument of disturbing such a form of faith. But at the same time, I cannot think that the partial existence of such notions,—unauthorized fancies, as I believe them to be,—is to be deemed so sacred a thing, that we are not to be allowed to speak of our hope and belief in the Future in any other form.

The author of the *Vestiges* had spoken, complainingly, of 'the awakened and craving mind asking what science can do for us in explaining the great ends of the Author of Nature, and our relations to him, to good and evil, to life and to eternity.' On this the author of the *Indications of the Creator* remarked, that we shall be able to bear our ignorance the more patiently, if we have not looked to science for that which science cannot give. In the truth and importance of this reflexion on such occasions, I entirely agree.

AB. Since this Dialogue was first printed, the Critic has published his arguments on the Plurality of Worlds, with his name to the book. But his book is still mainly a series of vituperative expressions about your Essay. He offers you warnings as to the tendency of your speculations. He says, 'However sincere may be his piety, which we do not question, we tell him with confidence that his theories are replete with danger;' and so on.

Z. No doubt he means well. But is this *we* the language employed in the book with the Author's name to it?

AB. Yes: I suppose he has become so accustomed to the tone of the anonymous reviewer that he uses it unconsciously. The 'boldness of the anonymous writer,' which he speaks of, becomes the 'confidence' of the singular 'we.' But what do you say to his advice?

Z. I should wish to give him a few words of warning in return, which I would offer with a very sincere desire that he should not do mischief without intending it.

I would warn him that the doctrine that the planets are all inhabited like the earth, and that the stars are centers of inhabited systems like the Sun, is a part of that *Nebular Hypothesis* which he holds in horror.

I would warn him that the author of the *Vestiges of Creation*, who would reject my opinions, would readily accept his.

I would warn him that the habit of accusing men of irreverence because they do not assent to his explanation of the uses of every part of creation, is very unlikely to promote a reverential feeling in his readers.

I would warn him that the doctrine that there are innumerable worlds besides that human world in which God has revealed himself to man, however familiar it may have become to those who take their opinions from modern books of popular astronomy, is not congenial to those who derive their views from revelation.

And having said this, I hope I have done with the Critic and his criticism.

BB. But here is a Critic of another mood, who speaks with great praise and admiration of your Essay. And you will not value his praise the less, that he is a Critic who habitually takes a religious view of literary works.

Z. I certainly hoped for sympathy from such persons, and I have found it.

BB. He quotes with strong approval the passage in which (Chap. XI. Art. 34) you speak of the 'lumps which have flown from the potter's wheel of the Great Worker,' &c.

AA. That is curious, for your other Critic 'consigns it to a note' and exclaims at it.

Z. But does my favourable Reviewer find no faults?

BB. He taxes you with misquoting Professor Owen: but that you have already answered in talking to *T*. He says also that he himself, as early as 1844, had written to the effect that the Earth alone may be the inhabited part of the universe.

Z. I have no wish to lay any stress upon the originality of the views presented in the Essay. I now know that several years ago, (in 1849) Hugh Miller, in his *First Impressions of England*, (Chap. XVII.) presented an argument from Geology, very much of the nature of that which I have employed; and that the Rev. Mr Birks, in a little Tract published in 1850, urged the very insecure character of the doctrine that the planets and stars are inhabited. These coincidences with my views, I did not know till my Essay was not only written, but printed. As to myself, the views which I have at length committed to paper, have long been in my mind. The convictions which they involved grew gradually deeper, through the effect of various trains of speculation; and I may also say, that when I proceeded to write the Essay, the arguments appeared to me to assume, by being fully unfolded, greater strength than I had expected. But however that may be, be the arguments strong or weak, there they are, delivered in all sincerity and simplicity. *Liberavi animam meam.*

AB. Here is another critic, who, also looking at your speculations from a religious point of view, approves them in general, but is offended at what you say in your last chapter; of the possibility of a supernatural interposition, which might take place, to plant upon the earth the germ of a Divine Society, as at the Creation of man there was placed upon the earth the germ of human society; and which might introduce a period in which the earth should be tenanted by a more excellent creature than at present.

Z. But what is there in this, at which a religious person can be offended?

AB. He says that by this supposition, Christ's work is to all intents and purposes eliminated from this present world; and asks, Has not the Supernatural Interposition already taken place?

Z. That a Supernatural Interposition in order to found a Divine Society upon earth has taken place already, Revelation informs us; and I hoped that I had prevented the recognition of this fact in the Essay from being overlooked. But the question, whether our scientific views can in any manner illustrate this Interposition, still remains, and is not to be lightly disposed of. Is it not necessary that we should touch with the utmost caution and forbearance, upon any physical analogies regarded as having such a bearing? The Interposition having occurred, there may be no physical analogy which at all illustrates it: but if it be analogous to the fact that, in the history of the earth, there has been already once placed upon the earth a creature incomparably more excellent than its previous inhabitants, the question occurs, how are we, on this view, to interpret this greater excellence of the Creature, since the Interposition? What supposition must we make? Is man more excellent than before, because he is becoming better by a slow progression? Or is it because he is to be further exalted, by a Second Coming of the Divine Power? Or is man's elevation to be, not an exaltation of his terrestrial condition, but his transfer to a new state of existence? I have, in the Essay, put forth, however briefly, these three suppositions: and having done this, I thought that religious persons would see, that whatever place each of them may hold in the contemplations of one who believes that such a Divine Interposition has taken place, he could neither feel himself justified in pressing upon his readers any one of these views, nor in rejecting them all. Analogies between the government of the natural and of the spiritual world are to be dwelt upon with great reserve and caution: and perhaps can never be safely urged, except as negative arguments, to shew that what religion reveals is not contrary to natural probability. And therefore I did not venture further, than to intimate, that when we are taught, that *as we have borne the image of the earthy we shall also bear the image of the heavenly*, we may find, even in natural science, reasons for opening our minds to the reception of the cheering and elevating announcement.

AB. I think that what you have said in Article 19—21 of your last Chapter is sufficient to suggest this; and certainly, by attempting to enforce any of the alternative suppositions which you have mentioned, you would have given more offence to many readers, than you have done by leaving your religious readers to follow their own choice among them. And surely the way in which these suppositions are put is such as to shew, to any attentive reader, that so far from doubting whether a Divine Interposition have taken place, the thing you doubt is, whether it can receive any illustration from the doctrines of geology.

FURTHER READING

Three books give the full background to the debate over extraterrestrial life. Stephen J. Dick starts the story in his *Plurality of Worlds: The Origin of the Extraterrestrial Life Debate from Democritus to Kant* (Cambridge: Cambridge University Press, 1982). This takes us to the middle of the eighteenth century. Michael Crowe picks up the tale with *The Extraterrestrial Life Debate 1750–1900: The Idea of a Plurality of Worlds from Kant to Lowell* (Cambridge: Cambridge University Press, 1986). Finally, Dick reappears with *The Biological Universe: The Twentieth-Century Extraterrestrial Life Debate and the Limits of Science* (Cambridge: Cambridge University Press, 1996). These really are wonderful books and give a detailed account of the science and of the underlying metaphysical and religious concerns and worries.

For a general understanding of William Whewell, the place to start is with his two nineteenth-century biographers and letter collectors. His sister-in-law Mrs Stair Douglas edited the personal letters in *The Life and Selections from the Correspondence of William Whewell, D.D.* (London: Kegan Paul, 1881). The mathematician Isaac Todhunter discussed the professional work, and edited the scientific and related letters in *William Whewell D.D. Master of Trinity College Cambridge: An Account of his Writings with Selections from his Literary and Scientific Correspondence* (London: Macmillan, 1876). In our time, Menachem Fisch and Simon Schaffer edited an excellent collection, *William Whewell: A Composite Portrait* (Oxford: Oxford University Press, 1991), dealing with the many-sided interests of Whewell.

The plurality of worlds debate is dealt with fully in a long and detailed article by John H. Brooke, "A natural theology and the plurality of worlds: observations on the Brewster-Whewell debate," (*Annals of Science*, 34 (1977), 221–86). Brooke thinks it was *Vestiges* which spurred Whewell to set pen to paper. Michael Crowe, in the work mentioned above, also has a very detailed discussion of Whewell and his critics, giving a slightly different emphasis. Crowe is more inclined to think that rethinking revealed theology made Whewell write his *Essay*. As you will have seen from my introduction, I think there is merit in both of these theses. (Crowe gives a very full listing of Whewell's critics with all of the pertinent references, and the interested reader should turn to his book for this information.)

There are many many books on mid-Victorian science and the events leading up to the *Origin of Species*. In my own, *The Darwinian Revolution: Science Red in Tooth and Claw* (Chicago: University of Chicago Press, 1979; second edition, 1999), I take pains to show that Darwin's success had many causes and many effects. The plurality of worlds debate existed in its own right, but it also was a factor in what was to happen at the end of the decade. An excellent collection of modern thoughts about the debate and its implications for religious thought has been edited by Stephen J. Dick, *Many Worlds: The New Universe, Extraterrestrial Life and the Theological Implications* (Philadelphia and London: Templeton Foundation Press, 2000). In my *Can a Darwinian be a Christian? The Relationship between Science and Religion* (Cambridge: Cambridge University Press, 2001), I, too, discuss the plurality of worlds debate today and the ways in which it is entwined with Christian beliefs and commitments. I am not sure that William Whewell would care greatly for my discussion, and I know that Sir David Brewster would not like it at all.